Ecology of
TEHRI DAM

Ecology of TEHRI DAM

Authors
D.R. Khanna
Pramod Kumar

Editor
Vikas Singh

BIOTECH
2013
BIOTECH BOOKS®

ISBN 978-81-7622-251-8

Published by: **BIOTECH BOOKS**®
4762-63/23, Ansari Road, Darya Ganj,
New Delhi - 110 002
Phone: +91-011-23262132
E-mail: biotechbooks@yahoo.co.in

Printed at: **Chawla Offset Printers**
Delhi - 110 052

PRINTED IN INDIA

Preface

Life on earth is available in different forms with water in its basic structure. Availability of water on earth is abundant but major portion is salty water. In comparison the freshwater resources are very less and pressure on them is continuously increasing. Reservoirs are one of the major sources of freshwater on earth, no matter whether they are man-made or nature but counted a freshwater ecosystem affecting the health, economy and environment of local area. Due to there importance we need to study their quality and monitor them on regular basis so that we can make available our natural resources to our coming generations. Even in ancient times people are aware about the importance of freshwater. In the ancient literature we have plenty of literature available describing how the important steps have been taken up by the kings to conserve the ponds, lakes, talabs, etc. and to store rain water which otherwise went to oceans as run-off.

In this book, we are concerning about the fifth tallest dam in the world – Tehri Dam, situated on the height of with a height of 260 meters (855 feet). The study done during was one of the first study done in context of limnology and aquatic biology. Including this, the book presents the environmental changes happen after the completion of the reservoir including loss of biodiversity, human settlement etc. Through this book we have tried to determine the present limnological status of Tehri Dam, their conservation strategies etc.

D.R. Khanna
Pramod Kumar
Vikas Singh

Contents

Chapter 1
Introduction

"Tehri dam is built with our tears"- Sunderlal Bahuguna

One of the most important gifts given by nature for mankind is water, which plays a significant role in different vital and structural activities. For the growth and development of life support system, it is absolutely essential to make water available in the acceptable quantity, quality and at the required time and place. The Earth appeared like a precious blue stone set in the darkness of outer space to astronauts for the first time. If inhabitants of other solar systems were to visit this lovely, cool, blue planet of ours, they might just call it "Aqua" rather than "Terra" because of great abundance of its most unique feature–liquid water in streams, rivers, lakes and oceans. Therefore, earth is popularly called as "Blue planet". Different kinds of physico-chemical processes occur in solutions comprising water as the solvent. The simple molecule, H_2O, aggregates to form water which is the basic requirement for sustaining life (Mohanty, 2006).

Water covers more than 75 per cent area of the earth surface either of salt water or fresh water, in which fresh water is less than 2.6 per cent of the total. 4/5 of this fresh water is in the form of ice. The remaining amount really available is that in the form of surface and ground water. The surface water includes the sources like (*i*) Rain water (*ii*) River water (*iii*) Lake water and (iv) Sea water. The sources of underground water are (*i*) Spring water and (*ii*) Well water found in lakes, rivers, reservoirs and shallow ground water sources, this small amount of water is continually renewed by the restless and endless movement of water in the global cycle in its various forms, *viz*, ice, water vapour, rain drops and snow. (Mohanty, 2006). Water is very important for all living organisms. It has a great economical and social value ultimately affecting human health. It is essential for industrial development, irrigation, fisheries, human life and hydro electric generation etc.

The fresh water bodies can be classified in two categories viz. running water (lotic) and standing water (lentic). The science dealing with the study of physical, chemical and biological characteristics of fresh water is termed as "Limnology" derived from the Greek word "Limnos" meaning pool.

Water is the nucleus for life support systems on earth. Unfortunately today the drinking water is not safe because of various anthropological activities and population pressure. The burgeoning human population and the demands for the various consumptive uses of water is creating a severe stress on the water resources of our planet with increasing water shortage and rising water quality problems in many parts of the world, ensuring universal access to safe, sufficient and affordable water for all remains is an enormous challenge. Industrial effluents, human abuse, misuse and overuse of chemicals, pesticides, dyes and detergents are the important sources of pollutants. Beside this over exploitation of water bodies due to civilization also contributes in deterioration of water quality. It is quite essential that the natural environment of the water body should be conducive to the extent that water could be used for drinking purpose. The health and well-being of the human race is closely tied up with the quality of water used (Sharma *et al.*, 2005). Water quality is primarily based upon its physical, chemical and biological characteristics. According to a report given by UN and the Washington Bank in the 'State of planet' fresh water are identified as the most threatened out of all the ecosystems in the world.

In India 14 major river systems share about 83 per cent of the drainage basin. In other words Indian rivers are supposed to carry about 1,645,000 million meter cube of water annually (Rao, 1975). The Himalaya is the source of fresh water, so that they can be called "the cradle of major rivers" that supports the life of many ecosystems. The Himalayan Rivers receive about 20-30 per cent of their water from melted snow during the summer and monsoon season as reported by Deming (1975). The importance of water resources has increased with industrialization and growth of population, as these are needed for sustaining crop productivity, meeting the ever increasing needs of the domestic, industrial and power sector (Rautela *et al.*, 2002).

Sun is the source of energy, this energy support the life forms to sustain on earth. It is the only primary source of energy from which all the living organisms are associated directly or indirectly. Just like that, human is in continuous search for an energy resource to fulfill the energy requirement needed for their modern life style, to run their industries, electronic equipments etc. Humans tried to escape from this problem by the use of fossil fuels, but they were limited and non-renewable. The most important drawback is the pollution. Then hydroelectricity emerged as a best alternative, it is renewable, pollution free and is unlimited. It resulted in the construction of dams on the rivers. With the increase in population and also in energy requirement, the number of dams on rivers also increased all over the world. In many countries hydropower is the only natural energy source. Hydropower accounts for about 20 percent of electricity production and about 7 percent of total energy production in the world. Hydropower is clean and the production from existing projects can be increased without a harmful impact on the environment. This potential is at least six times greater than its current level of production. In other words, hydropower from dams is one of the key sources for providing energy for expanding development (ICOLD, 2000).

Dams play an important role in utilizing water resources. A dam is defined as an artificial barrier with the ability to impound water, wastewater or any liquid-borne material, for the purpose of storage or control of water (Perry County, Hazard profiles draft, 2008). Humans have started building dams and canals a long time ago and this way influenced the natural flow regime. In 2600 B.C. the first large dam (14 m high, 113 m crest length) was built at Sadd el Kafara in Egypt in the Garawi ravine facing Memphis. Dam construction and the dam and reservoir itself has variable impacts: social, economical, geophysical and impacts on water quality, climate, flora and fauna (Horvath and Munico, 1998). For more than 5,000 years, dams have provided people with a reliable source of the water they need to live. Dams have enabled people to collect and store water when it is plentiful and then use it during dry periods. Dams have been essential in establishing and supporting towns and farms as well as providing food by irrigation of cropland. Today, dams and reservoirs also help to control flood waters to protect people and property, keep rivers navigable, provide electricity from renewable energy for towns and factories, and provide recreational opportunities such as fishing, camping and water sports (ICOLD, 2000).

Today more than 40,000 large dams improve the living conditions of the world's population that continues to grow at a rate of more than 100 million people per year. Yet, about 1.5 billion people still do not have access to a reliable source of suitable drinking water and more than two dozen countries do not have enough water to properly sustain their populations. Today more than one billion people are malnourished or starving. In many countries, increased food production is only possible through improved irrigation, which frequently depends on diminishing groundwater resources. Therefore, more dams are needed now and into the foreseeable future to improve the management of existing surface water resources (ICOLD, 2000).

There has been increasing recognition by both dam proponents and dam opponents that the social impacts of dams are complex, and can be far-reaching. The impacts of dam construction include quantifiable economic impacts, in the form of the changed flow of economic costs and benefits. Social impacts of dam can be positive (*e.g.* improved welfare resulting from new access to irrigation water) or negative (*e.g.* resettlement, decline of a downstream fishery due to flood control). Social impacts can be direct (*e.g.* the cultural trauma of involuntary resettlement), or the result of a cascade, where environmental impacts generate economic impacts, and these in turn cause social impacts (for example the impacts of changes in a river's flooding patterns reducing fish populations downstream of a dam, affecting the economic return form fishing and causing increased levels of out-migration of male fish catchers). Social impacts can be local to the dam site (*e.g.* the negative impacts associated with resettlement), or in places that are far from the dam, where water or electricity are consumed (Adam, 2000). Dams disconnect rivers from their floodplains and wetlands and reduce the speed at which water flows in rivers. They impact on the migratory patterns of fish, and flood riparian habitats, such as waterfalls, rapids, riverbanks and wetlands (Dynesius and Nilsson, 1994).

Large dams play an important role in social and economic development, as they provide water for agriculture and hydroelectricity generation. Half of the world's large dams were built primarily for irrigation (WCD, 2000), but beyond the dam itself irrigation systems utilized worldwide have water use efficiencies of only 38 per cent (UN, 2003).

Hydropower currently provides 19 per cent of the world's total electricity supply, with relatively low greenhouse gas emissions, and as a secure source of water supply, dams also play an important role with 12 per cent of large dams designated for water supply. More than 45,000 large dams are operational in over 150 countries (WCD, 2000).

The Himalayan terrain provides cost-effective conditions for dam construction deep gorges with wide valleys on the upstream (Rautela *et al.*, 2002). The Bhagirathi and its tributaries predominantly drain over greywackes, shales, slates, carbonates, quartzites, granites, gneisses, schist and calcsilicates (Chakrapani and Veizer, 2005). Tehri dam on India's River Bhagirathi, the main tributary of the Ganga River, is one of the world's largest most controversial hydroelectric project. It is under construction since 1978. Tehri dam is the fifth tallest dam in the world. This hydro-project is a joint venture of Uttar Pradesh Govt. and Govt. of India (IRN, 2002).

On Bhagirathi River there are many hydro-projects which are either under operation, under construction or proposed. Total number of hydro-projects on Bhagirathi river is 13 out of which two named Maneri Bhali I (90 MW) and Tehri (1000 MW) is operational, five named Loharinag Pala (600 MW), Pala Maneri I (480 MW), Maneri Bhali II (304 MW), Koteshwar (400 MW) and Kotli Bhel IA (195 MW) are under construction and seven named Karmoli (140 MW), Jadhganga (50 MW) Gangotri (55 MW), Bhaironghati I (380 MW), Bhaironghati II (65 MW), Harsil (210 MW) and Tehri PSS (1000 MW) are proposed (SANDRP, 2008). The location of Bhagirathi River and its hydro projects are shown in Map 1.1.

Salient features of Tehri Dam Project

Tehri dam has been constructed by the Tehri Hydro Development Corporation (THDC) with the government of India by Russian technical aid. With the height of 260.50 meters it is highest in Asia and ranks fifth highest earth and rock fill dam in the world. The dam side is surrounded by mountains in three sides having height of 1000-2000 meters above sea level. The dam has been constructed approx. 1.5 km down stream from Ganesh Prayag, the confluence of Bhagirathi and Bhilangana. This mega hydro electric project will generate 2400 MW electricity providing irrigation water for the area more than 6.6 million acres and municipal drinking water to large population in western UP, Uttarakhand and Delhi. The construction work on the project started in 1978 and is still going on. Generation of electricity from first turbine of 250 MW started on 17th July 2006 at 5.55 AM. It was formally inaugurated by central power minister Mr. Susheel Kumar Shinde on 30th July 2006 and devoted Tehri Project to nation.

First phase having four turbines of 250 MW each (250×4=1000) have been completed and all four turbines are generating electricity since 20th March 2007.

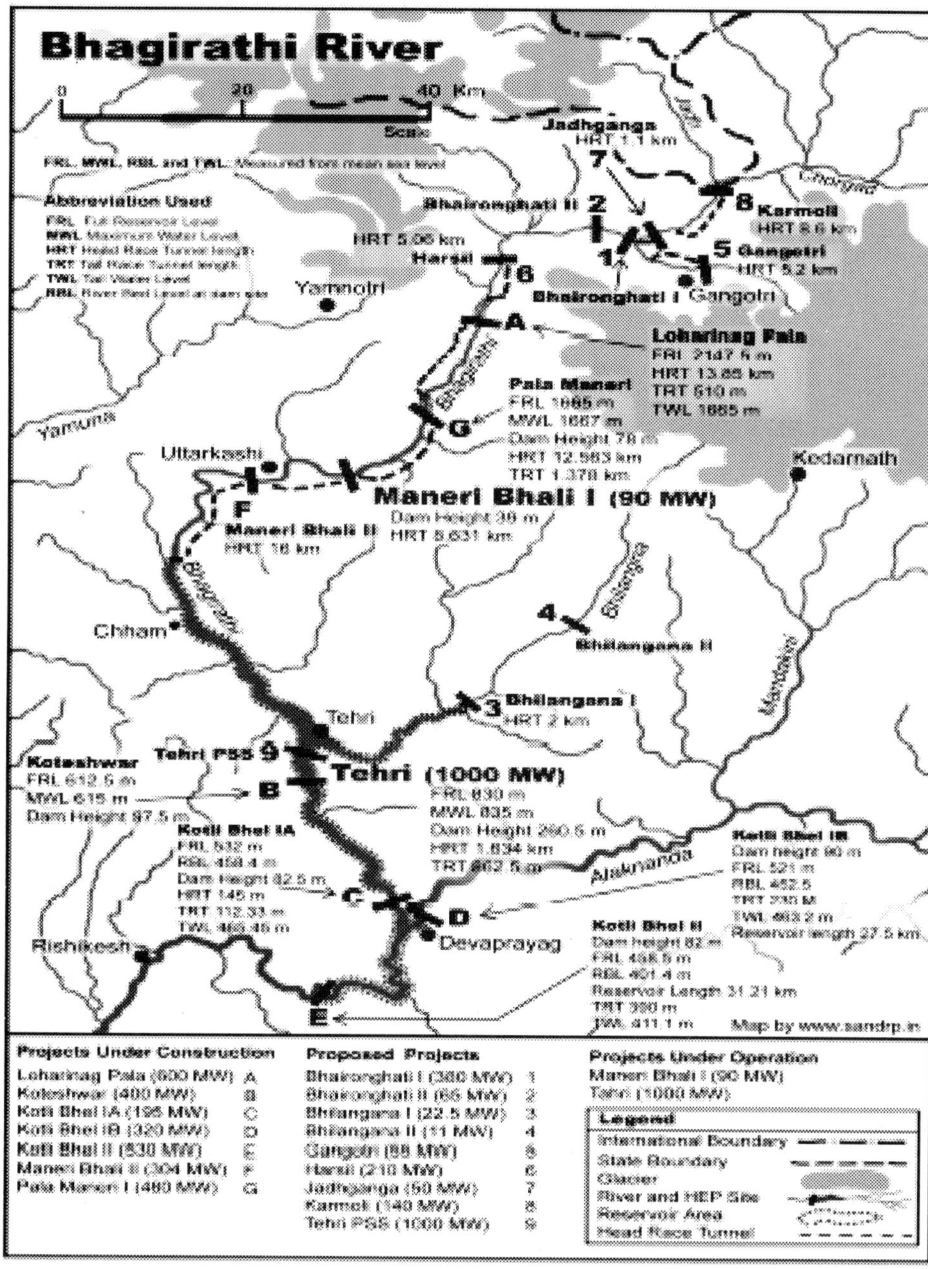

Map 1.1: Different Hydro-projects on Bhagirathi River (Courtesy: SANDRP, 2008)

Electricity produced by this project is being distributed in 9 states of north India. The second stage of 1000 MW and 400 MW Koteswar dam are under construction. These two together with first stage contribute total 2400 MW.

Catchment of Tehri Dam

The total catchments area upstream of Tehri dam is 6921.25 km which forms the water shed of river Bhagirathi and Bhilangana. Tehri dam has been constructed 1 km (approx.) down stream of the confluence of Bhagirathi and Bhilangana. The land use classification in Tehri dam catchments area assessed by UP forest department is as given below.

Agricultural area	88124 ha (13 per cent)
Forest	214458 ha (31 per cent)
Reserve	100 ha
Civil Soyam	6358 ha blanks
Alpine blanks	117761 ha
Other blanks and scrubs	33744 ha
Rocky	35687 ha (5 per cent)
Permanent snow bound	192445 ha (28 per cent)
Submerged	9906 ha (1 per cent)
Total	692125 ha

River Bhagirathi is famous for its unique nature. Its water does not allow any pathological activities in it so remain fresh for a long time. We know that there is difference in flora and fauna of running and standing water ecosystems. Construction of huge dam affects the physico-chemical and biological nature of river ecosystem. Therefore it is purposeful to carry out investigation to study the impact of Tehri dam on aquatic ecology of river Bhagirathi at Tehri Garhwal.

Physiography of the Study Area

The study area is located in Garhwal Himalaya which is an important zone of Himalaya and a part of new state Uttarakhand in India. It encompasses two districts Tehri and Uttarkashi. This area is very rich in biodiversity and the entire region of Garhwal Himalaya is rich in water resources in terms of major fluvial systems of the Ganges, Yamuna and their tributaries.

Various big River valley projects have been constructed and various more are under construction. Tehri Hydro-electric project is one of them. For present study following five sampling stations were selected. The total stretch, of these five stations is approximately 35 km (Map 1.2).

1. Dharanshu (30°36′41" North–78°19′05" East)
2. Chinyalisaur (30°33′05" North–78°20′54" East)
3. Bhaldiyana (30°28′47" North–78°24′21" East)
4. Koti (30°24′26" North–78°27′32" East)
5. Dam outlet near Chopra village (30°21′10" North–78°29′06" East)

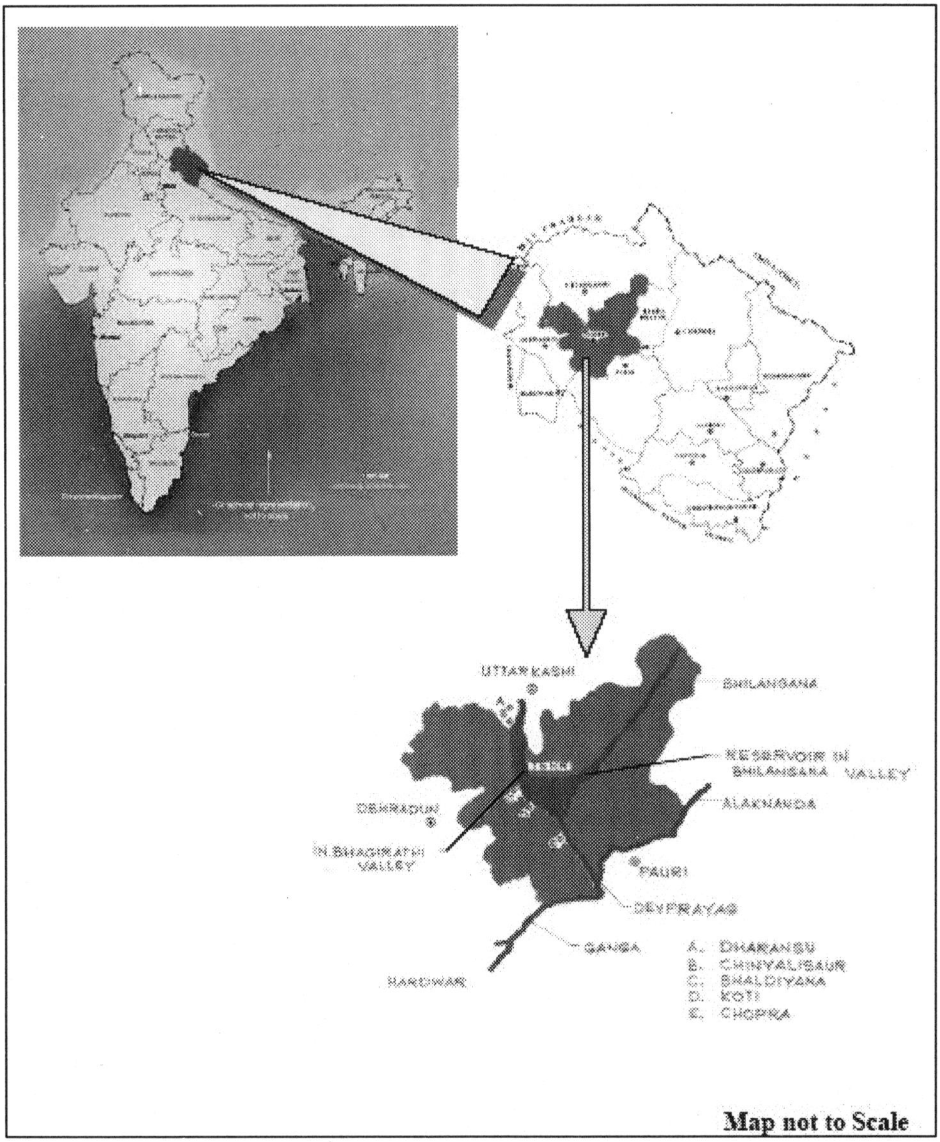

Map 1.2: Map of Tehri Garhwal District Showing Reservoir and Sampling Sites

Sampling Site I (Dharanshu)

Dharanshu is located at the distance of approx. 30 km from the district head quarters of Uttarakashi towards Tehri. It is located at the bank of river Bhagirathi. Here a perennial small tributary, Dharanshu Gadhera meets with Bhagirathi. The river bed consists of rocks, large and small stones, pebbles, and sand. It is the upstream of Bhagirathi, where water current is very high. The basic source of pollution is the domestic sewage that is discharged into the river.

Plate 1.1: Sampling Site I (Dharanshu)

Sampling Site II (Chinyalisaur)

Chinyalisaur is located at the distance of approximately 5 km from sampling site I towards Tehri. Sampling was done near Nagun bridge where Nagun Gad, a perennial small tributary meets with Bhagirathi. It is the starting point of lake (reservoir). In the months of winter water level of lake does not reach up to

Plate 1.2: Sampling Site II (Chinyalisaur)

Chinyalisaur. Due to glaciation amount of water decrease in upstream of river Bhagirathi, on the other hand electricity requirement increases, to fulfill increased electricity requirement more water is released from reservoir so water level in reservoir falls in these days. In winter lake reaches up to Bhaneti and some times up to Chham but not up to Chinyalisaur.

Sampling Site III (Bhaldiyana)

Bhaldiyana is located approximately 15 km from Chinyalisaur through reservoir and 8 km from Uttarkashi-Chamba highway. Most of the routes, roads and bridges to reach Bhaldiyana are submerged in reservoir; the only route to reach here is through a new road constructed from Chamba Uttarakashi road. Here THDC has provided boat for local people to cross the reservoir. It is middle part of reservoir. Illegal fishing from reservoir also has been observed at this sampling site during course of study.

Plate 1.3: Sampling Site III (Bhaldiyana)

Sampling Site IV (Koti)

Koti is located near dam and is approximately 10 km from sampling site III (Bhaldiyana). It is the place where the water of Bhagirathi and Bhilangana is mixed in reservoir. It is the widest part of the lake having Old Tehri, Dobata, Raja ka darbar, famous Ghantaghar, Koti, Tapsthali of Swami Ramteerth and many other historical monuments submerged in its water. Before the construction of dam there was Ganesh Prayag here in Old Tehri.

Sampling Site V (Dam Outlet Near Chopra Village)

Sampling of water coming out from dam was done near zero point bridge, which is the nearest approachable place for this purpose. It is located at the distance of

Plate 1.4: Sampling Site IV (Koti)

Plate 1.5: Sampling Site V (Dam Outlet Near Chopra Village)

approximately 5 km from sampling site IV (Koti). It is the place where maximum uncertainty of current and amount of water discharge from reservoir was found. Fluctuation in water discharge from reservoir is so high that some times river Bhagirathi is converted in to small nala. It is the place where cremation is commonly done by locals which adds pollution. The river bed consists of rocks, stones, pebbles and sand.

Chapter 2
Review of Literature

From the earliest evolution of hominid species around the lakeshores of northern Kenya to the development of main civilization on the banks of certain major rivers, human history can generally be considered to be water centered. The early important civilization developed and flourished on the bank of major rivers such as Nile, Euphrates, Tigris and Indus. Human history can, in fact, be written in terms of interactions and inter relations between man and water.

Much has been learned during the last several decades about the adverse effects of dams on the physical, chemical, and biological characteristics of rivers (Poff *et al.*, 1997 and Poff and Hart (2002). Attempts have been made by some workers in India and abroad to study the water quality of different rivers measuring various physico-chemical and biological parameters. Several workers both from India and abroad studied various aspects of water quality. Kofoid (1903) studied the plankton of River Illinois. Birge and Judey (1911) worked on physico-chemical and biological characteristics of Lake Wiscosin. Allen (1920) studied the plankton of San Joaquin River and its tributaries in California. Hora (1921) observed the fish and fisheries of Manipur with some observations on those of the Naga Hills. Richardson (1921) reported various changes in the composition of bottom and shore fauna of River Illinois and its connecting lakes. Atkin and Harris (1924) observed seasonal changes and the helioplankton of fresh water ponds. Galtsoff (1924) studied the limnology of upper Mississippi. Clarke (1924) analyzed the composition of the river and the lake waters of United States. Greenfield (1925) observed the chemical and biological parameters of Illinois River. Biswas (1928) had done the comparative study of the Chilka Lake and the Calcutta Salt lakes.

Pruthi (1933) studied the hydrology of a pond in Indian Museum compound at Calcutta, he studied the bionomics of fresh water of India. Eddy (1934) studied plankton communities in Illinois River. Garner *et al.* (1936) made a survey on chemical and biological characteristics of River Holme. Hutchinson (1937) has observed the ecology of fishes of Kashmir and Indian Tibet. Hora (1940) has observed some problems in migration of fishes because of dams. Chandler (1940) made limnological studies of Western Lake Erie. Hora (1942) had observed the effect of dams on the migration of the Hilsa fish in Indian waters. Berner (1951) made a study about the limnology of the lower Missouri River. Chacko *et al.* (1953) had given a report on survey of flora, fauna and fisheries of Pulicat Lake, Madras.

Ray (1955) studied plankton ecology of River Hoogly at Patna. Chacko and Srinivasan (1955) studied the hydrobiology of the major rivers of the Madras state of south India. Griffith (1955) analyzed the plankton yield in relation to certain physico-chemical parameters of Lake Michigen. Das and Srivastava (1956) studied the plankton of a fish tank in Lucknow. Talling (1957) investigated the phytoplankton population as a compound photosynthetic system. Oliff (1959) analyzed the hydrobiology of Tugela River. Chakarbarty *et al.* (1959) made a survey on the plankton and physico-chemical conditions of Jamuna River at Allahabad. Holden and Green (1960) studied the hydrology and plankton of River Sokoto. Singh (1960) studied the chlorophyceae of Kumaon Himalaya. Abraham (1962) had done the survey on hydrobiology and fisheries of the Cooun River. Zafer (1964) made studies on the ecology of algae in certain fishpond in Hyderabad. Michael (1964) correlated the plankton with physico-chemical parameters in three different ponds. Vyas (1968) studied the phytoplankton ecology of Pichhola Lake, Udaipur.

Talling (1965) investigated the photosynthetic activity of phytoplankton in east African lakes. Baloni and Sarkar (1965) made some observation on pollution of River Yamuna at Okhla Waterworks intake Delhi. Impacts of dam construction in the ecology of rivers have been studied by Cristofano (1965) and Bagnold (1966). Lakshminarayana (1965) studied the phytoplankton of the River Ganges in Varanasi. Pahwa and Mehrotra (1966) observed fluctuation in plankton in relation to hydrological conditions of River Ganga. Saxena *et al.* (1966) observed pollution of the Ganga near Kanpur. Imevbore (1967) studied the hydrology and plankton of Eleyele reservoir. Venkateshvarlu and Jayanti (1968) studied the hydro-biology of River Sabarmati to evaluate water quality. Verma and Shukla (1969) studied pollution in a perennial stream Khala by sugar factory effluent near Laksar, Saharanpur (U.P.). Baqai and Ishrat (1973) has done work on seasonal fluctuation of fresh water copepods of Keenjhar Lake in Sindh and its correlation with physico-chemical factors. George (1973) studied the role of Aswan high dam in changing the fisheries of south eastern Mediterranean. Fish fauna of District Tehri, Uttarkashi, Pauri and Chamoli Garhwal have been studied and reported by Badola and Pant (1973), Badola (1975) and Badola and Singh (1977 a and b).

Jhingran (1975) published a book on fish and fisheries of India. The biological and chemical study of Manchar Lake was done by Baig and Khan (1976). Aggarwal *et al.* (1976) analyzed the hydro-biology of River Ganga at Varanasi. Kant and Anand (1978) investigated the inter relationships of phytoplankton and physical parameters in Mansar Lake in Jammu. Verma (1978) worked upon pollution status of few rivers of western Uttar Pradesh. Badola (1979) studied ichthyofauna of Garhwal. Venkateshwarlu and Menon (1979) gave a list of fishes in River Ganga and its

Branches. Crayton and Sommerfield (1979) worked on composition and abundance of phytoplankton in tributaries of lower Colorado River. Sehgal (1980) studied the limnology of lake Sruinsar with reference to zooplankton and fishes. Jana *et al.* (1980) seasonally observed the effect of environmental factors on net plankton in two tropical fish ponds. Paasche (1980) studied physiological ecology of phytoplankton. Zingde *et al.* (1980) studied water quality of River Damanganga, Gujarat. Badola and Singh (1981) worked on hydro-biology of River Alaknanda. Rhee and Gotham (1981) studied the effects of environmental factors on phytoplankton growth. Badola and Singh (1981) described the fish and fisheries of River Alaknanda.

Arunachalam *et al.* (1982) has studies the seasonal variation of zooplankton of the Veli Lake. Singh and Dobriyal (1982) published the first paper on the occurrence of *Botio geto* in the River Alaknanda of Garhwal Himalaya. Tirath and Bhargava (1982) studied the water quality of Gomati River. The limnological study of Sampna reservoir at Betul District of Madhya Pradesh was done by Unni (1982). Mathur (1982) published a laboratory manual for water and waste water testing. Khan *et al.* (1983) studied the diurnal variations of physico-chemical factors and planktonic organisms in Jos plateau, West Africa. Raina *et al.* (1984) made pollution study of River Jhelum. Chattopadhaya *et al.* (1984) made a study on the pollution status of River Ganga at Kanpur. Verma *et al.* (1984) reported pollution of saprobic status of Eastern Kalindi. Reynolds (1984) studied ecology of fresh water phytoplankton. Adams (1985) studied downstream impacts of dams in Nigeria. Bhatt *et al.* (1985) studied ecology of the limnoflora in River Kosi of Kumaon. Bhowmick and Singh (1985) worked on the phytoplanktonic population in relation to physico-chemical parameters of River Ganga. Dobriyal (1985) studied ecology of limnofauna of Garhwal. Singh (1985) observed the limnology of Rihand reservoir, Uttar Pradesh with reference to the physical and chemical parameters of its water. Senger *et al.* (1985) studied distribution of algal flora in polluted and non polluted regions in Yamuna River at Agra. Sharma (1986) studied effect of physico-chemical factors on benthic fauna of Bhagirathi River at Garhwal Himalaya. Plankton community structure of Badua reservoir, Bihar was observed by Verma and Munshi (1987). Reddy and Venkateshwarlu (1987) studied River Tungbhadra near Kurnool (A.P.). Dobriyal and Singh (1988) reported ichthyofauna variation in two hill streams of Garhwal. Singh *et al.* (1988) worked upon pollution at Sangam Allahabad. Das (1989) wrote a hand book in limnology and water pollution.

Nautiyal (1990) explained the ecology of Ganga river system in the upland of Garhwal Himalayan environment. Jakher *et al.* (1990) had done the comparative study of the limnology of Sambhar and Didwana Lakes of Rajasthan. Mir and Kachroo (1990) made limnological studies on Kashmir Lakes on the dynamics of phytoplankton in Dal Lake. Singh and Ahmed (1990) worked on comparative study of the phytoplankton of River Ganga and ponds of Patna. Voros and Padisak (1991) studied phytoplankton biomass and chlorophyll-a in some shallow lakes in Central Europe. Pathak (1991) studied some aspects of ecology of River Sarju and Gomati of Kumaon Himalayas.

Pearce (1992) worked upon rivers, dams and the coming world crisis. Khanna *et al.* (1993) studied about the plankton ecology of River Ganga at Chandi Ghat,

Haridwar. Khanna (1993) reported the pollution aspect of River Ganga at Haridwar. The knowledge about the distribution of phytoplankton in reservoirs and lakes of Indian subcontinent was given by Unni (1993). Pandey *et al.* (1993) studied on physico-chemical quality of River Kosi at Purnia, in order to assess the status of water quality of the Ganges. Saxena and Chauhan (1993) observed the physico-chemical aspects of pollution of River Yamuna at Agra.

Kuuppo (1994) studied annual variation in abundance and size of heterotrophic nonflagillates on the south west coast of Finland. Snoeijs (1994) worked upon the distribution of epiphytic diatom species composition diversity and biomass of microalgal hosts in Baltic Sea. Tilzer *et al.* (1994) studied the impact of phytoplankton on spectral water transparency in the Southern ocean. Khulbe *et al.* (1995) worked on ecological study on water molds of some rivers of Kumaon Himalayas. Sarkar and Karagoz (1995) worked upon sustainable development of hydroelectric power. Dwivedi *et al.* (1995) made a limnological investigation in relation to fish production of two ponds of Patna. Chopra and Rahman (1995) made a study on self purification of physico-chemical properties of Ganga canal water at Jwalapur. Joshi (1996) studied the hydro biological profile of River Sutlej in its middle stretch in Western Himalayas. Suravit (1996) studied the relation between phytoplankton and water quality in Rachapapa Dam reservoir, Surathanee province.

Sharma and Pandey (1998) investigated the pollution status of Ramganga River at Moradabad. Gopal and Zutshi (1998) has reviewed the past, present and future scenario of limnology. Yakubo *et al.* (1998) investigated the physico-chemical conditions and planktonic organisms of the lower reaches of the Nun River, Nigeria. Wagh (1998) has studied the hydrobiological parameters of Harsul Dam. Khanna *et al.* (1999) studied correlation between abiotic factors and planktonic population in River Ganga at Rishikesh. Khanna *et al.* (1999) studied fishes and their ecology of the River Suswa at Raiwala Dehradun. Badola and Badola (1999) gave observations on destructive fishing methods and recommendations to protect fish fauna in Garhwal Himalayas. Akbay *et al.* (1999) studied the seasonal distribution of large phytoplankton in Keban Dam reservoir. Peerapornpisal *et al.* (1999) has observed the water quality and phytoplankton in the Mae Kuang Udomtara reservoir, Chiang Mai, Thailand.

Gautam *et al.* (2000) studied diurnal variation in physico- chemical characteristics of Ganga water at Triveni Ghat, Rishikesh. Adams (2000) studied social impacts of large dams and report submitted to World Commission of Dams. Khanna *et al.* (2000) observed 20 genera of phytoplanktonic population of a pond. Khanna and Singh (2000) reported seasonal fluctuation in the plankton of Suswa River at Raiwala, Dehradun. Turaga (2000) studied the impact of dam in Narmada valley. Jain and Sharma (2000) worked on water quality of Rampur reservoir of Guna district M.P. Gautam *et al.* (2001) made an attempt to explain the pollution sources along the River Ganga from Rishikesh to Haridwar. Sharma and Jain (2001) worked on diurnal variation in some aspects of limnlogy of Gopalpura Tank of district of Guna M.P. Impact of dam construction in the ecology of rivers have been studied by Biswas and Tortajado (2001) and Mc. Cully (2001).

Khanna and Singh (2002) published a book on pond fish ecology and economics. Basu and Mitra (2002) studied on pollution of River Damodar at Burdwan with special references to metals, COD and pH. Jhajharia (2002) recognised 11 genera of zooplankton and 19 genera of phytoplankton during limnobiotic studies of Sekhawat pond, Jodhpur. Kumar and Sharma (2002) studied the water quality of River Krishni (part II biological characteristics and bioindicators). Kala and Sharma (2002) worked on effect of physico-chemical factors of phytoplankton population. Lazaras and Mishra (2002) worked on physico-chemical components of Machna River water and related fish ecology at Betul M.P. Mahajan *et al.* (2002) made an attempt to present the result of physico-chemical and biological investigations carried out on River Kunda. Patel *et al.* (2002) studied the hydrobiological observations of Virla reservoir at Khargone. Kumar and Sharma (2002) studied the water quality of River Krishni (part I physico-chemical characteristics). Raghavendra and Hosmani (2002) worked on hydrobiology of Mandakally Lake. Shanthi *et al.* (2002) deals with the ecological studies of Lake Singanallur with reference to physico-chemical characteristics and water pollution. Mruthunjaya and Hosmani (2002) worked on comparative account of physico-chemical parameters, enzymatic activity and plankton population in three water bodies of Mysore city.

Monichan and Sharma (2003) studied the distribution status of fish Saun at Etawah district U.P. Khanna and Bhutiani (2003) worked on limnological status of Satikund pond at Haridwar and they observed that phytoplanktonic population was dominated by diatoms and recorded 3 genera of zooplankton. Prasad and Singh (2003) studied the composition, abundance and distribution of phytoplankton and zooplankton benthos in a tropical water body. Selva Raj *et al.* (2003) studied on hydrology and fish species diversity of Thamiraparani River in Kanyakumari District. Chatterjee *et al.* (2003) worked on physico-chemical parameters of Loco tank a reservoir in Asanol Town. Parashar *et al.* (2003) assessed the impact of collective bathing and increased organic load on quality of Ganga canal water downstream from Har ki Pauri. Prasad *et al.* (2003) studied environmental state of River Ganges in Rishikesh Haridwar and its management. Khanna and Bhutiani (2003) worked on limnological characteristics of River Ganga at Hardwar. Shivakumar *et al.* (2003) made studies on water quality of the River Ambrampalayam, Coimbatore, Tamilnadu. Subramanyam and Kumar (2003) worked on evaluation of water quality of lake by pollution index method. Arjariya (2003) studied the physico-chemical profile and plankton diversity of Ranital Lake. Jena *et al.* (2003) studied on water quality parameters around Sagar Island. Juettner *et al.* (2003) assessed the water quality characteristics of Bagmati River in Kathmandu Valley.

Ariyadej *et al.* (2004) has observed the diversity of phytoplankton and its relationships to the physico-chemical environment in Banglang Reservoir, located on the Pattani River in Southern Thjailand. Sakalauskiene *et al.* (2004) has observed the effects on water quality of Nemunas River because of the hydro power plants. Banatwala *et al.* (2004) made limnological study of Sapna reservoir of Betul (M.P.). Khanna and Bhutiani (2004) worked on fish and their ecology of River Ganga at Gohri Ghat, Garhwal. Sharma (2004) studied impact of construction of roads for transportation in the area of Tehri Dam on endangered fish *Tor tor* and *Tor putitora*.

Pathak and Mudgal (2005) studied limnology and biodiversity of fish fauna in Virla reservoir M.P. India. Khanna and Bhutiani (2005) worked upon benthic fauna and its ecology of River Ganga from Rishikesh to Haridwar. Khanna *et al.* (2005) made physico-chemical analysis of river Panvdhoi at Saharanpur U.P. Pathak and Mudgal (2006) worked upon correlation of physico-chemical parameters in Virla reservoir (MP). Jadhav *et al.* (2006) has done the work on seasonal variation of physico-chemical characteristics of Sonkhed Dam water. Khanna *et al.* (2006) studied the water quality of River Suswa near Raiwala, Uttarakhand.

Dayananda *et al.* (2007) studied limnology of Gudavi wetlands Sorab Shimoga, Karnataka. Bahuguna *et al.* (2007) studied fecundity and sex ratio in *Puntius conchonius* (pisces cyprinidae) from Garhwal Himalaya. Keshre *et al.* (2007) studied physico-chemical parameters for a reservoir at Khandawa District (M.P.). Khare (2007) made hydrobiological studies in the upstreams of River Kunda at Khargone, Madhya Pradesh. Vishnoi *et al.* (2007) studied water quality of River Ganga in respect of physico-chemical characteristics of Shyampur, District Haridwar. Zade and Sitre (2007) made study for water quality assessment of a polluted urban lake with respect to zooplankton. Dutta *et al.* (2007) studied ecology of plankton in some paddy fields, near Nagri bus stand, Kathua (J&K state). Pandey *et al.* (2007) studied seasonal variation in physico-chemical and biological properties of River Panar at Bihar. Sharma *et al.* (2007) studied biodiversity of fish fauna of Kishanpura Lake Indore (M.P.). Verma, (2007) studied pond and riverine algae of Khargone Madhya Pradesh. Vishnoi *et al.* (2007) studied abiotic status of Song River and its relation to zooplankton and phytoplankton at Nepali farm Dehradun. Khanna *et al.* (2007) studied biotic and abiotic factors of Song River at Dehradun.

Chapter 3
Materials and Methods

Any change in the quality of water has been determined by analysis and ensuring the qualitative and quantitative change in the parameters (physico-chemical and biological parameters) of the water. In this thesis we used standard methods to analyze the different parameters studied.

The water quality of River Bhagirathi and Tehri Dam reservoir from Dharanshu to Dam outlet near Chopra village was monitored at five sampling sites on monthly basis, throughout the year from July 2006 to June 2008. The samples were collected between 7AM to 11AM. The samples were taken in Borosil glass bottles of 300 ml as well as in 2 liter capacity of plastic cans from each sampling site.

The collection of water samples and analysis was done using standard methods with the help of Welch (1948), Trivedi and Goel (1984), APHA-AWWA-WPCF (1998) and Khanna and Bhutiani (2004) for physico-chemical parameters. Biological parameters were analysed with the help of Jhingran (1975), Badola (1979), Khanna (1993) and APHA-AWWA-WPCF (1998).

Sampling Programme and Analysis

Bearing in mind the river characteristics and the water quality requirements, the following routine sampling and analysis programmes were devised:

- ☆ *Number of sites*: Five
- ☆ *Collection of samples*: Monthly
- ☆ *Time of sample collection*: Between 7:00 A.M. to 11:00 A.M.
- ☆ *On site analysis*: Temperature, pH and velocity
- ☆ *Analysis within 24 hours of sampling*: All physico-chemical and biological parameters taken under study

Chemicals

The chemicals used in the present investigation were of analytical grade and were supplied by E. Merck India, S.D. fine chemicals and B.D.H., India.

Instruments

Following instruments were used during the study period.

Sl.No.	Name of Instruments	Make/Manufactured by
1.	BOD incubator	Tasco
2.	Electronic balance	Roy Electronic
3.	Jackson turbidity meter	-
4.	COD Assembly	-
5.	Hot plate	Sunvic
6.	Hot air oven	Modern Scientifics
7.	Microscope	-
8.	Plankton collection net	-
9.	Sedgwick rafter counting cell	-
10.	Mantle heater	A.J. Enterprises
11.	pH meter	Tasco
12.	Refrigerator	Godrej
13.	Thermometer (°C)	-
14.	Distil water plant	Tanco.
15.	Glass fiber filter	-
16.	Dessicator	-
17.	Autoclave	-

Glasswares

All the glasswares used were of corning grade, manufactured by Borosil India Ltd.

Sampling Methods

(A) Cleaning and Sterilization of Glassware

All the glasswares and jerricans were cleaned or washed with soda and chromic acid and rinsed with double distilled water 2-3 times.

(*i*) Sample Bottles

300 ml capacity BOD bottles made of borosil were used for sample collection. They were washed with washing soda and chromic acid and rinsed with tap water

followed by distilled water; the neck and stopper were wrapped with butter paper with the help of rubber band. The bottles were then sterilized in an autoclave at 15-lbs./inch^2 pressure (121°C) for 15-25 minutes.

(*ii*) Pipettes

Pipettes of different volume size were washed and fitted with cotton plug at the upper end; these were then wrapped in butter paper and sterilized in an autoclave at 15-lbs./inch^2 pressure at 121°C for 15-25 minutes.

(*iii*) Test Tubes

Borosil test tubes were washed and then plugged with non-absorbent cotton wool, these were arranged in test tube racks and sterilization was done in an autoclave at 15-lbs./inch^2 pressure at 121°C for 15-25 minutes.

(*iv*) Other Glasswares

Such as conical flasks, beakers, measuring cylinders etc. were washed carefully and then sterilized in hot air oven at 160-180 °C for 2-3 hours.

(B) Collection of Samples

(*i*) For Determination of the Parameters

For the estimation of total solids, TDS, TSS, turbidity, COD, alkalinity, pH (hydrogen ion concentration), total hardness, and chlorides samples were taken from 2 feet below the surface water by keeping and opening jerricans below the surface water. Caps of cans were removed after dipping the can and also closed in the water after filling up of can.

(*ii*) For Dissolved Oxygen/Biochemical Oxygen Demand

Samples were collected in sterilized 300 ml capacity BOD bottles. The bottles were filled completely with sample water up to the rim and stopper was placed inside water only avoiding any kind of air bubble inside it. DO was immediately fixed by adding 2 ml each of manganous sulphate ($MnSO_4$) and alkaline potassium iodide at sampling spot, while the sample for BOD was incubated for 5 days in BOD incubator at 20 °C.

(*iii*) Sampling for Bological Analysis

For biological analysis of water, samples were collected with the help of plankton net attached with a collection tube at the base of net. For this a known amount of water was filtered through the plankton collection net and sample was collected inside collection tube. The sample was then transferred in sterilized tube of 50 ml capacity and preserved by adding one drop of lugol's solution or formaldehyde solution.

(C) Reagent Preparation

(*i*) Potassium Dichromate Solution 0.025N

Dissolve 12.259 gm of dried A.R. grade $K_2Cr_2O_7$ in distilled water to make one liter of solution. Dilute it 10 times to make 0.025N $K_2Cr_2O_7$ solution.

***(ii)* Ferrous Ammonium Sulphate (0.01N)**

Dissolve 39.2 gm of Fe $(NH_4)_2(SO_4)_2.6H_2O$ in distilled water, adding 20 ml of concentrated H_2SO_4 to make 1 liter of solution. Dilute it 10 times to make 0.01 N Fe $(NH_4)_2(SO_4)_2.6H_2O$ solution.

***(iii)* Ferroin Indicator**

Dissolve 1.485 gm of 1, 10- Phenonthroline and 0.695 gm of ferrous sulphate $(FeSO_4.7H_2O)$ in distilled water to make 100ml of solution.

***(iv)* Sulfuric Acid**

H_2SO_4 Concentrated (Sp. Gr.1.84)

***(v)* Mercuric Sulphate**

$HgSO_4$ Solid

***(vi)* Sodium Thiosulphate (0.025N)**

Dissolve 24.82 gm of $Na_2S_2O_3.5H_2O$ in boiled distilled water to make final volume to one liter. A pellet of NaOH or 0.4 gm borax was added as stabilizer. This is 0.1 stock N solution. Dilute 0.1N stock solution 4 times with boiled distilled water to prepare 0.025N solution. Keep it in brown glass stopper bottle.

***(vii)* Alkali Iodide Azide Solution**

(*a*) Dissolve 500 gm of NaOH or 700 gm of KOH and 150 gm of KI in distilled water to makes one liter of solution.

(*b*) Dissolve 10 gm of NaN_2 in 40 ml of distilled water mix both (a) and (b) solution.

***(viii)* Manganous Sulphate Solution**

Dissolve 100 gm of $MnSO_4.4H_2O$ in 200 ml of boiled distilled water and filter.

***(ix)* Starch Solution**

Dissolve 1 gm of starch in 100 ml of warm distilled water and add a few drops of formaldehyde solution.

***(x)* Sodium Hydroxide (0.05N)**

Dissolve 40 gm of NaOH in one liter boiled CO_2 free distilled water to prepare 1N NaOH stock solution. Dilute 50 ml of 1N NaOH stock solution to 1 liter distilled water. Standardized the solution using H_2SO_4.

***(xi)* Phenolphthalein Indicator**

Dissolve 0.5 gm of phenolphthalein in 50ml of 95 per cent ethanol and add 50 ml of distilled water. Add slowly 0.05N CO_2 free NaOH until solution turns faintly pink.

***(xii)* Methyl Orange Indicator**

Dissolve 0.5 gm methyl orange in 100 ml of distilled water.

***(xiii)* Hydrochloric Acid (0.1N)**

Dilute 12N concentrate HCl (sp.gr.1.18) to 12 times (8.34–100ml) to prepare 1.0N HCl. It is further diluted ten times to make 0.1N HCl (100–1000ml).

(xiv) EDTA Solution (0.01M)

Dissolve 3.723 gm of disodium salt of EDTA in distilled water to prepare one liter of solution.

(xv) Buffer Solution

(a) Dissolve 16.9 gm of ammonium chloride (NH_4Cl) in 143 ml of concentrated ammonium hydroxide (NH_4OH).

(*b*) Dissolve 1.179 gm of disodium EDTA and 0.78 gm of $MgSO_4.7H_2O$ in 50 ml distilled water.

Mix (*a*) and (*b*) solution and make final volume to 250 ml using distilled water.

(xvi) Eriochrome Black-T Indicator

Mix 0.5 gm of EBT with 100 gm of NaCl (A.R) and grind.

(xvii) Sodium Hydroxide Solution (1N)

Dissolve 40 gm of NaOH in one liter distilled water.

(xviii) Silver Nitrate, (0.02N)

Dissolve 3.40 gm of dried $AgNO_3$ (A.R) in distilled water to make one liter of solution and kept in dark bottle.

(xix) Potassium Chromate (5 per cent)

Dissolve 5 gm of K_2CrO_4 in 100 ml of distilled water.

(xx) Ethyl alcohol, 95 per cent

(xxi) Lugol's solution

Dissolve 10gm of KI and 5 gm of iodine crystals in 20 ml of distilled water and then add 50 ml distilled water in which 5 gm anhydrous sodium acetate has been dissolved.

(D) Analytical Methods

Samples were analyzed for the following parameters:

(i) Physico-chemical Parameters

(*a*) Water temperature

(*b*) TS

(*c*) TDS

(*d*) TSS

(*e*) Velocity

(*f*) Turbidity

(*g*) DO

(*h*) COD

(*i*) BOD

(*j*) Alkalinity

(*k*) pH

(*l*) Total hardness
(*m*) Chlorides

(*ii*) Biological Parameters

(*a*) Plankton
1. Zooplankton
2. Phytoplankton
3. Planktonic biomass

(*b*) Fish diversity

Methodology

The collection of water samples and analysis was done with the help of Welch (1948), Trivedy and Goel (1984), APHA-AWWA-WPCF (1998) and Khanna and Bhutiani (2005) for physico-chemical parameters. Biological parameters were analysed with the help of Jhingran (1975), Badola (1979), Khanna (1993) and APHA-AWWA-WPCF (1998).

(*a*) Temperature

Centigrade calibrated thermometers were used to record the temperature of surface water for measuring water temperature. Thermometer was dipped into water up to a desirable depth for 5-7 minutes and reading was taken.

(*b*)Total Solids (TS)

Solids refer to matter suspended or dissolved in water or wastewater. Solids may affect water or effluent quality adversely in a number of ways. Total Solid is a term applied to the material residue left in a vessel after evaporation of a sample and its subsequent drying in an oven at a defined temperature.

Procedure

A 50ml washed and dried beaker was taken and weighted. 20ml of unfiltered sample was taken in that beaker and sample was evaporated over hot plate/water bath and then dried in an oven at 105 ±1°C for one hour. The final weight of the beaker was taken after drying of sample.

Calculation

$$\text{Total solids (mg/l)} = \frac{(A - B) \times 1000 \times 1000}{V}$$

where,

A = Final weight of beaker in gm
B = Initial weight of beaker in gm
V = Volume of sample taken in ml

(c) Total Dissolved Solids (TDS)

Total solids in any sample can be represented by dissolved organic and inorganic matter and may be estimated by evaporating the filtered sample through standard filter and weighing the residue left.

Calculation

$$\text{Total dissolved solids (mg/l)} = \frac{(\text{A-B}) \times 1000 \times 1000}{\text{Volume of sample}}$$

where,

A = Final weight of dish with sample

B = Previous weight of dish without sample

(d) Total Suspended Solids

Total suspended solids in sample are the retained material on Whattman no. 42 filter paper after filtration of well mixed sample. TSS was determined by taking difference between the total solids and total dissolved solids.

Calculation

$$\text{TSS} = \text{TS} - \text{TDS}$$

(e) Velocity

The speed of water current was measured by the surface float method. The rate of flow of the uppermost stratum of the water in the river was measured by releasing the pre-weighted light piece of wood in different portions of the river. The time taken by the wood piece to cover a known distance was calculated by the given formula:

Calculation

$$V = \frac{d}{t}$$

where,

V = Velocity (meter/second)

d = Distance (meter)

t = Time (second)

(f) Turbidity

Clarity of water is important in producing product destined for human consumption and in many manufacturing operations. The clarity of a natural body of water is an important determinant of its condition and productivity.

Turbidity in water is caused by suspended and colloidal matter such as clay, silts, finely divided organic and inorganic matter, plankton and other microscopic organisms. Turbidity is an expression of the optical property that causes light to be scattered and absorbed rather than transmitted with no change in direction or flux level through the sample. The turbidity of water was measured with the help of Jackson's Candle turbidity meter. It consists of three parts (a) Candle (b) Holder (c) Calibrated glass tube

Procedure

Collected samples were shaked thoroughly in jerricans and were slowly transferred into calibrated glass tube of Jackson's candle turbidity meter until the flame of the candle disappears and the light path was noted in centimeters (measured from inside the bottom of the glass tube). The reading (cm) was then converted in to Jackson turbidity unit (JTU) using conversion table.

(g) Dissolved Oxygen (DO)

Dissolved oxygen (DO) level in natural water and wastewater depend on the physical, chemical and biochemical activities in the water body. The analysis for DO is a key test in water pollution and waste treatment process control.

DO is the measure of oxygen concentration present in a given water sample. The concentration of oxygen reflects whether the process undergoing is aerobic or anaerobic.

Principle

The manganous sulphate reacts with the alkali (KOH, NaOH) to form a white precipitate of manganous hydroxide, which in the presence of oxygen, gets oxidized to a brown colour compound. In the strong acid medium manganic ions are reduced by iodide ions which get converted into iodine equivalent to the original concentration of oxygen in the sample. The iodine can be titrated against thiosulphate using starch as an indicator.

Certain oxidizing and reducing materials may effectively interfere with the determination of oxygen by converting iodide ions to iodine or vice-versa. The azide modification removes the interference substance specially nitrite, which is destroyed by sodium azide (NaN_2). That's why azide modification method is used to determine dissolved oxygen instead of Winkler's iodometric method.

Procedure

Fill sample water in 300 ml capacity BOD bottle up to the rim and stopper is placed inside water body avoiding any kind of air bubble inside it. Add 2 ml each of $MnSO_4$ and alkaline iodide azide solution using separate pipettes. A precipitate was appeared. The bottle was kept for some time after mixing thoroughly. At this stage the DO was fixed 2 ml of Concentrated H_2SO_4 was added to dissolve the precipitate by shaking it. The part or whole of the content was titrated against sodium thio-sulphate using starch as an indicator. At end point blue colour changes to colourless.

Calculation

When whole contents have been titrated

$$\text{Dissolved oxygen (mg/l)} = \frac{(\text{ml} \times \text{N}) \text{ of titrant} \times 1000 \times 8}{V_1 - V}$$

When only a part of content has been titrated

$$\text{Dissolved oxygen (mg/l)} = \frac{(\text{ml} \times \text{N}) \text{ of titrant} \times 1000 \times 8}{V_2 (V_1 - V) / V_1}$$

where,

V_1 = Volume of sample bottle after placing stopper

V_2 = Volume of part of the contents titrated

V = Volume of $MnSO_4$ and KI added.

(*h*) Chemical Oxygen Demand

Chemical oxygen demand (COD) is defined as the amount of specified oxidant that reacts with the sample under controlled conditions. COD is widely used to characterize the organic strength of wastewater and pollution of natural water.

It is the measure of oxygen consumed during the oxidation of the oxidizable organic matter by a strong oxidizing agent.

Principle

The sample is refluxed with $K_2Cr_2O_7$ and H_2SO_4 in presence of mercuric sulphate, to neutralize the effect of chlorides and silver sulphate, which acts as a catalyst. Excess potassium dichromate ($K_2Cr_2O_7$) is titrated against ferrous ammonium sulphate using ferroin as an indicator. The amount of $K_2Cr_2O_7$ used is proportional to the oxidizable organic matter present in the sample.

Procedure

20 ml of sample was taken in a 250 ml capacity COD flask; slowly add 10 ml of $K_2Cr_2O_7$ and a pinch of $HgSO_4$ and Ag_2SO_4 were added. Add slowly 30 ml of concentrated H_2SO_4 and reflux condenser was connected to the COD flask, reflux for 2 hours on hot plate. Now remove the flask, cooled and make the final volume to 140 ml in flask using distilled water, take suitable amount of sample in conical flask add 2-3 drops of ferroin indicator and titrate against 0.01N ferrous ammonium sulphate.

Run a blank with distilled water using same quantity of chemicals.

Calculation

$$\text{COD (mg/l)} = \frac{(\text{B-A}) \times \text{ N of Ferrous ammonium sulphate} \times 1000 \times 8}{\text{ml of sample}}$$

where,

A = ml of titrant with sample

B = ml of titrant with blank

(*i*) Biochemical Oxygen Demand (BOD)

The biochemical oxygen demand (BOD) determination is an empirical test in which standardized laboratory procedures are used to determine the relative oxygen requirement of wastewater effluents, and polluted waters.

The test measure the molecular oxygen utilized during a specified incubation period for the biochemical degradation of organic matter (carbonaceous demand) and the oxygen used to oxidize inorganic material such as sulfides and ferrous ions. In other words biological oxygen demand, (BOD) represents the amount of oxygen required for the microbial decomposition of the organic matter in the water. The BOD procedure, which is used extensively in monitoring water quality and biodegradation of waste material, is designed to determine how much oxygen microorganisms consume during oxidation of the organic matter in the sample.

Principle

The principle of the method involves, measuring the difference of the oxygen concentration between the samples after incubating it for 5 days at 20°C.

Procedure

Samples were collected in two BOD bottles of 300 ml capacity from each site. One set of the bottle were kept in BOD incubator at 20°C for 5 days and the DO constant was determined in another set immediately. After the compilation of 5 days incubation, the DO of first set was determined and BOD was calculated by means of given formula:

Calculation

$$\text{BOD (mg/l)} = D_0 - D_5$$

where,

D_0 = Initial DO in the sample

D_5 = DO after five days incubation

(*j*) Alkalinity

Alkalinity of water is its acid- neutralizing capacity. It is the sum of all the titrable bases.

Principle

Total alkalinity, carbonates and bicarbonates can be estimated by titrating the sample with a strong acid (HCl or H_2SO_4), first to pH 8.3 using phenolphthalein as an indicator and then further to pH between 4.2 and 5.4 with methyl orange.

Procedure

Take100 ml of sample in a conical flask and add 2 drops of phenolphthalein indicator, if solution remain colourless then phenolphthalein alkalinity (PA) =0 *i.e.*

absent and total alkalinity was determined, if solution turns pink, titrate it against 0.1 N HCl until color disappears this is phenolphthalein alkalinity (PA). Add 2-3 drops of methyl orange indicator to the same sample and titrate against 0.1N HCl until yellow color changes to pink at the end point this is total alkalinity (TA).

Calculation

$$\text{Alkalinity (mg/l)} = \frac{(\text{ml} \times \text{N}) \text{ of HCl} \times 1000 \times 50}{\text{Volume of Sample}}$$

(*k*) pH

Measurement of pH is one of the most important and frequently used tests in water chemistry. pH is the hydrogen ion activity or concentration in the given water sample. pH equal to negatively $\log_{10}$ of hydrogen ion concentration

$$\text{pH} = -\log_{10}(\text{H}^+)$$

pH as defined by Sorenson (1909) is $-\log_{10}(\text{H}^+)$; it is the "intensity" factor of acidity. pH of water sample was measured by the use of pH strips, but the accurate method is the electrometric method. The sample was analyzed with the help of electronic digital pH meter consisting of a reference and indicator electrode.

Procedure

Calibrate the electrode system against standard buffer solution of known pH. Rinse the electrode using distilled water, now dip the electrode in sample water and note the reading on digital screen.

(*l*) Total Hardness

Water hardness was understood to be a measure of the capacity of water to precipitate soap. Soap is precipitated chiefly by the calcium and magnesium ions present.

Principle

If a small amount of dye such as Eriochrome black T [EBT] is added to an aqueous solution containing calcium and magnesium ions at a pH of 10.0 ±0.1, the solution becomes wine red. If EDTA is added as a titrant, the calcium and magnesium will be complexed and solution turn from wine red to blue, marking the end point of titration.

Procedure

Take 100 ml of sample in a conical flask and add a pinch of EBT indicator and one ml of buffer solution, the solution turns wine red, titrate it against EDTA until wine red color change to blue.

Calculation

$$\text{Hardness (mg/l)} = \frac{\text{ml of EDTA} \times 1000}{\text{ml of sample}}$$

(m) Chloride

The chloride concentration is higher in wastewater than in raw water because sodium chloride (NaCl) is a common article of diet and passes unchanged through the digestive system.

Principle

Silver nitrate reacts with chloride to form very slightly soluble white precipitate of AgCl. At the end point when all the chlorides get precipitated, free silver ions react with chromate to form silver chromate of reddish brown colour.

Procedure

Take 50 ml of sample in a conical flask and add 2 ml of potassium chromate to it, titrate the solution against silver nitrate until persistent red tinge appears.

Calculation

$$\text{Chloride (mg/l)} = \frac{(\text{ml} \times \text{N}) \text{ of } AgNO_3 \times 1000 \times 35.5}{\text{ml of sample}}$$

Biological Examination

Plankton

The term 'plankton' refers to those microscopic aquatic forms having little or no resistance to currents and living free floating and suspended in natural water. The phytoplankton (microscopic algae) occurs as unicellular, colonial, or filamentous forms. The zooplankton in fresh water comprises principally protozoan, rotifers, cladocerans and copepods.

Collection of Plankton

The plankton of river water was collected by fine plankton collection net. A known amount of water was filtered through plankton collection net and concentrated in desired quantity inside collection tube. The sample was preserved in Lugol's solution.

Procedure

1 ml of concentrate was taken and placed in Sedgwick rafter counting cell a glass slip was placed over it avoiding any kind of bubble. Calculation of the organisms was done by applying the following formula:

Calculation

$$\text{Plankton (unit/l)} = \frac{(\text{a} \times 1000)\ \text{C}}{\text{L}}$$

where,

a = Average number of plankton in one small counting chamber of Sedgwick rafter counting cell

C = ml. of plankton concentrate

L = Volume of original water filtered in liter.

Planktonic Biomass

In our study planktonic biomass estimates are major concerns in biological parameters to determine the impact of dam on aquatic ecology of Bhagirathi. We have used gravimetric method for the calculation of planktonic biomass.

Principle

Biomass is a quantitative estimate of the total mass of living organism within a given area or volume. It may include the mass of a population (species biomass) or a community biomass but gave no information on community structure or function. The most accurate methods for the estimation of biomass are dry weight, ash free dry weight or volume of living organism (APHA, 1998).

Procedure

Determine dry weight by placing 100 mg wet concentrated sample in a clean, ignited, and tared porcelain crucible and dry at 105 °C for 24 hour. Alternatively, filter a known volume of sample through a pre rinsed dried and pre weighted glass-fiber filter. Cool sample in a dessicator and weigh. Obtain ash weight by igniting the dried sample at 500 °C for 1 h. Cool, rewet ash with distilled water, and bring to constant weight at 105 °C. The ash is rewetted to restore water of hydration of clays and other minerals; this may amount to as much as 10 per cent of weight lost during incineration. The ash-free dry weight is the difference between the dry weight and the weight of the ash residue after ashing. The ash-free dry weight is preferred to dry weight to compare mixed assemblages. The ash content may constitute 50 per cent or more of the dry weight in phytoplankton having inorganic structures, such as the diatoms. In other forms the ash content is only about 5 per cent of dry weight.

Fish Diversity

Fishes were collected by using jal, fatiyala (triangular net) and fine clothes. Fishes were preserved in 5 per cent formaline and identified with the help of Badola (1979) and Jhingran (1975).

Statistical Methods

(*i*) Standard Deviation

Standard deviation was calculated to find out the fluctuation from the average value by the following formula:

$$S=\sqrt{\frac{\Sigma x^2}{N}} \text{ or } S=\sqrt{\frac{\Sigma x^2}{N-1}}$$

(*ii*) Coefficient of Correlation (r)

Coefficient of correlation (r) was calculated with the help of an electronic calculator to find out the relationship (+ve or –ve) between different ecological parameters. It is calculated by the following formula:

$$r_{x,y} = \frac{\Sigma x \times y}{\sqrt{\Sigma x^2 \times \Sigma y^2}}$$

Chapter 4
Observations

The importance of water resources has increased with industrialization and growth of population, as these are needed for sustaining crop productivity, meeting the ever increasing needs of the domestic, industrial and power sector. (Rautela *et al.*, 2002). River flow regime is coherent with the surrounding environment, anthropogenic factors, first of all the hydropower development may destabilize a complex river environment system and cause unpredictable consequences (Zdankus *et al.*, 2008). Dams are therefore being constructed for harnessing water resources (Rautela *et al.*, 2002).

The quality of natural water bodies impacts those using or living within those water bodies. It is therefore necessary to make use of technical observation of the various aspects of water quality. In particular, reference to the different parameters of physical, chemical and biological quality needs to be made.

4.1. Physico-chemical Parameters

The physico-chemical parameters were studied in upstream of the river Bhagirathi, Tehri dam reservoir and down stream of reservoir. Parameters analyzed during the study were water temperature (°C), total solids (mg/l), total dissolved solids (mg/l), total suspended solids (mg/l), turbidity (JTU), velocity (m/s), dissolved oxygen (mg/l), chemical oxygen demand (mg/l), biochemical oxygen demand (mg/l), alkalinity (mg/l), pH, total hardness (mg/l) and chloride (mg/l).

(The time taken for doing whole study is from July 2006 to June 2008, which are presented in the Tables 4.1 to 4.20 and in Figures 4.1 to 4.60).

Sampling Site I (Dharanshu)

The results of physico-chemical parameters for the sampling site I are presented in the Tables 4.1 to 4.4 and Figures 4.1 to 4.12.

Water Temperature (°C)

The water temperature at this sampling site was found minimum 6.49°C ±0.29 in the month of January in 2007-2008 and maximum 18.80°C ±0.25 in the month of September of 2006-2007. The average value in the year 2006-2007 was observed 12.94°C ±4.20 and in the year 2007-2008 it was recorded 12.72°C ±4.11.

Total Solids (mg/l)

The total solids at this sampling site were recorded minimum 157.29 mg/l ±1.61 in the month of April in 2006-2007 and maximum 3069.91 mg/l ±2.43 in the month of August in 2006-2007. The average value in the year 2006-2007 was found 1011.72 mg/l ±868.11 and in the year 2007-2008 it was found 985.03 mg/l ±832.01.

Total Dissolved Solids (mg/l)

The total dissolved solids at this sampling site were found minimum 71.45 mg/l ±1.33 in the month of April in 2006-2007 and maximum 429.97 mg/l ± 1.15 in the month of August in 2007-2008. The average value in the year 2006-2007 was recorded 161.73 mg/l ±112.40 and in the year 2007-2008 it was observed 155.57 mg/l ±112.55.

Total Suspended Solids (mg/l)

The total suspended solids at this sampling site were observed minimum 85.84 mg/l ±2.11 in the month of April in 2006-2007 and maximum 2641.24 mg/l ±2.07 in the month of August in 2006-2007. The average value in the year 2006-2007 was observed 849.99 mg/l ±758.56 and in the year 2007-2008 it was observed 826.46 mg/l ±724.62.

Velocity (m/s)

The velocity at this sampling site was found minimum 0.49 m/s ±0.03 in the month of March in 2007-2008 and maximum 2.30 m/s ±0.12 in the month of August in 2007-2008. The average value in the year 2006-2007 was observed 1.16 m/s ±0.56 and in the year 2007-2008 it was observed 1.25 m/s ±0.65.

Turbidity (JTU)

The turbidity at this sampling site was recorded minimum 0.00 JTU ±0.00 in the month of March of 2006-2007 and in the April month of both the years 2006-2007 and 2007-2008. The maximum turbidity was found 362.70 JTU ±1.55 in the month of August in 2006-2007. The average value in the year 2006-2007 was observed 65.83 JTU ±110.06 and in the year 2007-2008 it was noted 58.42 JTU ±91.18.

Dissolved Oxygen (mg/l)

The dissolved oxygen at this sampling site was observed minimum 6.90 mg/l ±0.21 in the month of August in 2006-2007 and maximum 11.90 mg/l ±0.17 in the month of January in 2007-2008. The average value in the year 2006-2007 was observed 9.68 mg/l ±1.45 and in the year 2007-2008 it was recorded 10.13 mg/l ±1.17.

Chemical Oxygen Demand (mg/l)

The COD at this sampling site was recorded minimum 2.16 mg/l ±0.16 in the month of January in 2007-2008 and maximum 9.20 mg/l ±0.16 in the month of August in 2006-2007. The average value in the year 2006-2007 was observed 5.30 mg/l ±2.07 and in the year 2007-2008 it was noted 4.60 mg/l ±1.82.

Table 4.1: Monthly Variation in Physical Parameters of River Bhagirathi at Sampling Station I Dharanshu (Upstream running water) during 2006–2007

Month	*Temp. (°C)*	*Total Solids (mg/l)*	*TDS (mg/l)*	*TSS (mg/l)*	*Velocity (m/s)*	*Turbidity (JTU)*
July	15.50±0.38	1946.00±2.48	320.85±1.24	1625.15±2.22	1.95±0.15	180.50±1.15
August	17.50±0.24	3069.91±2.43	428.67±1.52	2641.24±2.07	2.10±0.22	362.70±1.55
September	18.80±0.25	1735.36±1.49	223.51±1.60	1511.85±2.45	1.90±0.15	135.30±1.08
October	16.50±0.57	1120.41±1.89	192.16±1.52	928.25±2.34	1.50±0.21	35.10±1.46
November	11.40±0.76	818.24±2.18	104.32±1.76	713.92±2.54	1.10±0.11	19.50±1.87
December	10.10±0.84	636.17±2.45	97.98±1.04	538.19±2.16	0.85±0.10	4.20±0.85
January	7.46±0.46	361.11±1.49	86.57±1.87	274.54±2.48	0.87±0.18	3.20±0.22
February	6.50±0.48	257.62±1.37	79.38±1.67	178.24±2.47	0.67±0.15	2.30±0.12
March	8.30±0.55	169.12±1.75	74.27±1.58	94.85±2.34	0.53±0.04	0.00±0.00
April	11.90±0.29	157.29±1.61	71.45±1.33	85.84±2.11	0.57±0.05	0.00±0.00
May	16.10±0.36	827.57±2.11	122.29±1.05	705.28±1.87	0.80±0.11	18.40±1.24
June	15.20±0.46	1041.81±1.88	139.32±1.01	902.49±1.74	1.10±0.24	28.80±1.57
Average± SD	12.94±4.20	1011.72±868.11	161.73±112.40	849.99±758.56	1.16±0.56	65.83±110.06

Table 4.2: Monthly Variation in Physical Parameters of River Bhagirathi at Sampling Station I Dharanshu (Upstream running water) during 2007–2008

Month	*Temp. (°C)*	*Total Solids (mg/l)*	*TDS (mg/l)*	*TSS (mg/l)*	*Velocity (m/s)*	*Turbidity (JTU)*
July	15.01±0.25	1937.10±2.45	321.85±1.22	1615.25±2.04	1.93±0.10	182.00±1.18
August	17.25±0.37	2971.11±2.11	429.97±1.15	2541.14±1.99	2.30±0.12	284.00±1.15
September	18.15±0.54	1532.96±2.15	220.51±1.27	1312.45±1.88	2.20±0.11	121.00±1.51
October	16.22±0.38	1153.21±1.43	105.12±1.16	1048.09±1.76	1.70±0.10	57.10±1.55
November	11.11±0.18	808.54±1.58	104.82±1.54	703.72±2.08	1.40±0.14	9.80±0.89
December	9.89±0.46	637.47±1.04	97.88±1.08	539.59±2.17	0.87±0.07	4.90±0.78
January	6.49±0.29	330.71±1.03	90.57±1.17	240.14±1.76	0.85±0.08	4.30±1.46
February	7.15±0.45	275.28±1.11	84.34±1.09	190.94±1.49	0.67±0.05	3.30±1.01
March	8.30±0.25	176.71±0.24	76.22±1.07	100.49±1.57	0.49±0.03	3.40±1.05
April	11.71±0.16	167.09±0.52	72.45±1.11	94.64±2.26	0.53±0.12	0.00±0.00
May	16.10±0.22	816.97±0.15	125.79±1.29	691.18±2.13	0.81±0.01	9.40±1.58
June	15.20±0.40	1013.23±2.15	137.34±1.57	875.89±1.55	1.20±0.07	21.8±1.42
Average± SD	12.72±4.11	985.03±832.01	155.57±11.55	826.46±724.62	1.25±0.65	58.42± 91.18

Table 4.3: Monthly Variation in Chemical Parameters of River Bhagirathi at Sampling Site I (Dharanshu) during 2006–2007

Month	*DO (mg/l)*	*COD (mg/l)*	*BOD (mg/l)*	*Alkalinity (mg/l)*	*pH*	*Total Hardness (mg/l)*	*Chloride (mg/l)*
July	7.20±0.21	8.91±0.15	3.15±0.12	38.40±1.11	7.31±0.18	40.70±1.15	3.11±0.11
August	6.90±0.21	9.20±0.16	3.45±0.06	35.20±1.09	7.12±0.22	31.20±1.13	3.98±0.13
September	9.60±0.18	5.00±0.21	2.14±0.10	28.50±1.21	7.25±0.15	46.30±1.20	1.74±0.12
October	9.30±0.23	6.19±0.22	2.20±0.15	42.20±1.20	7.43±0.19	48.50±1.15	1.65±0.08
November	9.40±0.24	5.42±0.18	2.10±0.08	69.30±1.15	8.19±0.18	56.50±1.17	1.40±0.07
December	9.80±0.21	4.75±0.17	1.90±0.16	87.50±1.16	7.37±0.16	62.10±1.18	1.42±0.16
January	11.20±0.24	3.07±0.09	1.20±0.21	92.50±1.19	7.30±0.21	65.30±0.89	2.00±0.14
February	11.00±0.25	3.21±0.11	1.30±0.09	65.60±1.20	7.93±0.20	61.20±0.97	2.15±0.17
March	10.80±0.22	3.85±0.12	1.35±0.07	62.50±1.13	7.73±0.17	56.20±0.76	2.27±0.13
April	9.10±0.19	5.74±0.21	2.20±0.13	50.20±1.11	7.64±0.14	52.40±0.88	2.57±0.11
May	11.20±0.20	2.83±0.13	1.15±0.07	41.60±1.10	7.22±0.15	49.30±1.08	2.13±0.05
June	10.70±0.22	5.47±0.10	1.80±0.19	39.50±1.12	7.15±0.13	47.10±1.07	2.83±0.08
Average±SD	9.68±1.45	5.30±2.07	2.00±0.73	54.42±20.91	7.47±0.33	51.45±9.69	2.27±0.76

Table 4.4: Monthly Variation in Chemical Parameters of River Bhagirathi at Sampling Site I (Dharanshu) during 2007–2008

Month	*DO (mg/l)*	*COD (mg/l)*	*BOD (mg/l)*	*Alkalinity (mg/l)*	*pH*	*Total Hardness (mg/l)*	*Chloride (mg/l)*
July	8.10±0.14	8.12±0.18	2.80±0.11	31.40±1.23	7.10±0.08	45.00±1.26	3.18±0.25
August	9.20±0.23	5.75±0.09	2.15±0.09	27.50±1.05	7.24±0.11	41.20±1.45	4.22±0.21
September	8.70±0.08	6.18±0.14	2.40±0.15	49.20±1.65	7.35±0.24	32.20±1.23	2.01±0.20
October	9.60±0.18	5.27±0.11	1.90±0.13	45.20±1.23	7.17±0.09	49.30±0.96	1.80±0.15
November	10.10±0.16	4.47±0.06	1.70±0.04	41.20±1.56	7.58±0.16	55.30±1.17	1.60±0.11
December	11.30±0.12	2.50±0.10	1.00±0.08	69.30±1.98	7.76±0.13	58.20±0.84	2.13±0.08
January	11.90±0.17	2.16±0.16	0.80±0.09	92.50±2.15	7.90±0.05	63.20±1.47	2.30±0.14
February	11.10±0.24	3.07±0.17	1.26±0.15	89.50±1.15	7.57±0.08	61.50±1.25	2.10±0.05
March	10.30±0.10	4.52±0.07	1.60±0.09	61.50±0.98	7.29±0.03	59.10±1.42	2.50±0.09
April	9.50±0.26	5.59±0.19	2.00±0.15	59.30±1.26	7.54±0.11	54.20±1.33	2.20±0.10
May	11.50±0.22	2.21±0.06	0.95±0.03	53.50±1.13	7.34±0.09	51.50±0.76	2.10±0.05
June	10.20±0.17	5.30±0.09	1.75±0.08	35.20±0.53	7.13±0.02	46.30±0.51	3.23±0.18
Average±SD	10.13±1.17	4.60±1.82	1.69±0.61	54.61±21.08	7.41±0.26	51.42±9.11	2.45±0.74

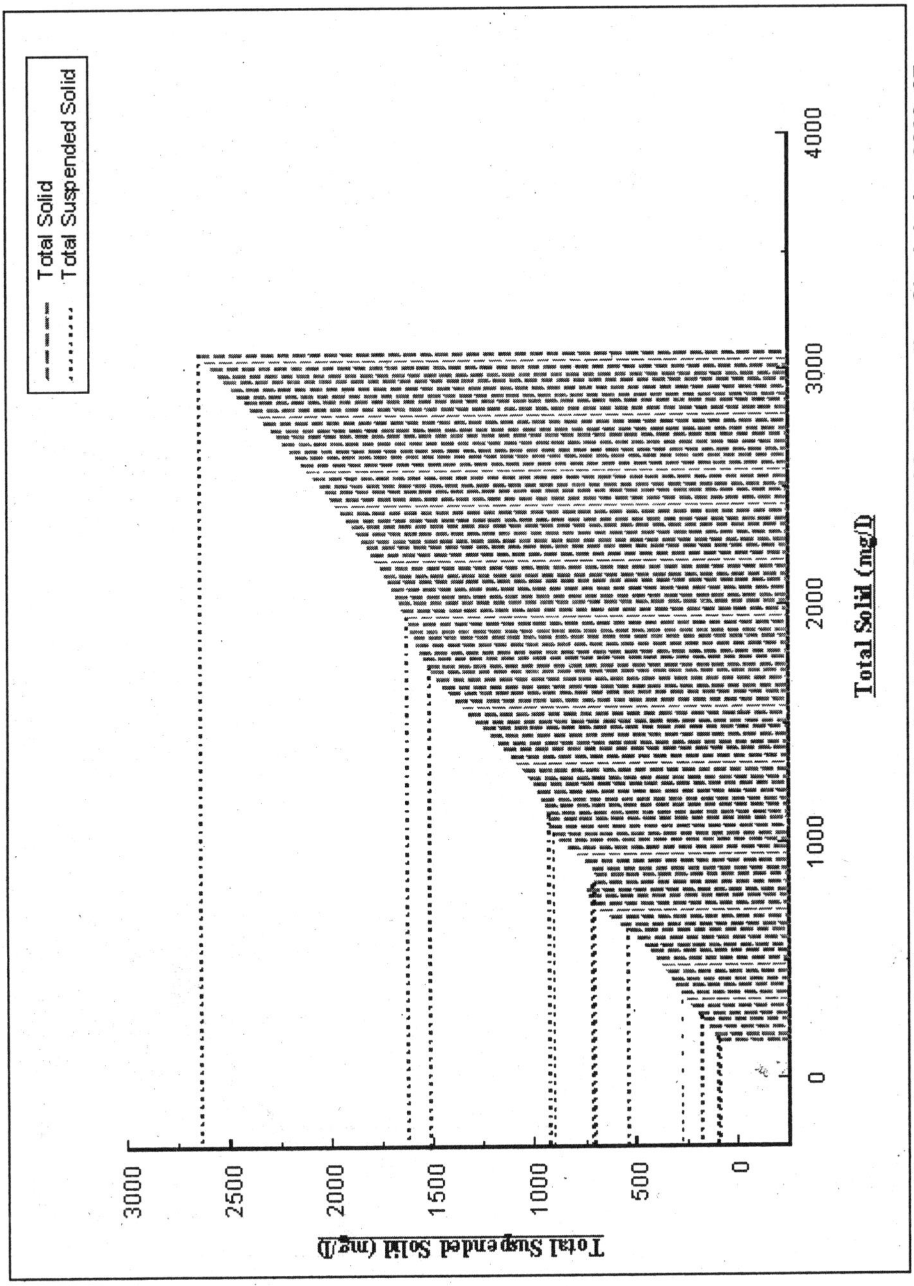

Figure 4.1: Graph Showing Quantitative Variation of TS and TSS at Sampling Site I during 2006–07

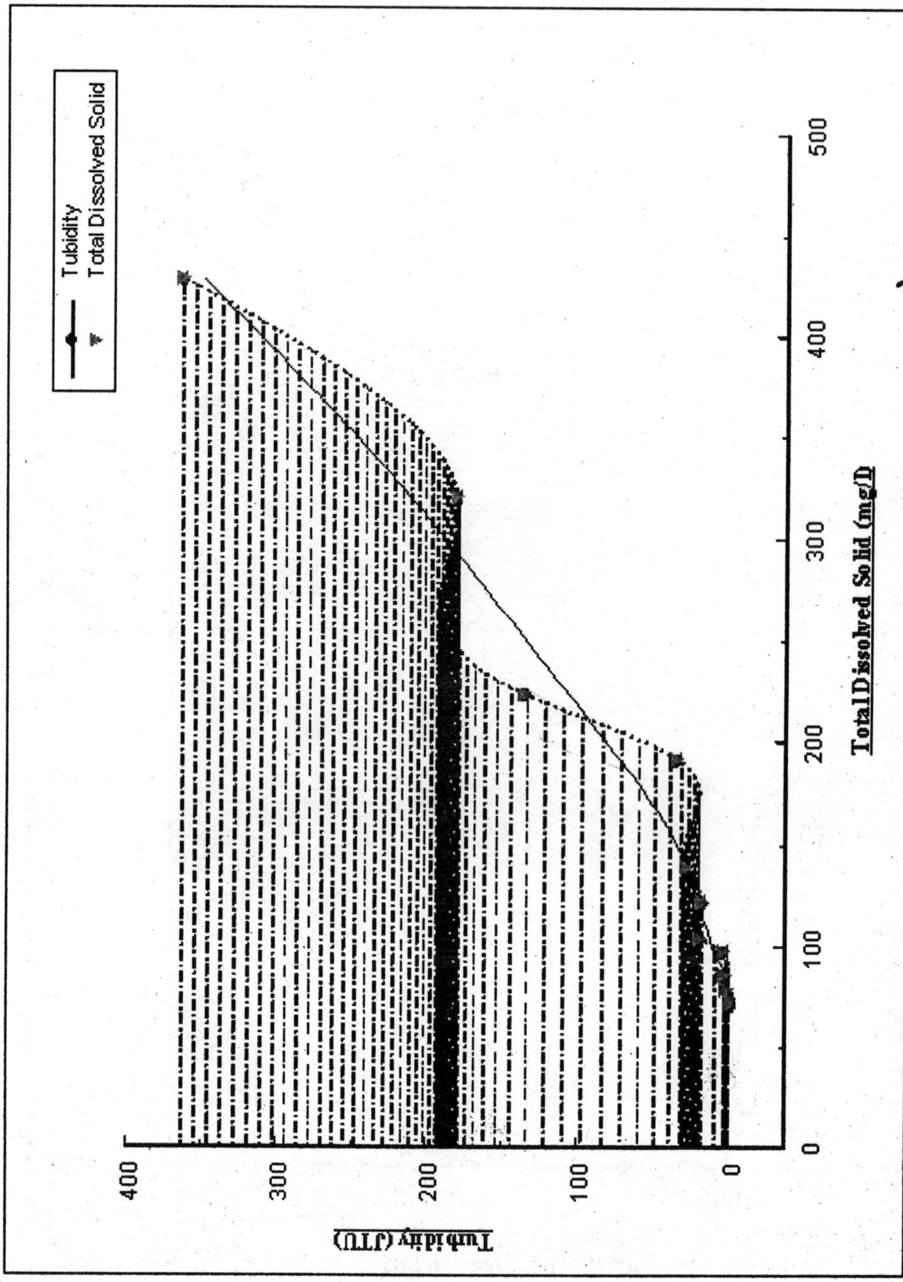

Figure 4.2: Graph Showing Quantitative Variation of Turbidity and TDS at Sampling Site I during 2006–07

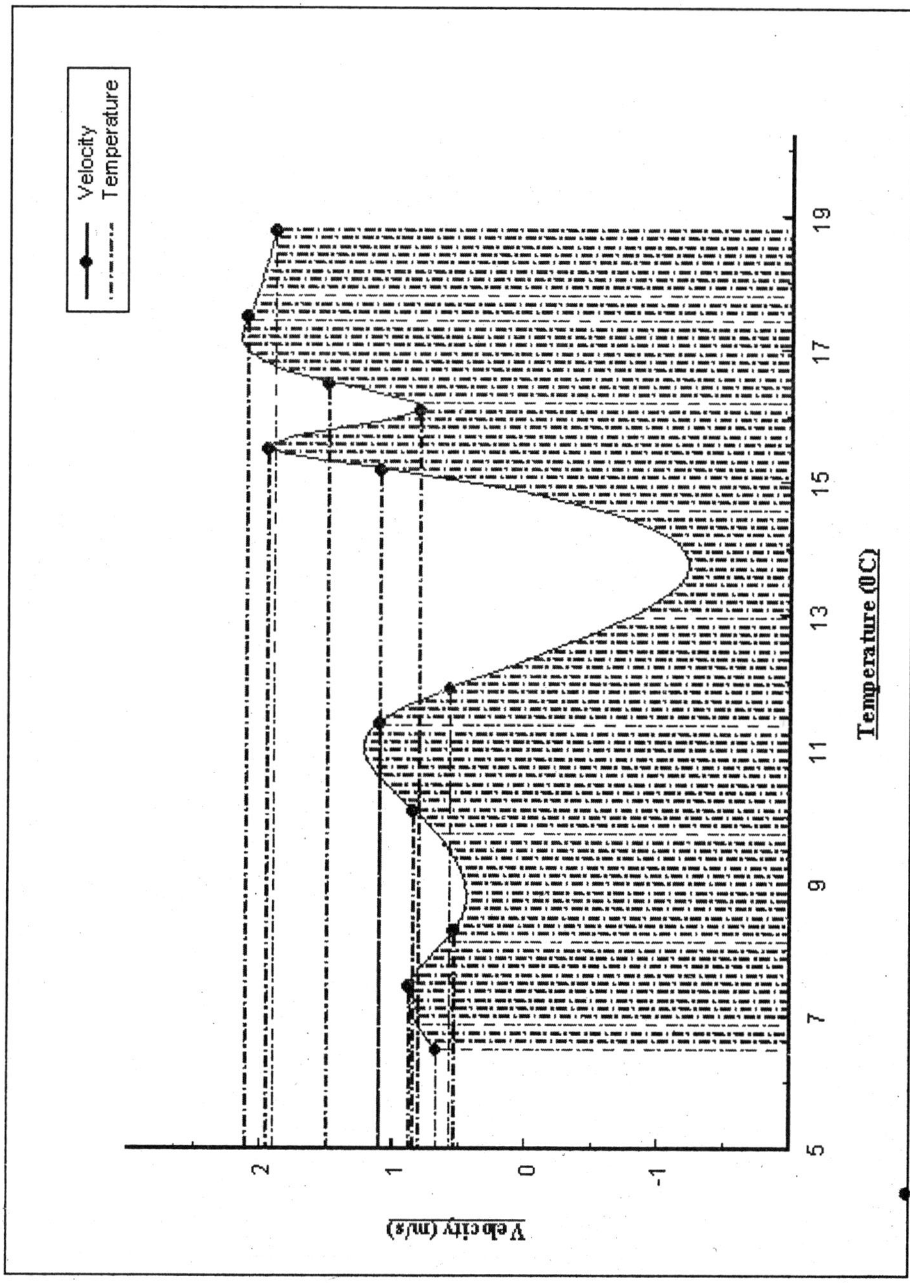

Figure 4.3: Graph Showing Quantitative Variation of Velocity and Temperature at Sampling Site I during 2006–07

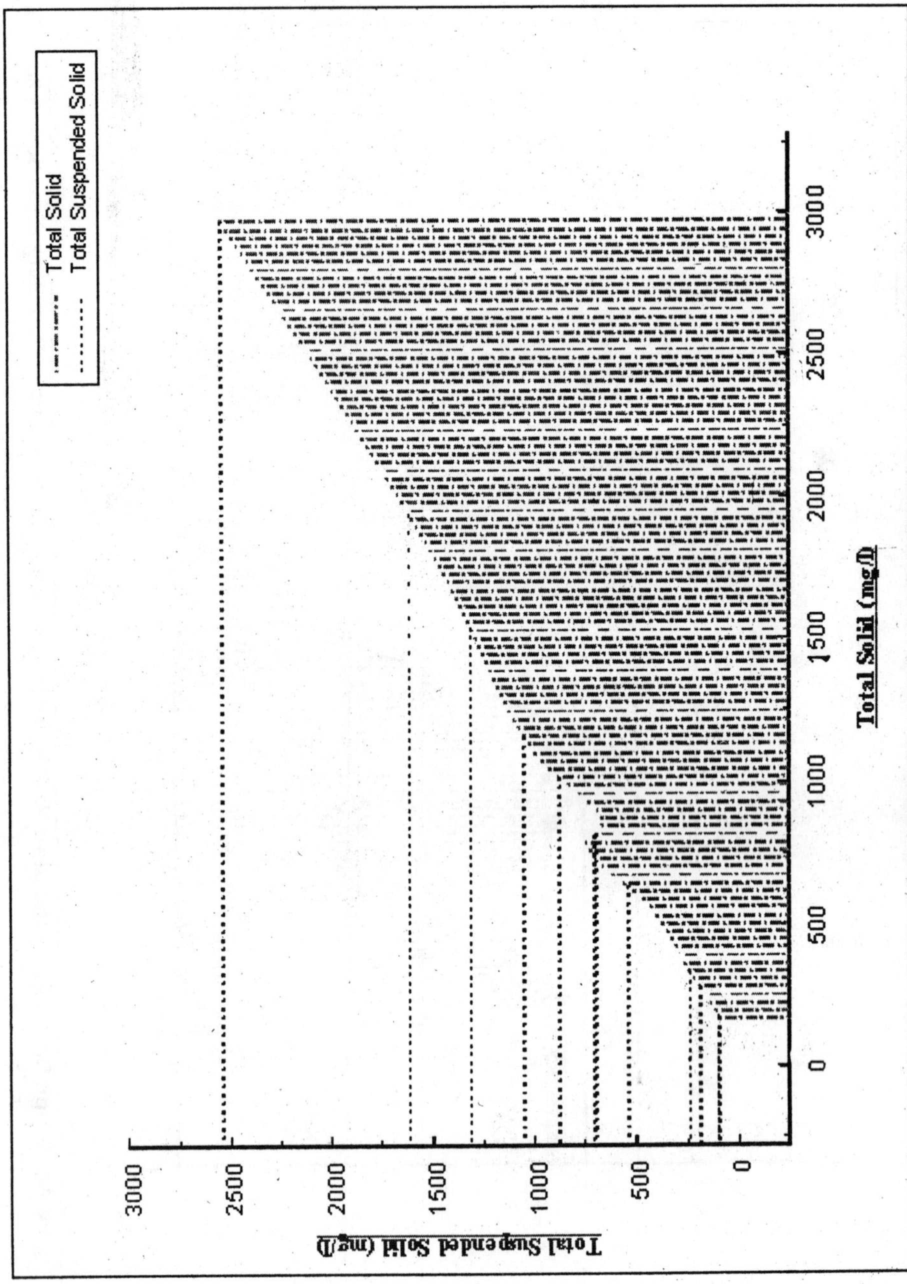

Figure 4.4: Graph Showing Quantitative Variation of TS and TSS at Sampling Site I during 2007–08

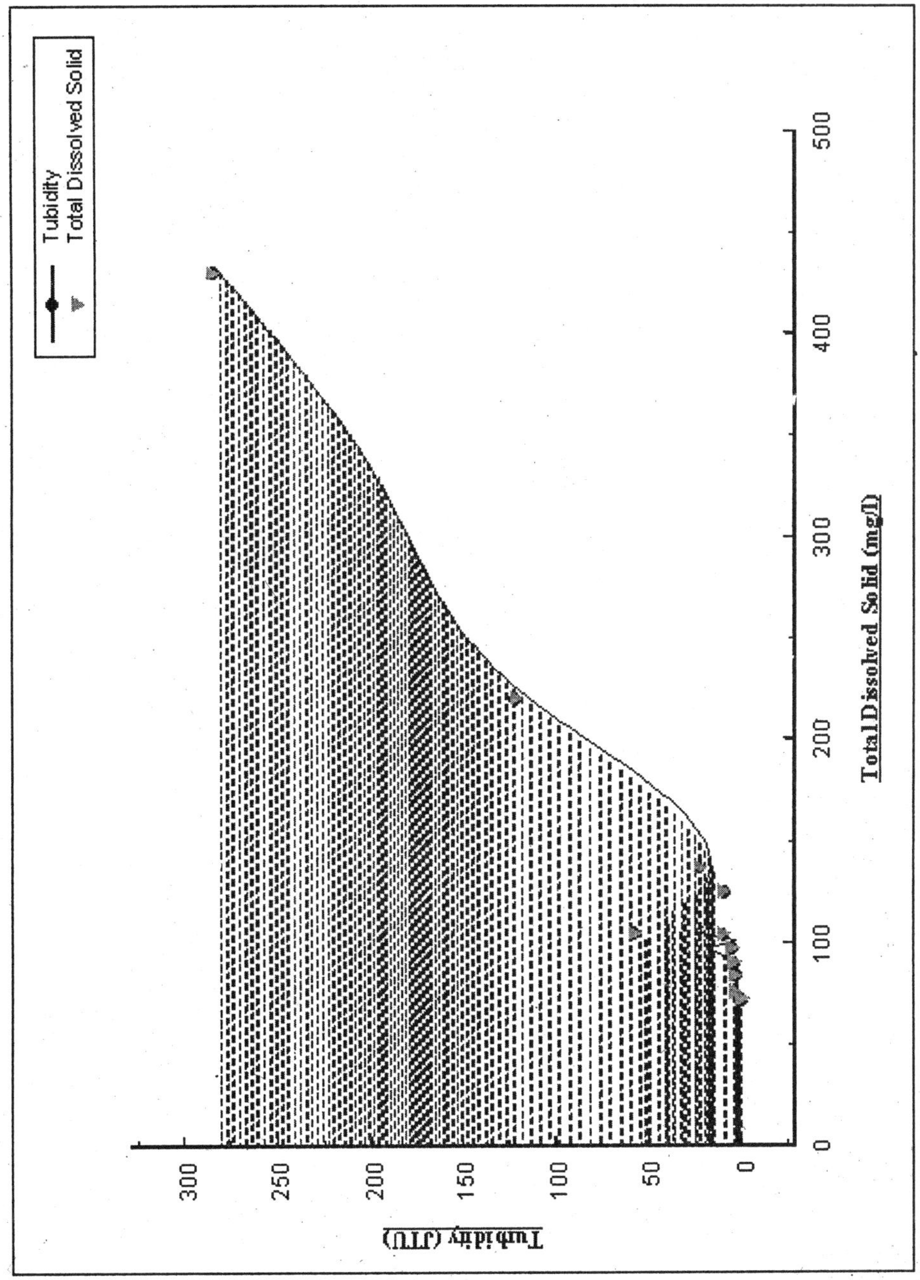

Figure 4.5: Graph Showing Quantitative Variation of Turbidity and TDS at Sampling Site I during 2007–08

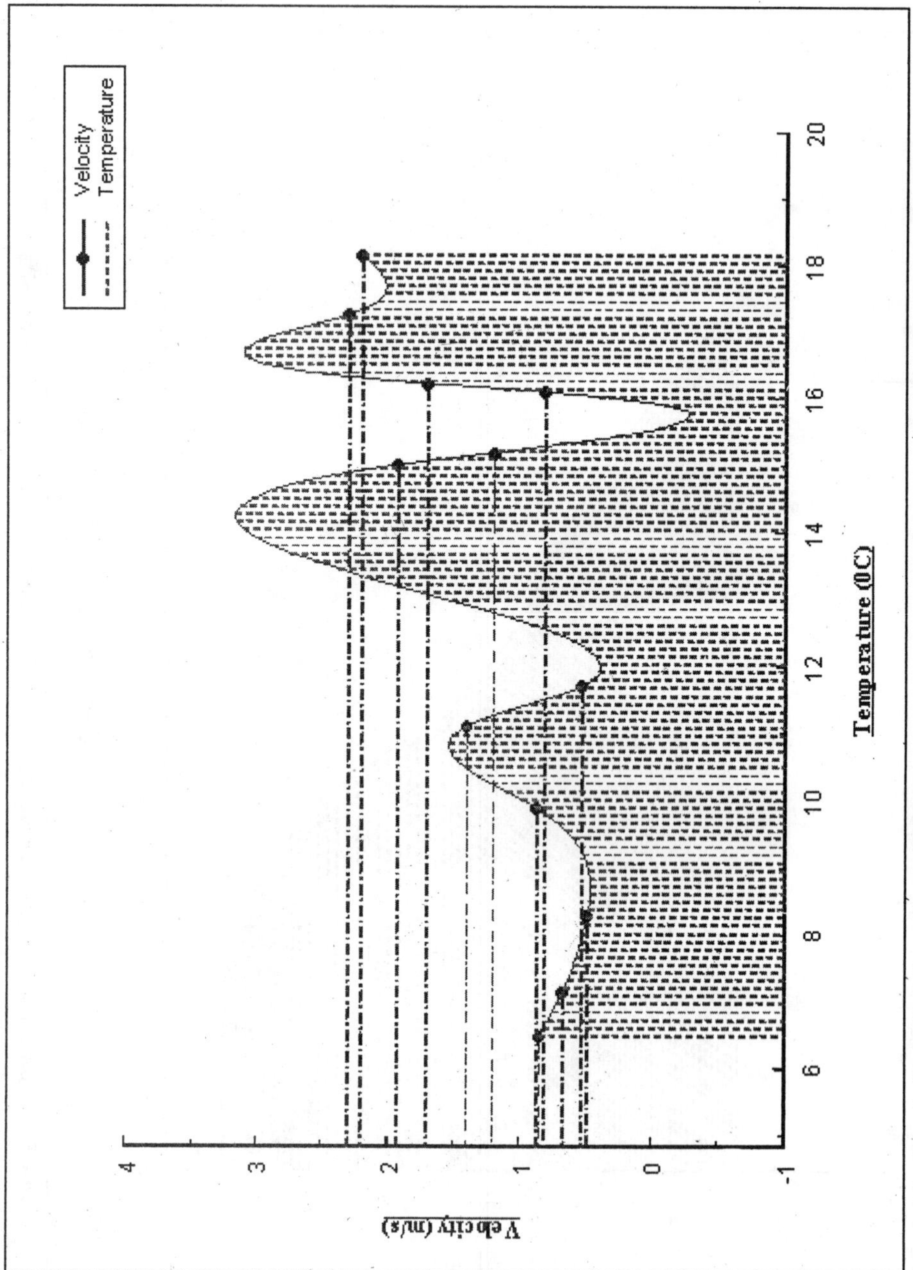

Figure 4.6: Graph Showing Quantitative Variation of Velocity and Temperature at Sampling Site I during 2007–08

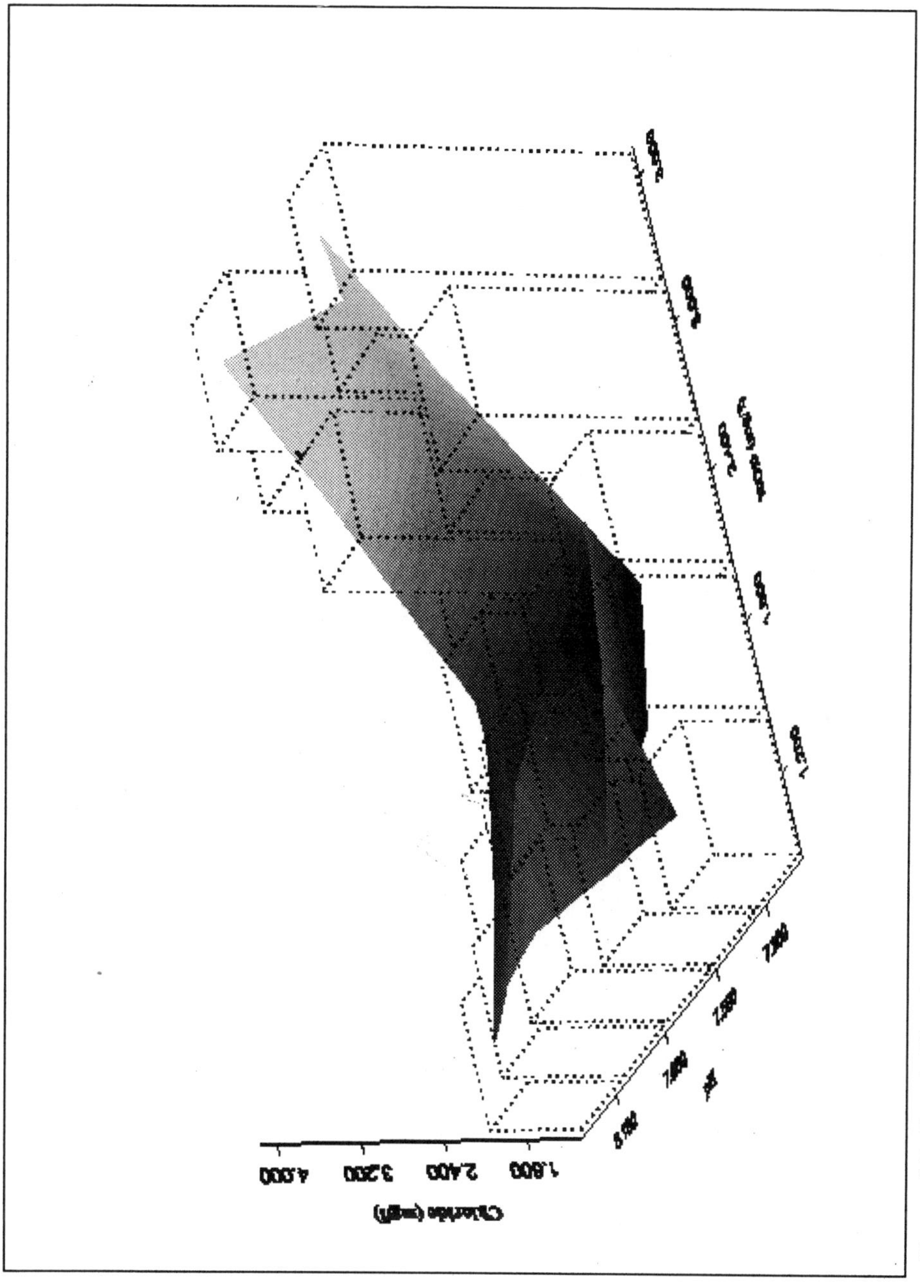

Figure 4.7: Graph Showing Quantitative Variation of Chloride, pH and BOD at Sampling Site I during 2006–07

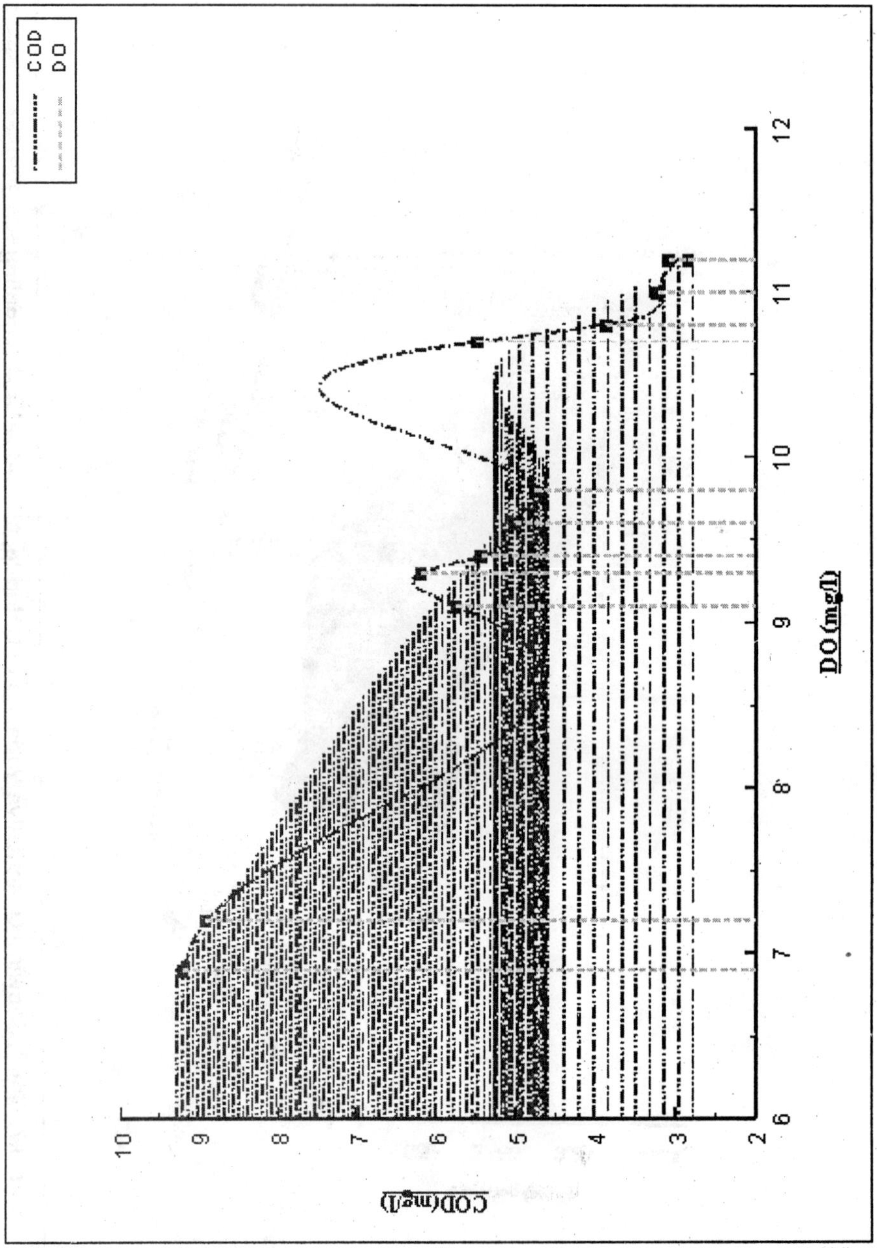

Figure 4.8: Graph Showing Quantitative Variation of COD and DO at Sampling Site I during 2006–07

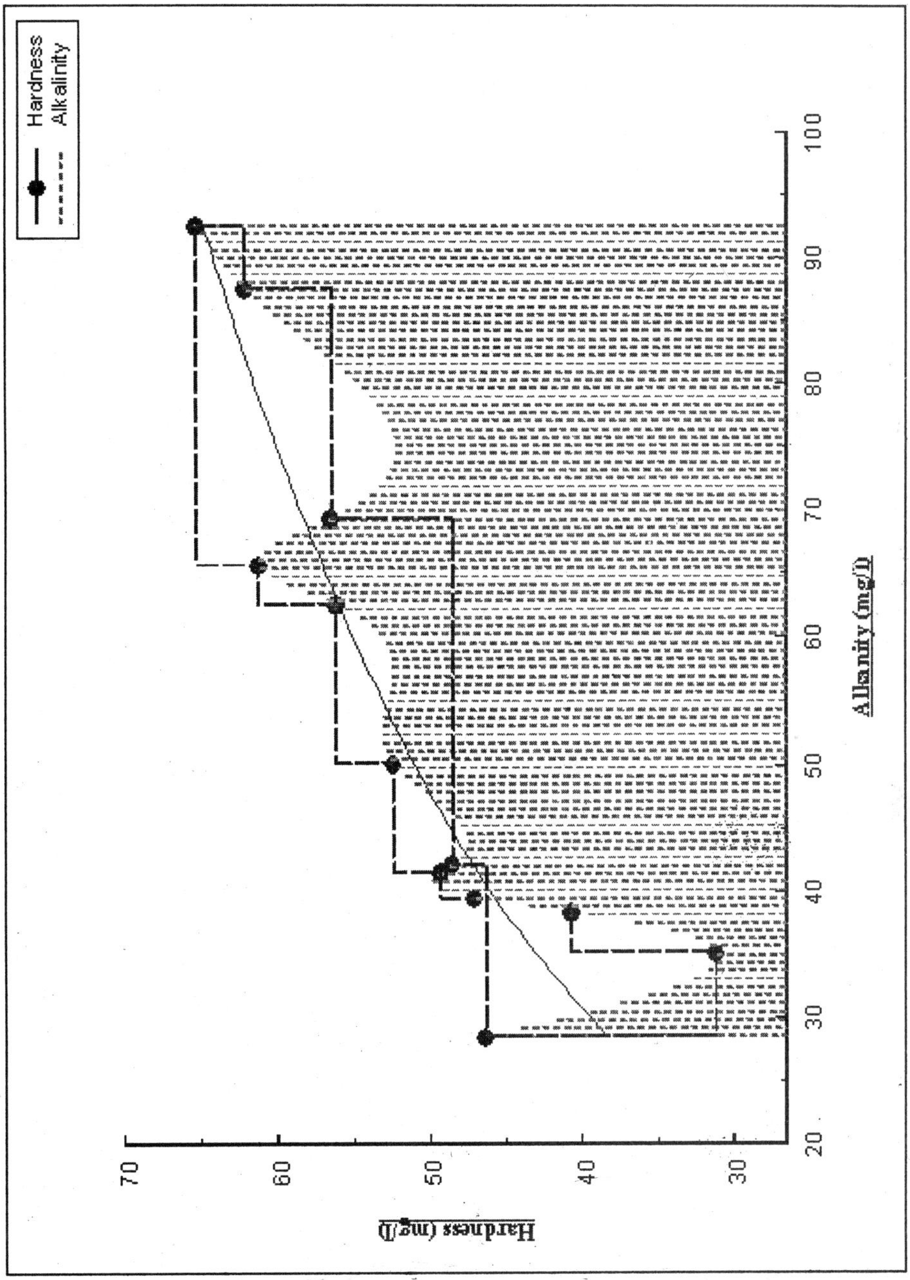

Figure 4.9: Graph Showing Quantitative Variation of Hardness and Alkalinity at Sampling Site I during 2006–07

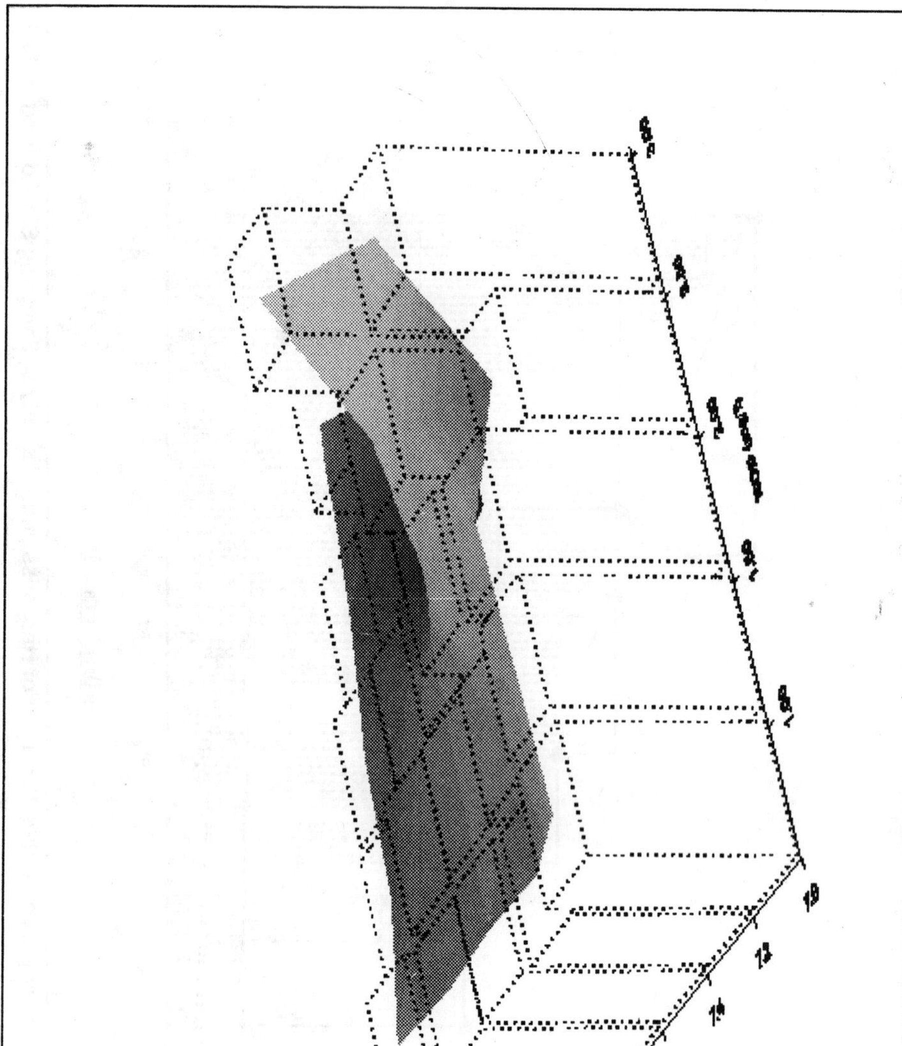

Figure 4.10: Graph Showing Quantitative Variation of Chloride, pH and BOD at Sampling Site I during 2007–08

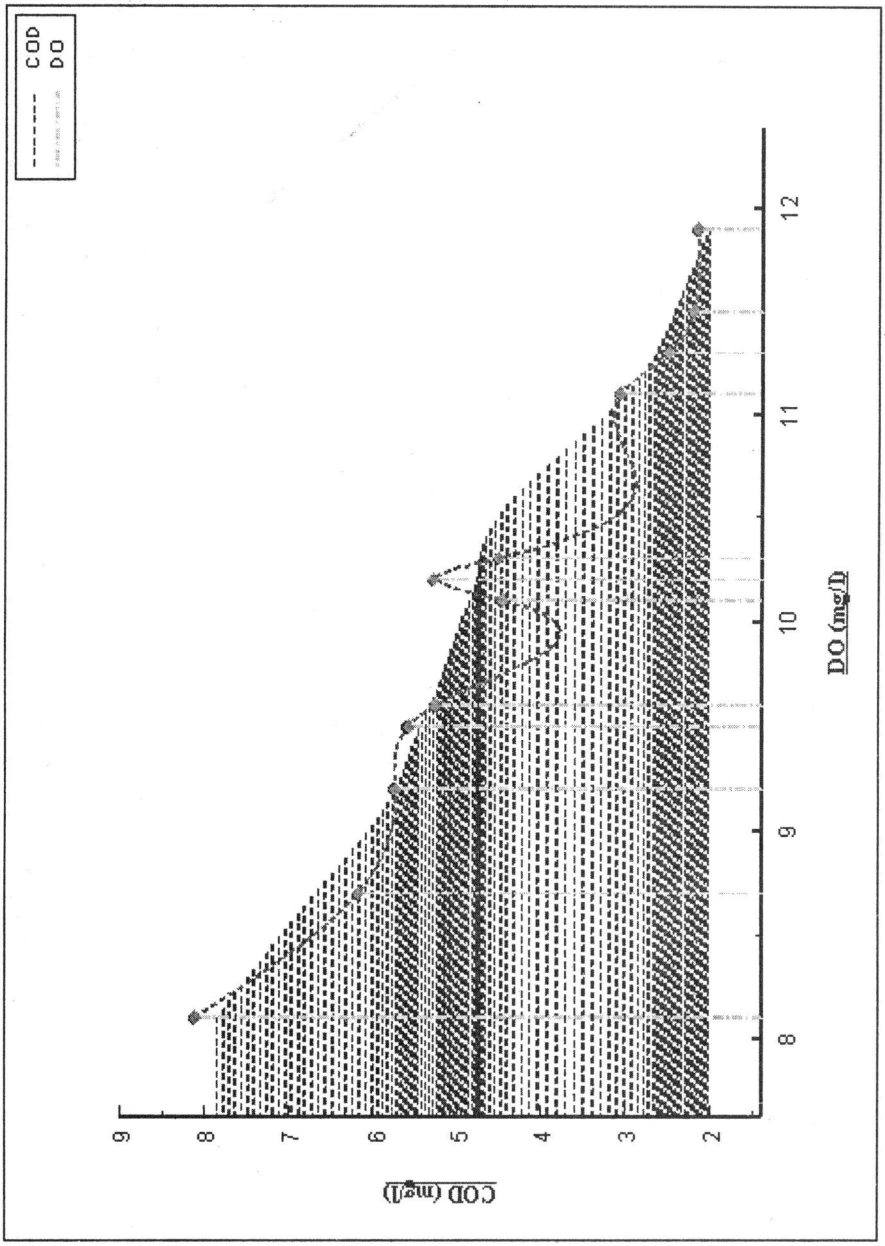

Figure 4.11: Graph Showing Quantitative Variation of COD and DO at Sampling Site I during 2007–08

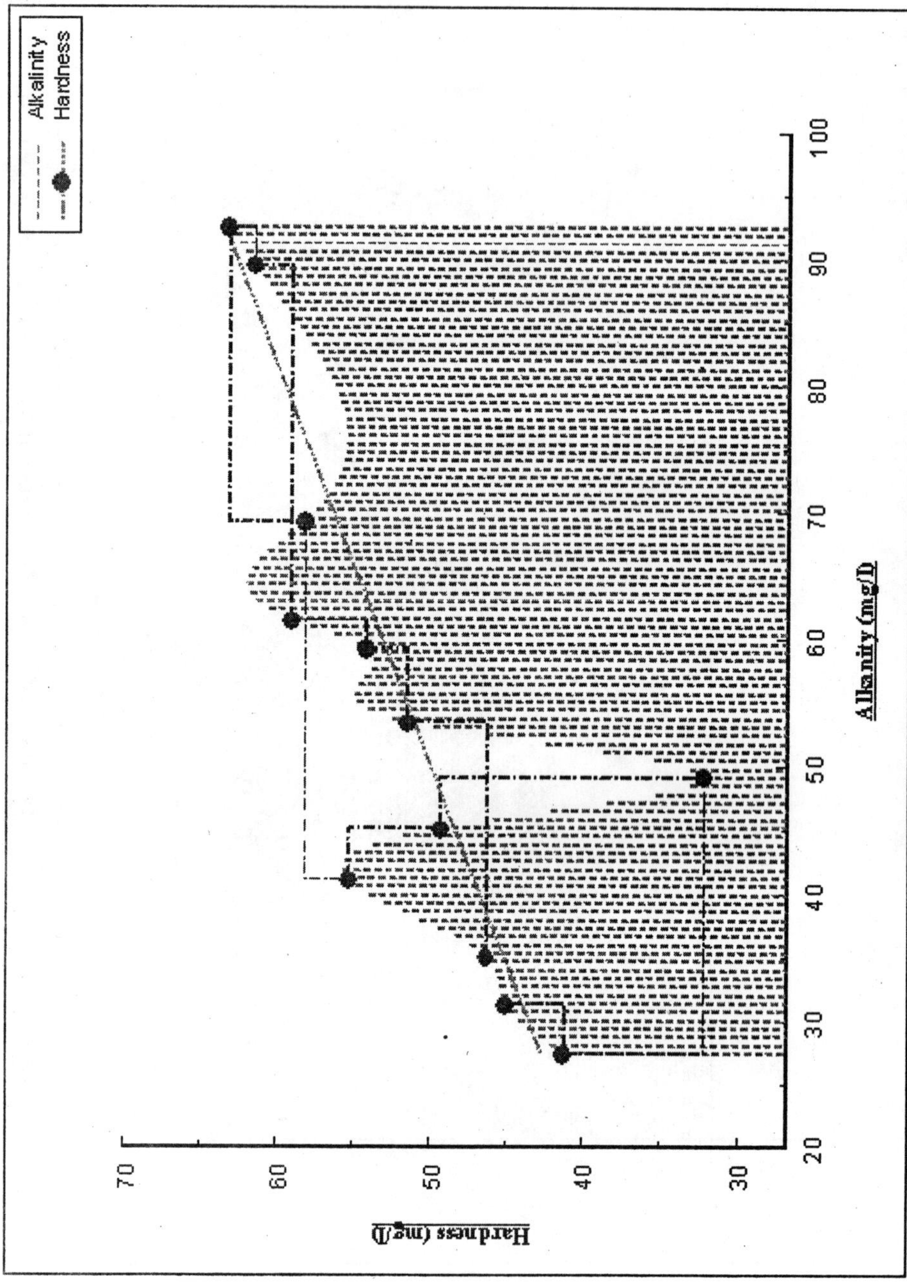

Figure 4.12: Graph Showing Quantitative Variation of Hardness and Alkalinity at Sampling Site I during 2007–08

Biochemical Oxygen Demand (mg/l)

The BOD at this sampling site was found minimum 0.80 mg/l ±0.09 in the month of January in 2007-2008 and maximum 3.45 mg/l ±0.06 in the month of August in 2006-2007. The average value in the year 2006-2007 was recorded 2.00 mg/l ±0.73 and in the year 2007-2008 it was observed 1.69 mg/l ±0.61.

Alkalinity (mg/l)

The alkalinity at this sampling site was observed minimum 27.50 mg/l ±1.05 in the month of August in 2007-2008 and maximum 92.50 mg/l ±1.19 in the month of January of both the years 2006-2007 and 2007-2008. The average value in the year 2006-2007 was observed 54.42 mg/l ±20.91 and in the year 2007-2008 it was recorded 54.61 mg/l ±21.08.

pH

pH at this sampling site was recorded minimum 7.10 ±0.08 in the month of July in 2007-2008 and maximum 8.19 ±0.18 in the month of November in 2006-2007. The average value in the year 2006-2007 was found 7.47 ±0.33 and in the year 2007-2008 it was observed 7.41 ±0.26.

Total Hardness (mg/l)

The total hardness at this sampling site was observed minimum 31.20 mg/l ±1.13 in the month of August in 2006-2007 and maximum 65.30 mg/l ±0.89 in the month of January in 2006-2007. The average value in the year 2006-2007 was observed 51.45 mg/l ±9.69 and in the year 2007-2008 it was recorded 51.42 mg/l ±9.11.

Chloride (mg/l)

The chloride at this sampling site was recorded minimum1.40 mg/l ±0.07 in the month of November in 2006-2007 and maximum 4.22 mg/l ±0.21 in the month of August in 2007-2008. The average value in the year 2006-2007 was noted 2.27 mg/l ±0.76 and in the year 2007-2008 it was found 2.45 mg/l ±0.74.

Sampling Site II (Chinyalisaur)

The results of physico-chemical parameters for the sampling site II are presented in the Tables 4.5 to 4.8 and Figures 4.13 to 4.24.

Water Temperature (°C)

The water temperature at this sampling site was observed minimum 6.90°C ±0.54 in the month of February in 2006-2007 and maximum 21.20°C ±0.68 in the month of September in 2007-2008. The average value in the year 2006-2007 was recorded 13.43 °C ±3.88 and in the year 2007-2008 it was found 13.79 °C ±4.52.

Total Solids (mg/l)

The total solids at this sampling site were recorded minimum 105.21 mg/l ±1.42 in the month of April in 2007-2008 and maximum 2331.4 mg/l ±2.05 in the month of August in 2006-2007. The average value in the year 2006-2007 was noted 878.88 mg/l ±669.49 and in the year 2007-2008 it was observed 712.37 mg/l ±592.10.

Table 4.5: Monthly Variation in Physical Parameters of Bhagirathi River at Sampling Station II (Chinyalisaur) during 2006–2007

Month	*Temp. (°C)*	*Total Solids (mg/l)*	*TDS (mg/l)*	*TSS (mg/l)*	*Velocity (m/s)*	*Turbidity (JTU)*
July	15.80±0.55	1925.12±2.15	314.20±1.35	1610.92±2.05	1.91±0.24	161.50±1.54
August	17.90±0.42	2331.40±2.05	421.10±1.25	1910.30±1.89	1.99±0.08	272.30±1.24
September	18.50±0.37	1321.54±2.00	207.15±1.24	1114.39±1.87	1.87±0.06	159.20±1.46
October	16.90±0.57	1087.28±1.85	151.87±1.13	935.41±1.58	1.52±0.54	17.50±0.45
November	11.90±0.98	745.25±1.76	97.01±1.11	648.24±1.25	1.08±0.12	8.10±0.21
December	11.80±0.76	534.21±1.36	95.75±1.51	438.46±1.03	0.87±0.11	4.50±0.13
January	7.85±0.57	321.43±1.46	90.18±1.46	231.25±1.04	0.81±0.16	3.10±0.16
February	6.90±0.54	245.57±1.78	89.49±1.27	156.08±1.23	0.61±0.08	2.10±0.14
March	8.90±0.66	150.22±1.99	78.25±1.05	71.97±1.55	0.45±0.14	0.50±0.10
April	12.20±0.38	141.10±1.28	75.42±1.02	65.68±1.51	0.50±0.16	0.00±0.00
May	16.80±0.85	797.57±2.07	114.23±1.05	683.34±2.09	0.75±0.02	9.40±0.45
June	15.70±0.89	945.78±2.11	132.15±1.11	813.63±2.45	0.98±0.09	16.40±0.87
Average± SD	13.43±3.88	878.88±669.49	155.57±103.54	723.31±570.95	1.11±0.54	54.55±86.90

Table 4.6: Monthly Variation in Physical Parameters of Bhagirathi River at Sampling Station II (Chinyalisaur) during 2007–2008

Month	Temp. (°C)	Total Solids (mg/l)	TDS (mg/l)	TSS (mg/l)	Velocity (m/s)	Turbidity (JTU)
July	15.10±0.88	1921.5±2.26	310.23±1.24	1611.27±2.13	1.81±0.13	170.10±0.21
August	20.20±0.45	1551.24±2.07	202.50±1.52	1348.74±2.15	0.59±0.05	155.50±1.05
September	21.20±0.68	1121.24±2.13	197.31±1.40	923.93±2.42	0.21±0.04	91.00±0.55
October	17.50±0.79	897.23±1.45	115.92±1.28	871.31±2.00	0.14±0.02	65.00±0.46
November	12.20±0.84	675.30±1.78	91.14±1.43	584.16±1.87	0.81±0.06	15.20±0.12
December	10.80±0.34	375.85±1.46	88.24±1.11	287.61±1.49	0.75±0.07	11.90±0.11
January	7.98±0.55	175.21±1.28	81.45±1.08	93.76±1.57	0.85±0.03	7.10±0.05
February	7.63±0.28	111.94±1.81	69.40±2.04	42.54±1.57	0.61±0.05	1.80±0.10
March	9.20±0.31	122.32±1.43	75.39±1.45	46.93±1.54	0.49±0.02	2.30±0.03
April	12.40±0.22	105.21±1.42	68.24±1.19	36.97±0.89	0.56±0.00	0.00±0.00
May	16.20±0.16	688.23±1.89	99.34±1.31	588.89±1.46	0.72±0.01	17.30±0.13
June	15.10±0.54	803.21±2.38	105.32±1.12	697.89±1.11	1.12±0.03	21.20±0.78
Average± SD	13.79±4.52	712.37±592.10	125.37±73.76	594.5±527.34	0.72±0.42	46.53±60.91

Table 4.7: Monthly Variation in Chemical Parameters of Bhagirathi River at Sampling Site II (Chinyalisaur) during 2006–2007

Month	*DO (mg/l)*	*COD (mg/l)*	*BOD (mg/l)*	*Alkalinity (mg/l)*	*pH*	*Total Hardness (mg/l)*	*Chloride (mg/l)*
July	8.10± 0.23	7.93± 0.35	2.70±0.06	37.22±1.2	7.20±0.03	43.25±1.38	3.56±0.15
August	8.55±0.15	6.57±0.23	2.50±0.05	38.35±1.35	7.13±0.05	38.68±1.15	3.84±0.13
September	9.20±0.11	5.40±0.13	2.10±0.11	27.45±0.76	7.12±0.09	31.50±1.42	1.70±0.09
October	8.60±0.05	7.26±0.16	2.65±0.15	56.51±1.22	7.15±0.10	50.85±1.35	1.58±0.05
November	10.50±0.17	3.90±0.11	1.45±0.06	58.66±1.32	7.65±0.15	49.15±0.68	1.56±0.09
December	9.00±0.16	5.87±0.21	2.35±0.03	49.65±1..40	7.77±0.16	61.30±1.25	1.27±0.12
January	10.30±0.13	4.54±0.10	1.90±0.07	86.27±1.75	7.50±0.13	62.65±0.90	1.75±0.10
February	10.50±0.11	3.92±0.06	1.55±0.11	48.56±0.86	7.65±0.17	61.51±1.82	1.60±0.09
March	10.70±0.06	3.6±0.13	1.30±0.16	46.05±0.58	7.80±0.06	56.62±1.35	2.02±0.03
April	8.80±0.08	6.18±0.11	2.25±0.19	41.25±0.48	7.29±0.04	54.00±1.46	2.41±0.05
May	9.80±0.10	4.48±0.16	1.85±0.13	42.34±0.21	7.12±0.00	51.37± 1.26	2.55±0.08
June	9.60±0.18	6.26±0.19	2.00±0.09	40.12±0.15	7.18±0.21	47.33±0.53	2.84±0.07
Average±SD	9.47±0.89	5.49±1.41	2.05±1.41	47.70±14.86	7.38±0.27	50.68±9.54	2.22±0.83

Table 4.8: Monthly Variation in Chemical Parameters of Bhagirathi River at Sampling Site II (Chinyalisaur) during 2007–2008

Month	*DO (mg/l)*	*COD (mg/l)*	*BOD (mg/l)*	*Alkalinity (mg/l)*	*pH*	*Total Hardness (mg/l)*	*Chloride (mg/l)*
July	8.50±0.25	7.36±1.1	2.60±0.16	31.23±0.29	7.18±0.06	43.97±1.35	3.80±0.16
August	9.70±0.15	5.15±0.12	1.90±0.05	29.50±0.32	7.25±0.05	43.23±1.25	3.40±0.06
September	9.10±0.26	5.30±0.15	2.15±0.09	40.35±0.48	7.00±0.15	49.25±1.3	2.60±0.09
October	9.80±0.23	4.94±0.29	1.75±0.15	39.23±0.15	7.10±0.16	47.62±0.56	1.30±0.18
November	10.30±0.46	3.81±0.24	1.45±0.11	55.30±0.24	7.46±0.10	61.20±0.25	1.20±0.11
December	10.50±0.48	4.17±0.26	1.60±0.06	57.30±0.16	7.75±0.06	66.10±0.14	2.10±0.15
January	11.00±0.12	3.26±0.3	1.20±0.05	77.50±0.22	7.89±0.08	67.25±0.38	2.40±0.03
February	11.20±0.06	2.63±0.11	1.00±0.05	81.20±0.27	7.35±0.03	61.51±0.23	1.90±0.08
March	10.50±0.16	4.08±0.22	1.50±0.06	52.22±0.35	7.30±0.05	58.25±0.67	2.70±0.14
April	9.20±0.04	5.95±0.09	2.18±0.16	35.50±0.16	7.40±0.08	56.15±0.11	2.90±0.16
May	10.00±0.19	4.20±0.11	1.78±0.03	37.50±0.11	7.21±0.09	52.29±0.26	2.50±0.19
June	10.20±0.24	4.92±0.06	1.70±0.07	47.60±0.18	7.15±0.03	51.21±0.15	3.00±0.21
Average±SD	10.00±0.79	4.64±1.26	1.73±0.44	48.70±16.92	7.34±0.26	54.84±8.17	2.48±0.78

Total Dissolved Solids (mg/l)

The total dissolved solids at this sampling site were observed minimum 68.24 mg/l ±1.19 in the month of April in 2007-2008 and maximum 421.10 mg/l ±1025 in the month of August 2006-2007. The average value in the year 2006-2007 was found 155.57 mg/l ±103.54 and in the year 2007-2008 it was recorded 125.37 mg/l ±73.76.

Total Suspended Solids (mg/l)

The total suspended solids at this sampling site were found minimum 36.97 mg/l ±0.89 in the month of April in 2007-2008 and maximum 1910.3 mg/l ±1.89 in the month of August in 2006-2007. The average value in the year 2006-2007 was found 723.31 mg/l ±570.95 and in the year 2007-2008 it was recorded 594.50 mg/l ±527.34.

Velocity (m/s)

The velocity at this sampling site was recorded minimum 0.14 m/s ±0.02 in the month of October in 2007-2008 and maximum 1.99 m/s ±0.08 in the month of August in 2006-2007. The average value in the year 2006-2007 was observed 1.11 m/s ±0.54 and in the year 2007-2008 it was noted 0.72 m/s ±0.42.

Turbidity (JTU)

The turbidity at this sampling site was observed minimum 0.00 in the month of April in 2006-2007 and 2007-2008 and maximum 272.34 JTU ±1.24 in the month of August 2006-2007. The average value in the year 2006-2007 was noted 54.55 JTU ±86.90 and in the year 2007-2008 it was recorded 46.53 JTU ±60.91.

Dissolved Oxygen (mg/l)

The dissolved oxygen at this sampling site was recorded minimum 8.10 mg/l ±0.23 in the month of July in 2006-2007 and maximum 11.20 mg/l ±0.06 in the month of February in 2007-2008. The annual average value in the year 2006-2007 was observed 9.47 mg/l ±0.89 and in the year 2007-2008 it was noted 10.00 mg/l ±0.79.

Chemical Oxygen Demand (mg/l)

The COD at this sampling site was observed minimum 2.63 mg/l ±0.11 in the month of February in 2007-2008 and maximum 7.93 mg/l ±0.35 in the month of July in 2006-2007. The average value in the year 2006-2007 was recorded 5.49 mg/l ±1.41 and in the year 2007-2008 it was found 4.64 mg/l ±1.26.

Biochemical Oxygen Demand (mg/l)

The BOD at this sampling site was found minimum 1.00 mg/l ±0.05 in the month of February in 2007-2008 and maximum 2.70 mg/l ±0.06 in the month of July in 2006-2007. The average value in the year 2006-2007 was observed 2.05 mg/l ±1.41 and in the year 2007-2008 it was observed 1.73 mg/l ±0.44.

Alkalinity (mg/l)

The alkalinity at this sampling site was observed minimum 27.45 mg/l ±0.76 in the month of September in 2006-2007 and maximum 86.27 mg/l ±1.75 in the month of January in 2006-2007. The average value in the year 2006-2007 was noted 47.7 mg/l ±14.86 and in the year 2007-2008 it was recorded 48.70 mg/l ±16.92.

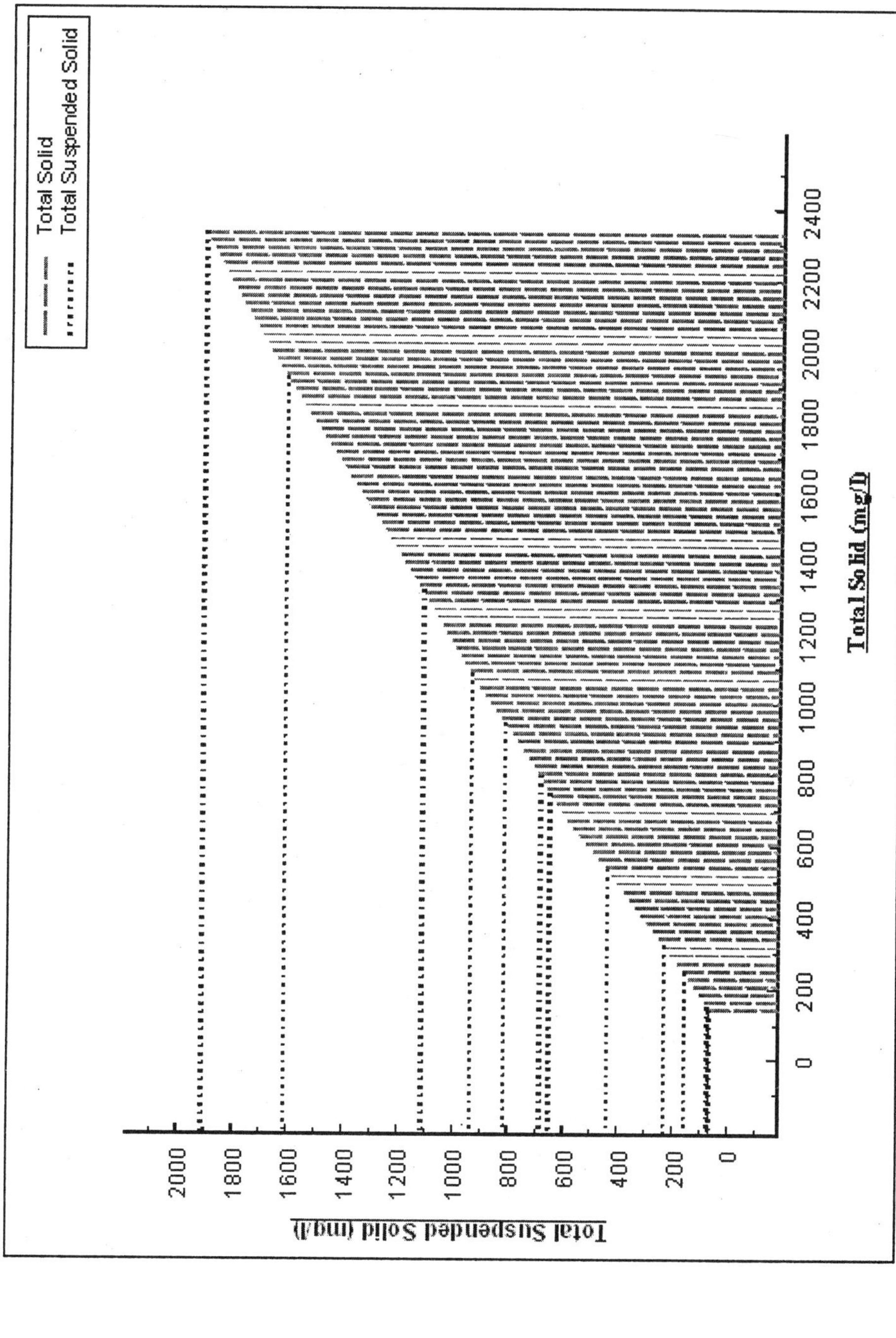

Figure 4.13: Graph Showing Quantitative Variation of TS and TSS at Sampling Site II during 2006–07

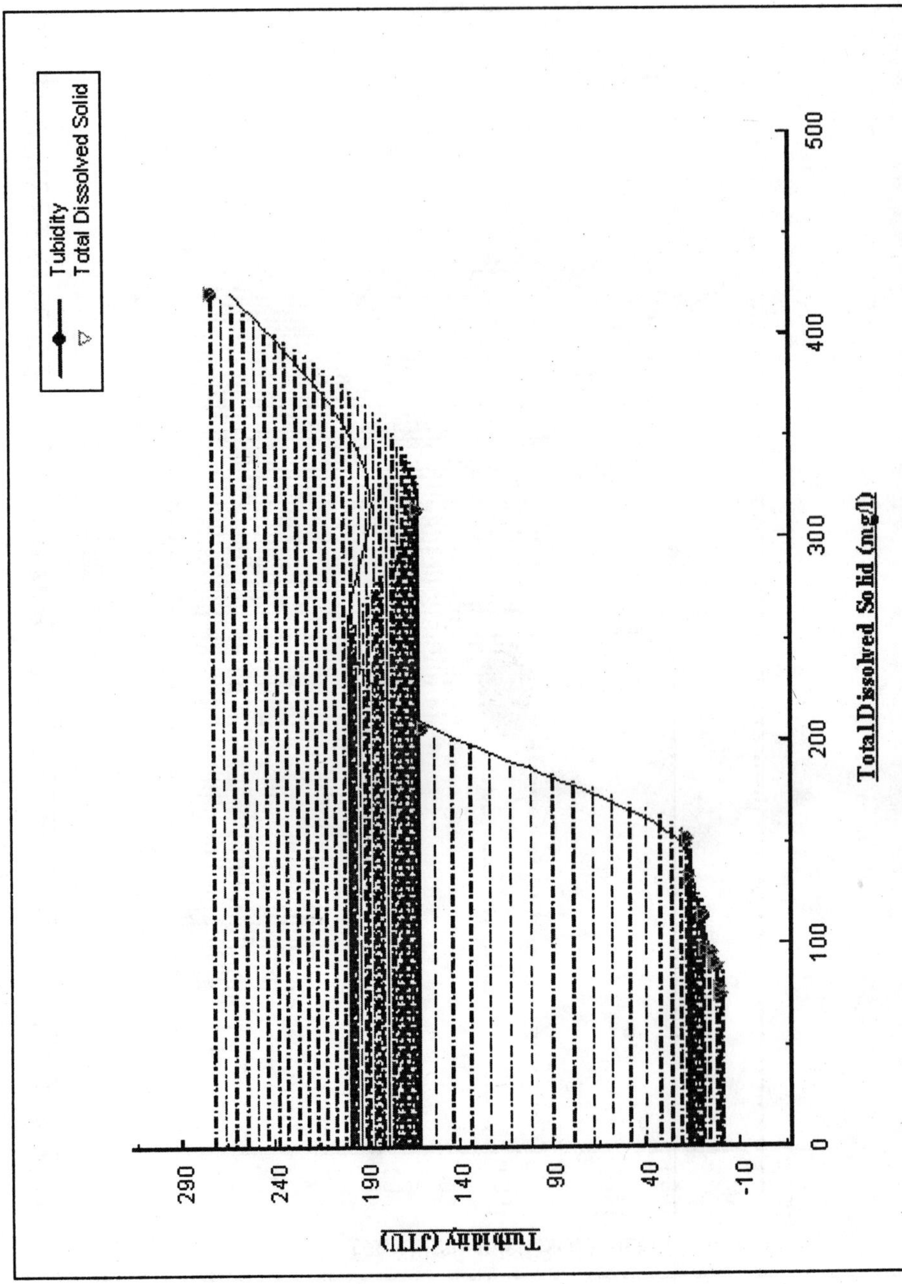

Figure 4.14: Graph Showing Quantitative Variation of Turbidity and TDS at Sampling Site II during 2006–07

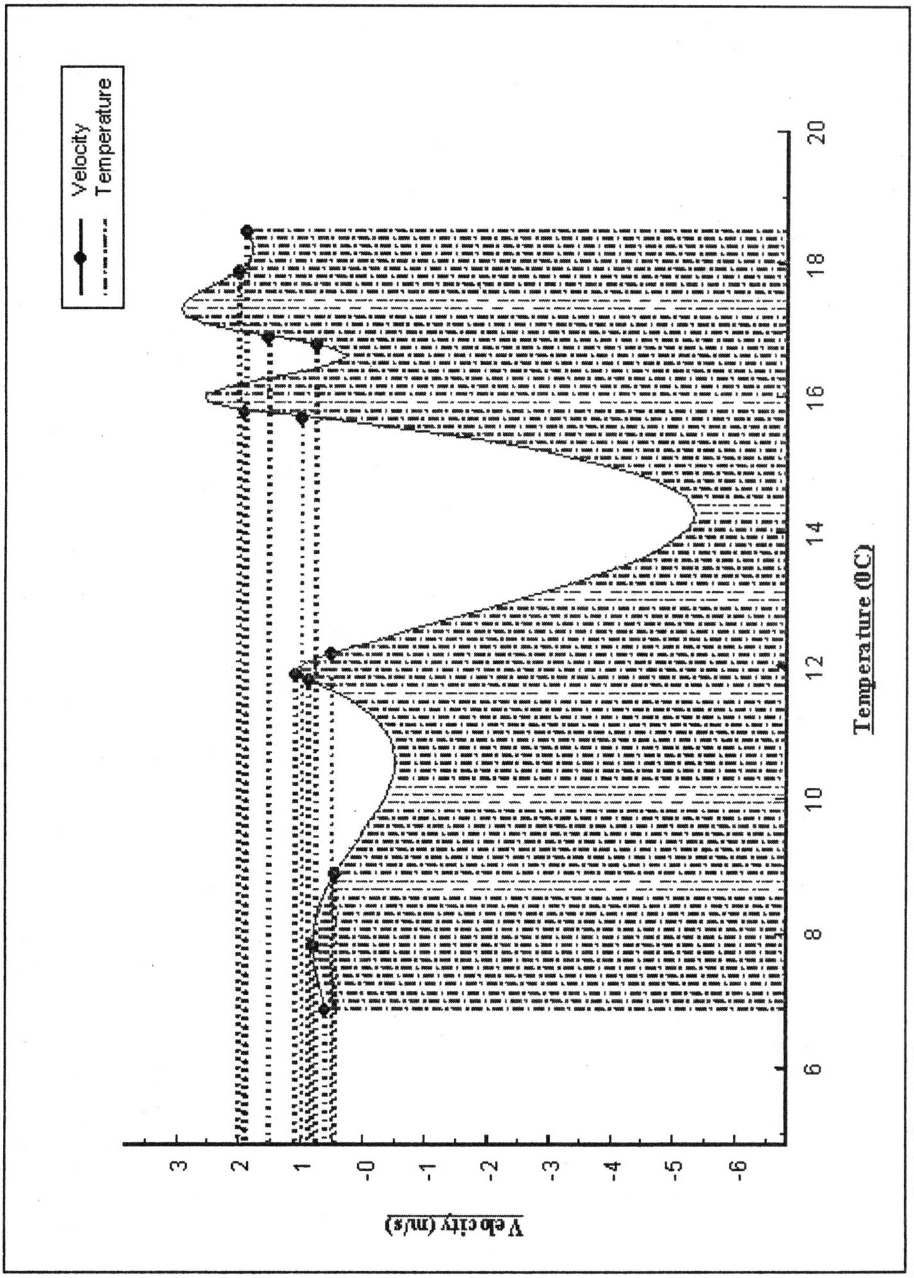

Figure 4.15: Graph Showing Quantitative Variation of Velocity and Temperature at Sampling Site II during 2006–07

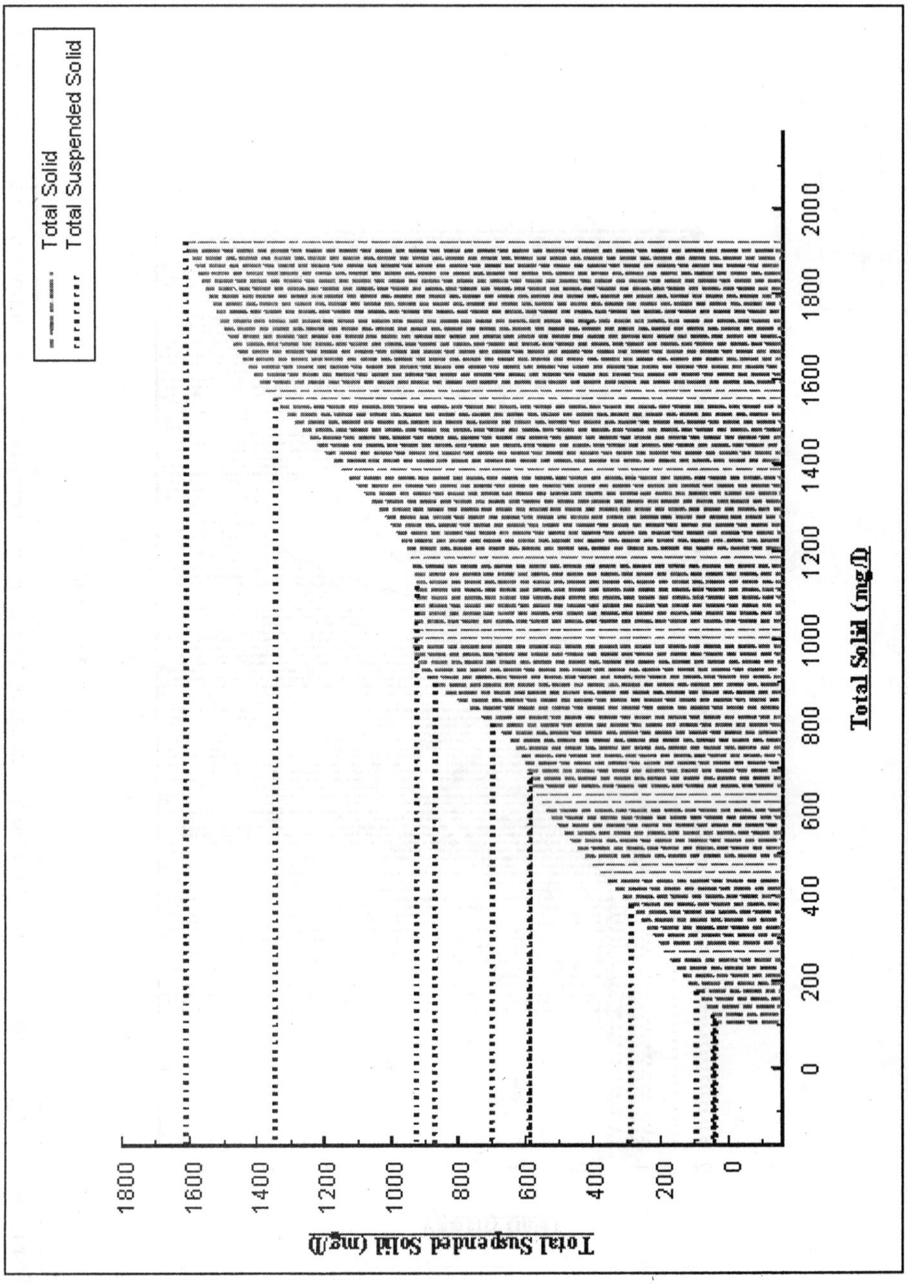

Figure 4.16: Graph Showing Quantitative Variation of TS and TSS at Sampling Site II during 2007–08

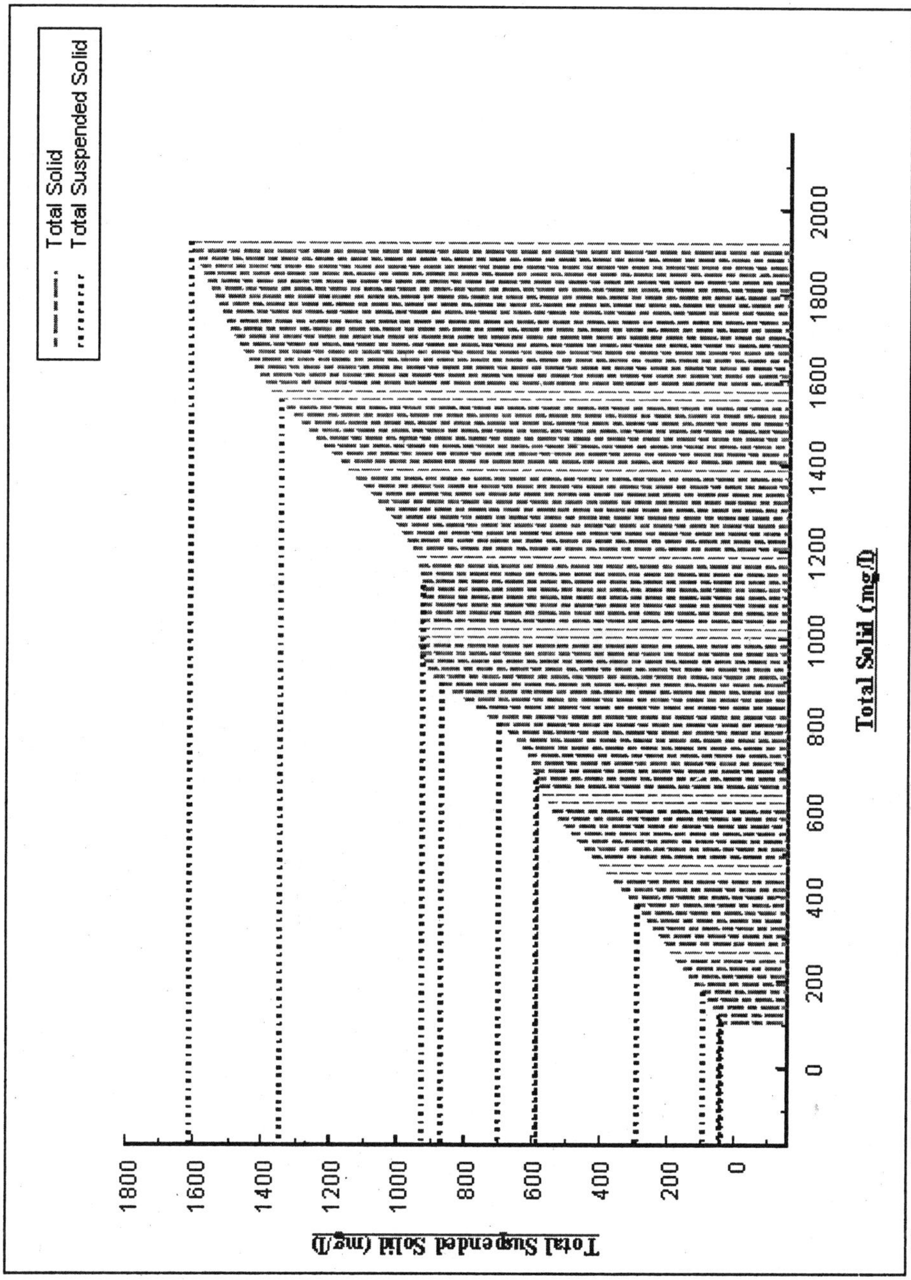

Figure 4.17: Graph Showing Quantitative Variation of Turbidity and TDS at Sampling Site II during 2007–08

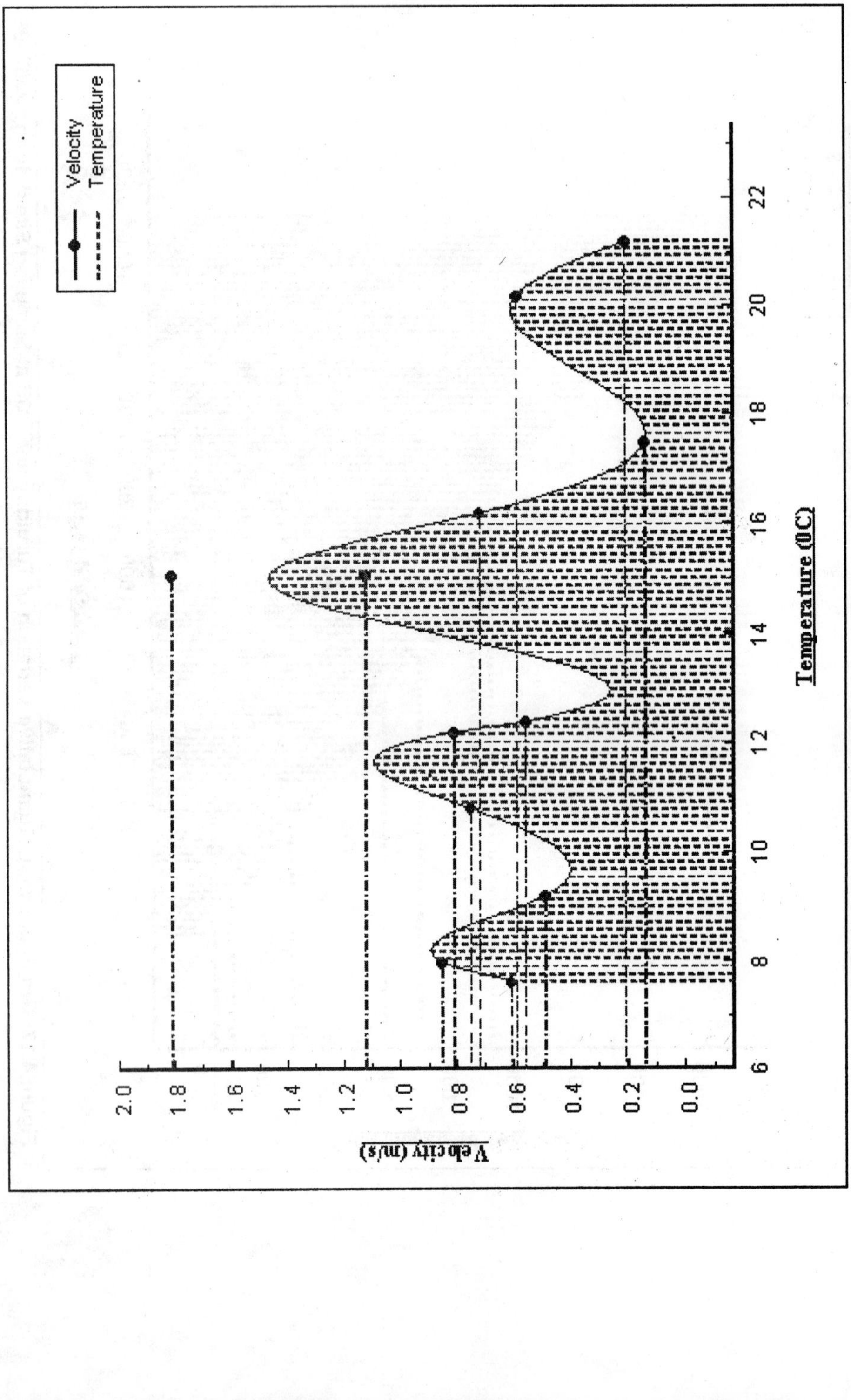

Figure 4.18: Graph Showing Quantitative Variation of Velocity and Temperature at Sampling Site II during 2007–08

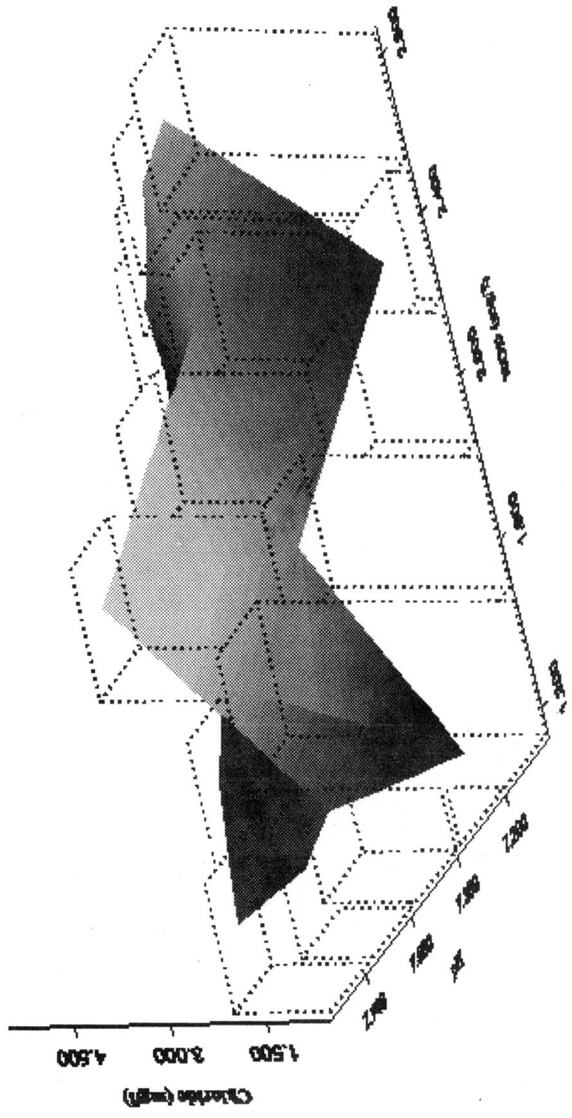

Figure 4.19: Graph Showing Quantitative Variation of Chloride, pH and BOD at Sampling Site II during 2006–07

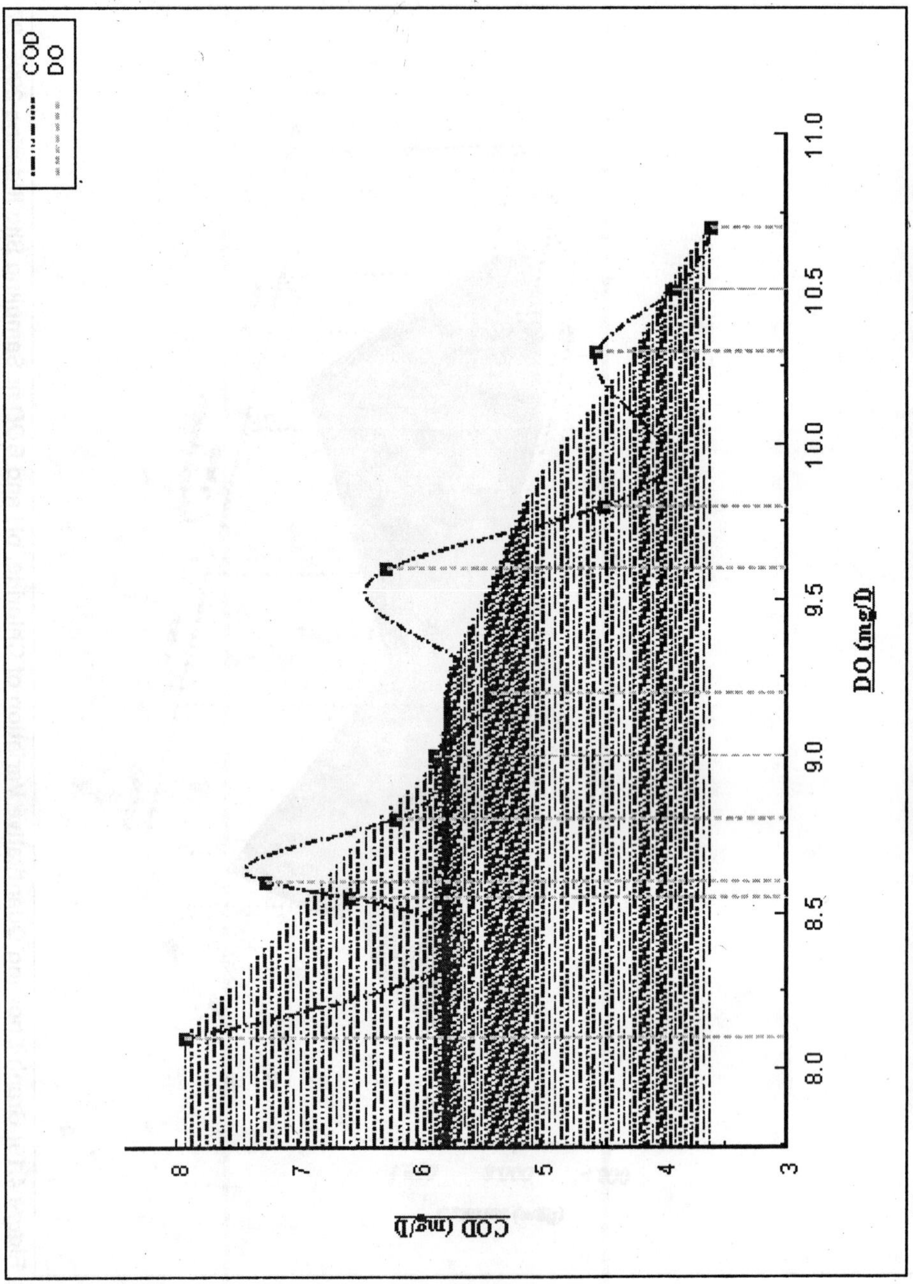

Figure 4.20: Graph Showing Quantitative Variation of COD and DO at Sampling Site II during 2006–07

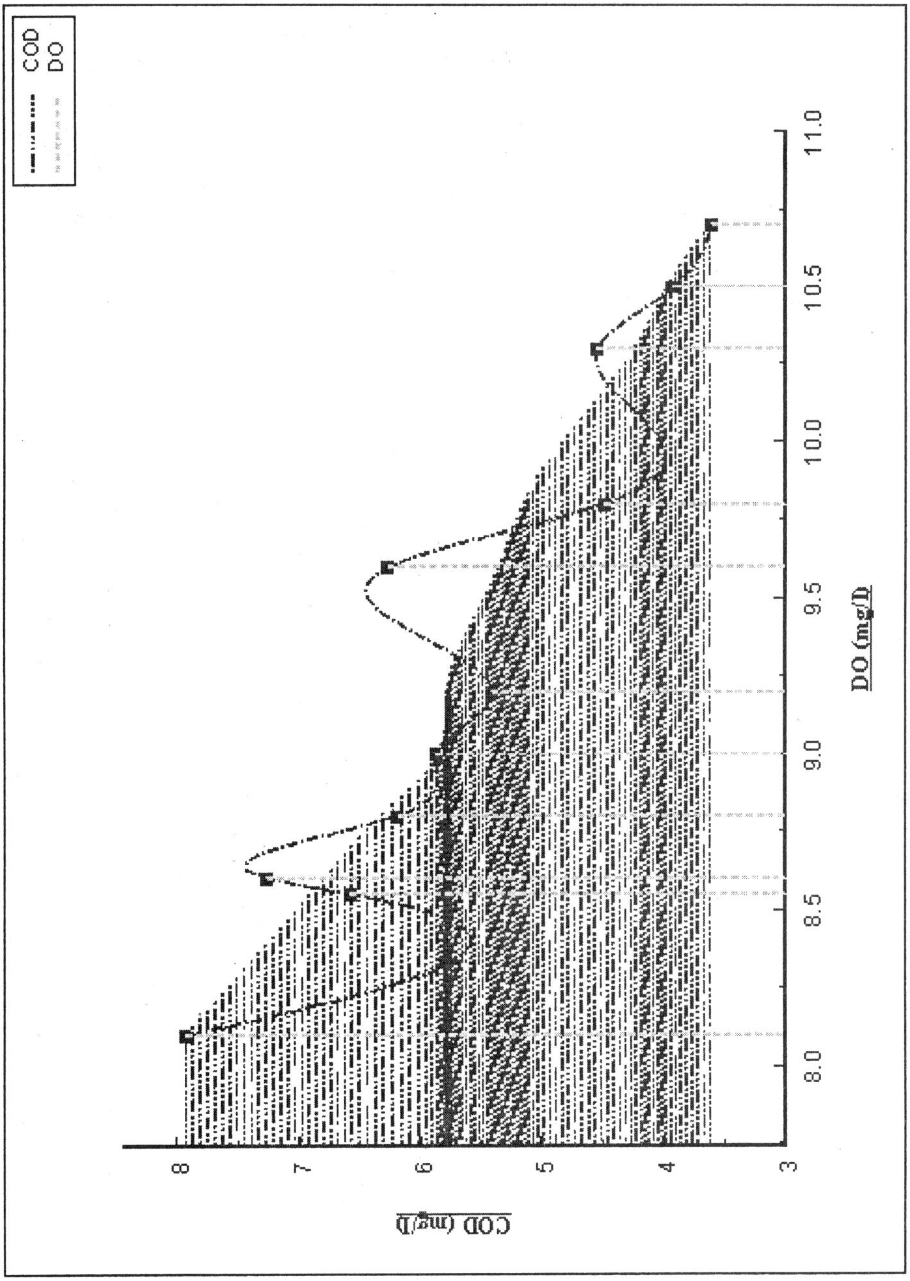

Figure 4.21: Graph Showing Quantitative Variation of Hardness and Alkalinity at Sampling Site II during 2006–07

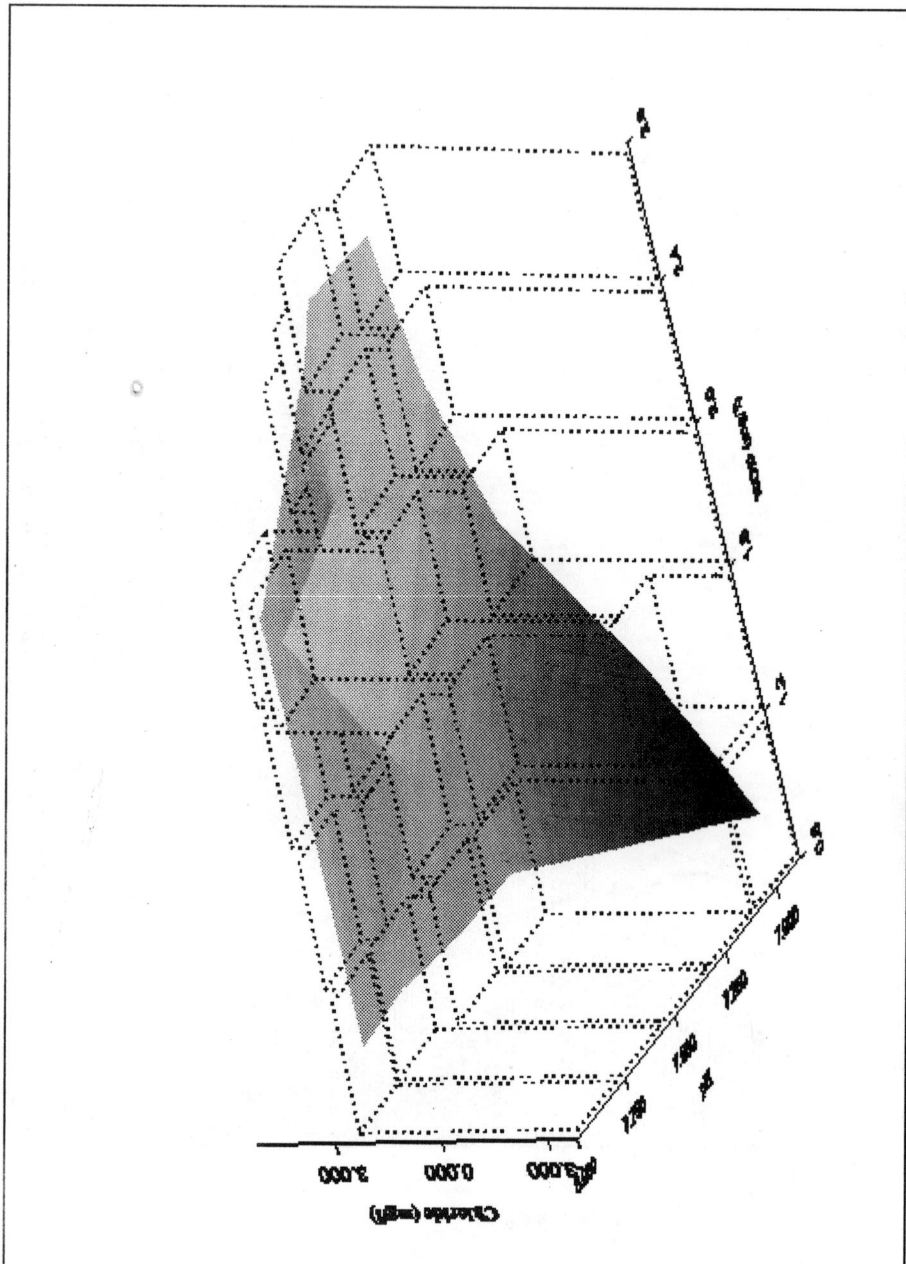

Figure 4.22: Graph Showing Quantitative Variation of Chloride, pH and BOD at Sampling Site II during 2007–08

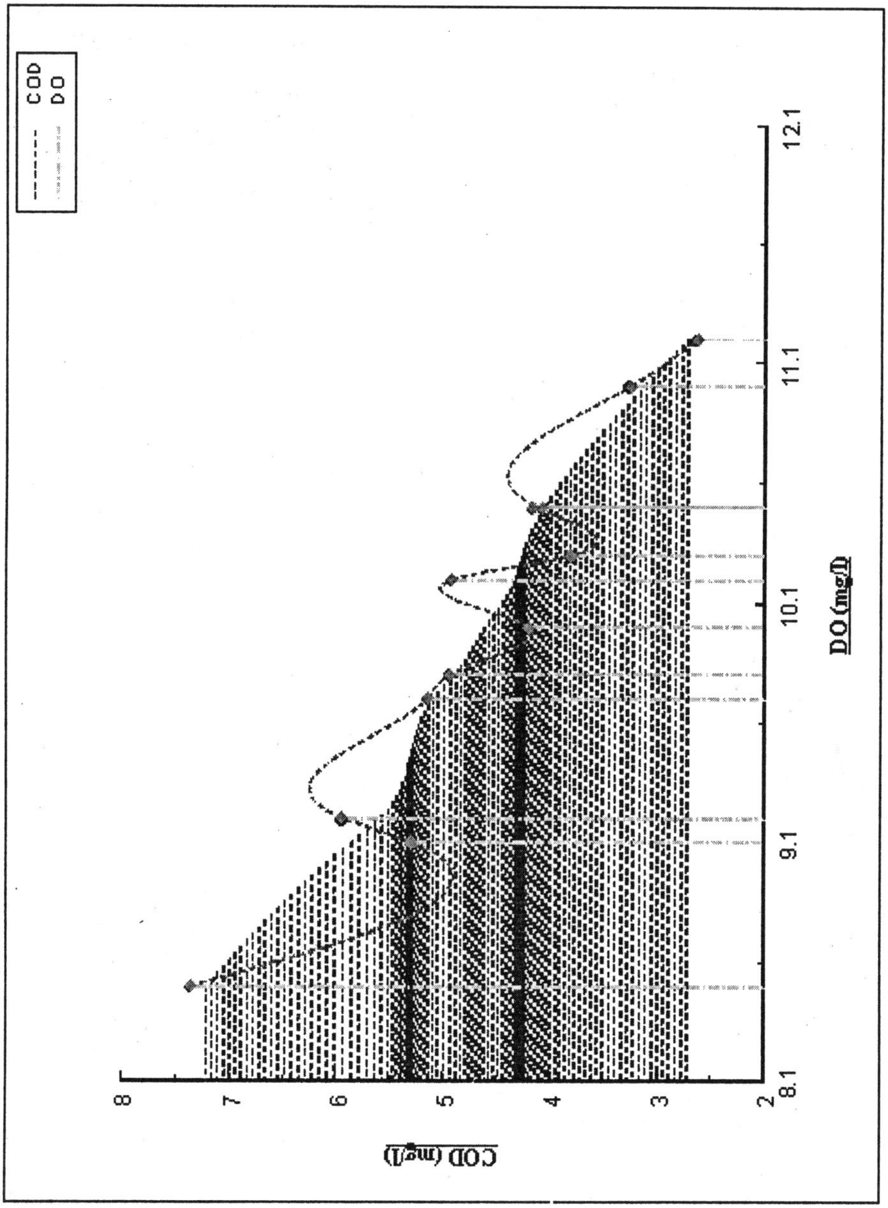

Figure 4.23: Graph Showing Quantitative Variation of COD and DO at Sampling Site II during 2007–08

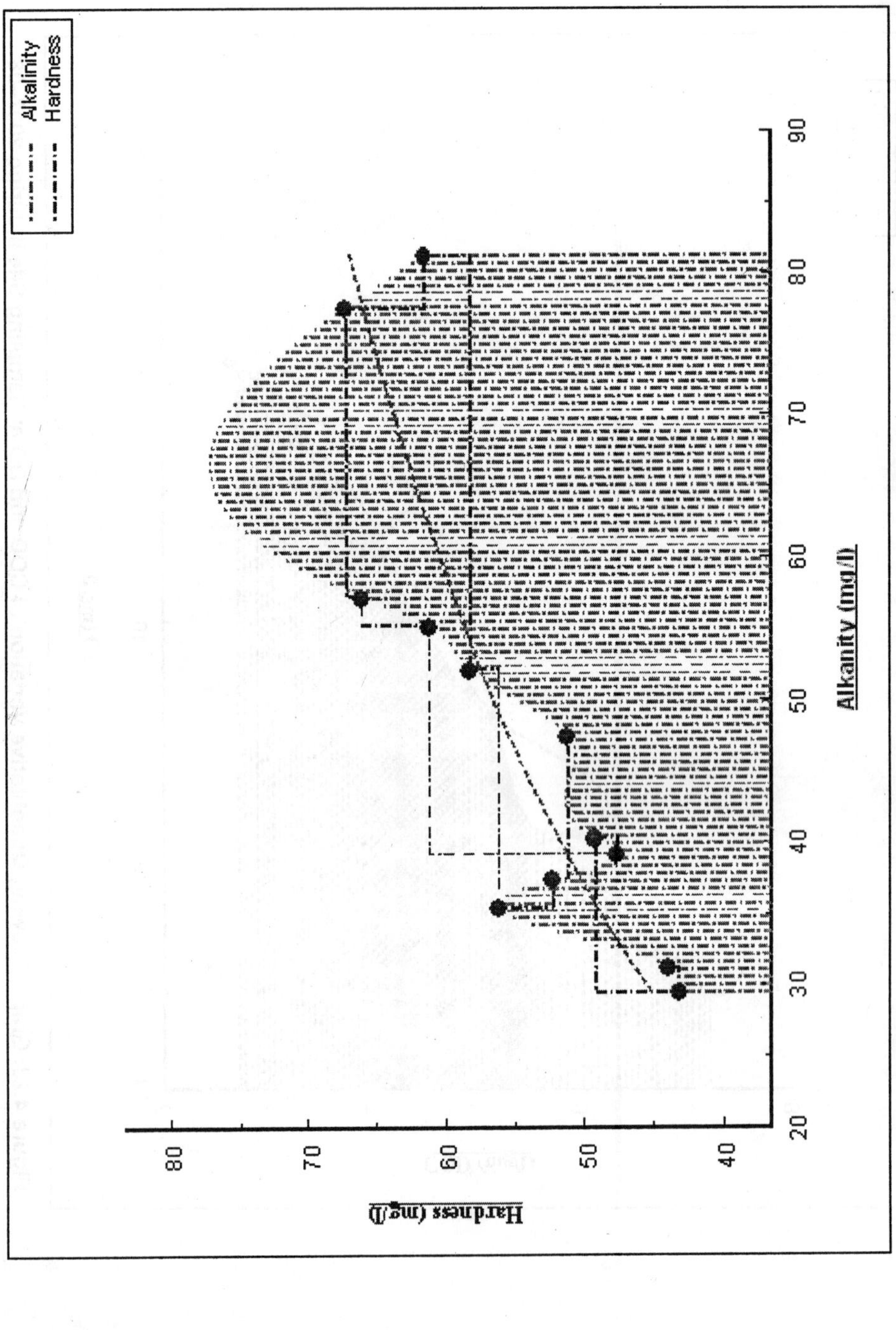

Figure 4.24: Graph Showing Quantitative Variation of Hardness and Alkalinity at Sampling Site II during 2007–08

pH

pH at this sampling site was found minimum 7.00 ±0.15 in the month of September in 2007-2008 and maximum 7.89 ±0.08 in the month of January in 2007-2008. The average value in the year 2006-2007 was noted 7.38 ±0.27 and in the year 2007-2008 it was observed 7.34 ±0.26.

Total Hardness (mg/l)

The total hardness at this sampling site was recorded minimum 31.50 mg/l ±1.42 in the month of September in 2006-2007 and maximum 67.25 mg/l ±0.38 in the month of January in 2007-2008. The average value in the year 2006-2007 was observed 50.68 mg/l ±9.54 and in the year 2007-2008 it was noted 54.84 mg/l ±8.17.

Chloride (mg/l)

The chloride at this sampling site was observed minimum 1.20 mg/l ±0.11 in the month of November in 2007-2008 and maximum 3.84 mg/l ±0.13 in the month of August in 2006-2007. The average value in the year 2006-2007 was noted 2.22 mg/l ±0.83 and in the year 2007-2008 it was observed 2.48 mg/l ±0.78.

Sampling Site III (Bhaldiyana)

The results of physico chemical parameters for the sampling site III are presented in the Tables 4.9 to 4.12 and Figures 4.25 to 4.36.

Water Temperature (°C)

The water temperature at this sampling site was observed minimum 9.20 °C ±0.25 in the month of January in 2007-2008 and maximum 29.30 °C ±0.24 in the month of July in 2006-2007. The average value in the year 2006-2007 was recorded 21.01 °C ±4.89 in the year 2007-2008 it was recorded 21.60 °C ±6.26.

Total Solids (mg/l)

The total solids at this sampling site were found minimum 89.50 mg/l ±1.36 in the month of February in 2007-2008 and maximum 1429.21 mg/l ±2.01 in the month of July in 2006-2007. The average value in the year 2006-2007 was observed 552.94 mg/l ±504.17 and in the year 2007-2008 it was recorded 337.40 mg/l ±323.11.

Total Dissolved Solids (mg/l)

The total dissolved solids at this sampling site were observed minimum 53.45 mg/l ±1.23, in the month of February in year 2007-2008 and maximum 190.32 mg/l ± 1.24 in the month of July in year 2006-2007. The average value in the year 2006-2007 was observed 89.42 mg/l ±42.50 and in the year 2007-2008 it was recorded 67.81 mg/l ±12.53.

Total Suspended Solids (mg/l)

The total suspended solids at this sampling site were recorded minimum 36.05 mg/l ±1.01 in the month of February in year 2007-2008 and maximum 1238.89 mg/l ± 2.16 in the month of July in year 2006-2007. The average value in the year 2006-2007 was observed 463.52 mg/l ±464.51 and in the year 2007-2008 it was noted 269.59 mg/l ±311.53.

Table 4.9: Monthly Variation in Physical Parameters of Tehri Reservoir at Sampling Station III (Bhaldiyana) during 2006–2007

Month	*Temp. (°C)*	*Total Solids (mg/l)*	*TDS (mg/l)*	*TSS (mg/l)*	*Velocity (m/s)*	*Turbidity (JTU)*
July	29.30±0.24	1429.21±2.01	190.32±1.24	1238.89±2.16	0.00±0.00	75.10±1.35
August	25.80±0.34	1310.57±2.15	158.45±1.21	1152.12±2.12	0.00±0.00	71.20±1.34
September	23.10±0.31	1121.05±2.00	118.34±1.11	1002.71±2.13	0.00±0.00	37.50±1.23
October	23.50±0.34	1021.24±1.89	97.32±1.05	923.92±1.85	0.00±0.00	31.00±1.85
November	21.10±0.11	681.55±1.84	82.43±1.12	599.12±1.44	0.00±0.00	25.40±1.23
December	17.50±0.21	299.21±1.87	67.50±1.16	231.71±1.75	0.00±0.00	12.70±0.85
January	12.40±0.15	185.50±1.46	63.20±1.24	122.30±1.43	0.00±0.00	8.10±0.79
February	11.90±0.16	151.54±1.54	61.43±1.20	90.11±1.62	0.00±0.00	7.50±0.46
March	19.50±0.08	115.21±1.52	59.24±1.13	55.97±1.21	0.00±0.00	4.20±0.05
April	21.50±0.19	96.50±1.13	56.94±1.01	39.56±1.35	0.00±0.00	1.10±0.03
May	22.20±0.24	101.21±1.11	57.30±1.02	43.91±1.48	0.00±0.00	6.00±0.09
June	24.30±0.22	122.50±1.12	60.54±1.03	61.96±1.38	0.00±0.00	6.10±0.50
Average± SD	21.01±4.89	552.94±504.17	89.42±42.50	463.52±464.51	0.00±0.00	23.83±24.64

Table 4.10: Monthly Variation in Physical Parameters of Tehri Reservoir at Sampling Station III (Bhaldiyana) during 2007–2008

Month	Temp. (°C)	Total Solids (mg/l)	TDS (mg/l)	TSS (mg/l)	Velocity (m/s)	Turbidity (JTU)
July	28.10±0.88	945.00±1.78	93.50±1.05	851.50±2.51	0.00±0.00	37.00±1.21
August	27.60±0.75	871.63±1.68	82.38±1.01	789.25±2.13	0.00±0.00	41.10±1.32
September	26.50±0.79	675.50±1.49	79.96±1.16	595.54±2.00	0.00±0.00	27.30±1.16
October	25.80±0.29	489.47±1.94	71.97±1.24	417.50±1.15	0.00±0.00	19.10±0.46
November	21.20±0.61	207.50±1.52	60.11±1.13	147.39±1.87	0.00±0.00	12.50±0.78
December	17.50±0.54	108.35±1.67	64.85±1.16	43.50±1.46	0.00±0.00	2.50±0.42
January	9.20±0.25	91.84±1.12	54.18±1.28	37.66±1.13	0.00±0.00	1.50±0.41
February	11.20±0.46	89.50±1.36	53.45±1.23	36.05±1.01	0.00±0.00	0.90±0.10
March	21.30±0.37	97.94±1.54	56.34±1.20	41.60±1.06	0.00±0.00	1.10±0.46
April	20.10±0.48	108.05±1.67	58.23±1.09	49.82±1.24	0.00±0.00	1.30±0.11
May	24.50±0.16	122.47±1.22	68.43±1.06	54.04±1.04	0.00±0.00	7.50±0.08
June	26.30±0.34	241.55±1.41	70.34±1.52	171.21±1.00	0.00±0.00	12.20±0.07
Average± SD	21.60±6.26	337.4±323.11	67.81±12.53	269.59±311.53	0.00±0.00	11.17±12.16

Table 4.11: Monthly Variation in Chemical Parameters of Tehri Reservoir at Sampling Site III (Bhaldiyana) during 2006–2007

Month	*DO (mg/l)*	*COD (mg/l)*	*BOD (mg/l)*	*Alkalinity (mg/l)*	*pH*	*Total Hardness (mg/l)*	*Chloride (mg/l)*
July	7.40±0.15	9.00±0.35	3.20±0.16	34.20±0.16	7.14±0.03	40.15±0.25	2.84±0.29
August	8.50±0.16	6.46±0.26	2.40±0.11	28.50±0.11	7.27±0.08	36.13±0.24	2.70±0.25
September	7.50±0.13	7.62±0.24	3.10±0.19	54.30±0.15	7.18±0.12	45.23 ±0.68	1.63±0.09
October	9.50±0.21	5.97±0.12	2.08±0.16	40.10±0.13	7.61±0.15	56.62±0.15	1.56±0.18
November	10.70±0.16	3.85±0.09	1.43±0.08	78.40±0.09	7.48±0.11	61.25±0.49	1.42±0.26
December	11.70±0.09	4.80±0.13	1.94±0.06	85.30±0.18	7.92±0.16	66.35±1.24	1.24±0.11
January	12.00±0.18	2.10±0.06	0.76±0.07	91.50±0.12	7.40±0.05	68.50±1.49	1.70±0.18
February	11.10±0.25	3.35±0.15	1.25±0.15	79.30±0.21	7.73±0.03	65.21±0.94	1.63±0.09
March	11.50±0.16	3.18±0.24	1.08±0.13	65.80±0.32	7.53±0.08	58.75±1.28	1.06±0.05
April	10.30±0.22	4.63±0.26	1.68±0.07	59.30±0.26	7.59±0.17	55.13±1.53	2.13±0.14
May	9.90±0.18	4.32±0.13	1.75±0.11	39.50±0.13	7.10±0.15	50.30±0.65	2.31±0.05
June	9.60±0.12	5.86±0.11	1.97±0.05	45.20±0.18	7.12±0.13	47.25±0.24	2.08±0.14
Average±SD	9.98±1.56	5.10±1.98	1.89±0.74	58.45±21.48	7.42±0.27	54.24±10.52	1.86±0.56

Table 4.12: Monthly Variation in Chemical Parameters of Tehri Reservoir at Sampling Site III (Bhaldiyana) during 2007–2008

Month	*DO (mg/l)*	*COD (mg/l)*	*BOD (mg/l)*	*Alkalinity (mg/l)*	*pH*	*Total Hardness (mg/l)*	*Chloride (mg/l)*
July	8.90±0.15	6.64±0.05	2.22±0.06	39.40±1.21	7.12±0.21	45.23±1.54	1.90±0.12
August	9.50±0.18	5.26±0.15	2.04±0.05	30.10±0.95	7.18±0.14	40.15±1.15	1.50±0.06
September	9.30±0.25	5.55±0.13	2.14±0.10	55.30±1.42	7.15±0.13	37.79±0.65	1.80±0.09
October	9.70±0.26	5.26±0.21	1.86±0.13	60.20±0.85	7.22±0.19	47.56±0.24	1.60±0.11
November	10.50±0.15	4.13±0.26	1.52±.0.14	70.50±0.56	7.61±0.15	58.12±0.35	1.10±0.16
December	10.10±0.13	4.40±0.19	1.76±0.06	81.50±0.22	7.50±0.09	61.75±0.14	1.20±0.05
January	11.20±0.09	3.22±0.17	1.18±0.03	95.20±0.16	7.80±0.03	62.15±0.25	1.80±.0.18
February	10.90±0.17	3.29±0.13	1.27±0.18	89.20±1.58	7.71±0.04	59.20±0.24	1.20±0.04
March	9.60±0.24	5.55±0.20	1.98±0.20	51.20±1.13	7.42±0.13	58.18±1.19	1.60±0.13
April	8.90±0.05	6.19±0.10	2.33±0.00	45.30±0.54	7.15±0.05	57.13±0.63	1.75±0.02
May	10.40±0.24	3.37±0.16	1.50±0.05	45.30±0.25	7.14±0.11	51.63±0.94	1.90±0.05
June	9.30±0.11	6.77±0.08	2.17±0.14	40.30±0.21	7.23±0.05	51.25±0.25	2.24±0.09
Average±SD	9.86±0.76	4.97±1.27	1.83±0.38	58.63±21.06	7.35±0.25	52.51±8.30	1.63±0.34

Velocity (m/s)

The velocity of surface water at this sampling site was recorded nil throughout the year during the period of study.

Turbidity (JTU)

The turbidity at this sampling site was recorded minimum 0.90 JTU ± 0.10, in the month of February in year 2007-2008 and maximum 75.12 JTU ±1.35 in the month of July year 2006-2007. The average value in the year 2006-2007 was observed 23.83 JTU ±24.64 and in the year 2007-2008 it was noted 11.17 JTU ±12.16.

Dissolved Oxygen (mg/l)

The dissolved Oxygen at this sampling site was found minimum 7.40 mg/l ±0.15, in the month of July in year 2006-2007 and maximum 12.00 mg/l ±0.18 in the month of January in year 2006-2007. The average value in the year 2006-2007 was observed 9.98 mg/l ±1.56 and in the year 2007-2008 it was recorded 9.86 mg/l ±0.76.

Chemical Oxygen Demand (mg/l)

The COD at this sampling site was observed minimum 2.10 mg/l ±0.06, in the month of January in year 2007-2008 and maximum 9.00 mg/l ±0.35 in the month of July in year 2006-2007. The average value in the year 2006-2007 was found 5.10 mg/l ±1.98 and in the year 2007-2008 it was observed 4.97 mg/l ±1.27.

Biochemical Oxygen Demand (mg/l)

The BOD at this sampling site was recorded minimum 0.76 mg/l ±0.07, in the month of January in year 2006-2007 and maximum 3.20 mg/l ±0.16 in the month of July in year 2006-2007. The average value in the year 2006-2007 was recorded 1.89 mg/l ±0.74 and in the year 2007-2008 it was observed 1.83 mg/l ±0.38.

Alkalinity (mg/l)

The alkalinity at this sampling site was observed minimum 28.50 mg/l ±0.11, in the month of August. in year 2006-2007 and maximum 95.20 mg/l ±0.16 in the month of January in year 2007-2008. The average value in the year 2006-2007 was observed 58.45 mg/l ±21.48 and in the year 2007-2008 it was noted 58.63 mg/l ±21.06.

pH

pH at this sampling site was observed minimum 7.10 ±0.15 in the month of

May in year 2006-2007 and maximum 7.92 ±0.16 in the month of December in year 2006-2007. The average value in the year 2006-2007 was observed 7.42 ±0.27 and in the year 2007-2008 it was recorded 7.35±0.25.

Total Hardness (mg/l)

The total hardness at this sampling site was observed minimum 36.13 mg/l ±0.24, in the month of August in year 2006-2007 and maximum 68.50 mg/l ±1.49 in the month of January in year 2006-2007. The average value in the year 2006-2007 was noted 54.24 mg/l ±10.52 and in the year 2007-2008 it was noted 52.51 mg/l ±8.30.

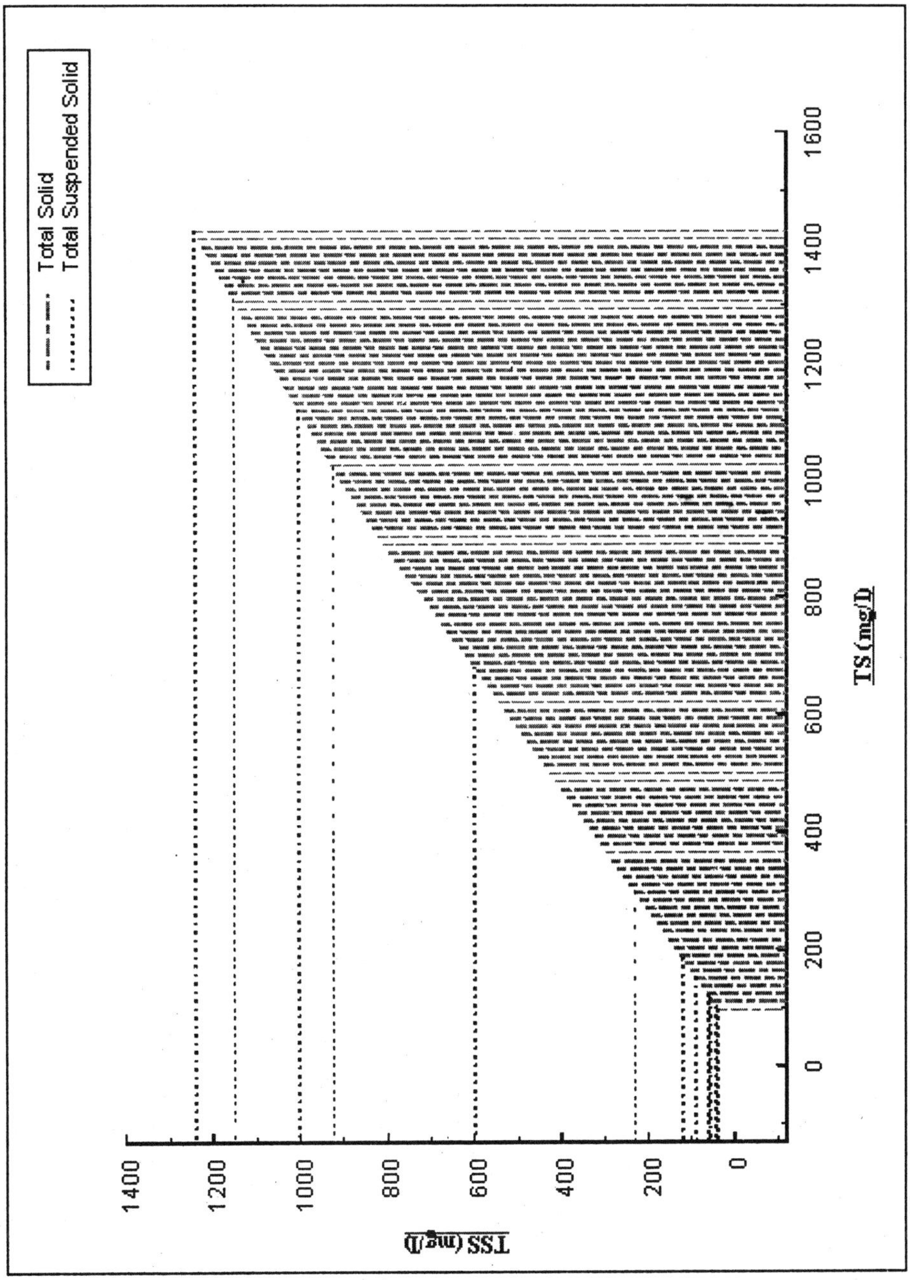

Figure 4. 25: Graph Showing Quantitative Variation of TS and TSS at Sampling Site III during 2006–07

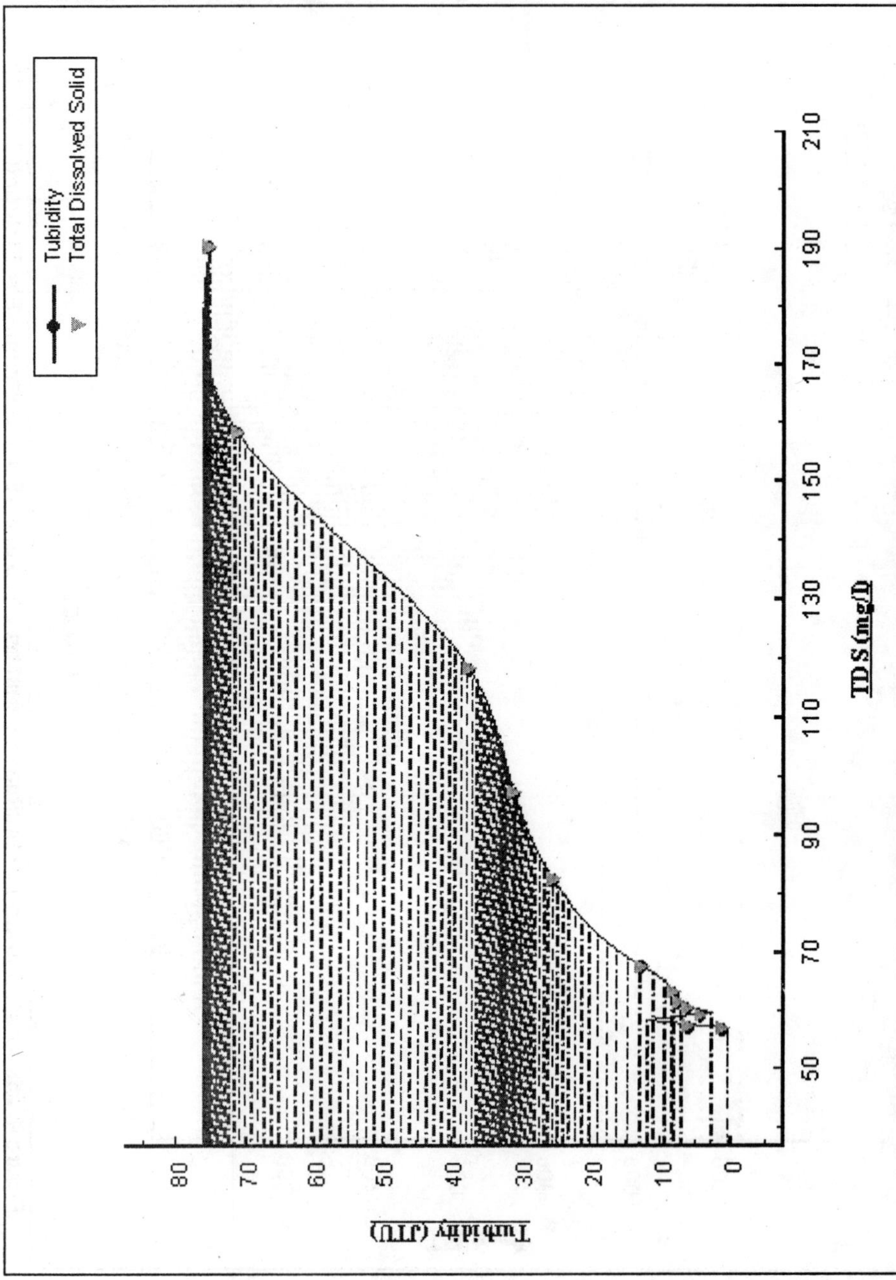

Figure 4.26: Graph Showing Quantitative Variation of Turbidity and TDS at Sampling Site III during 2006–07

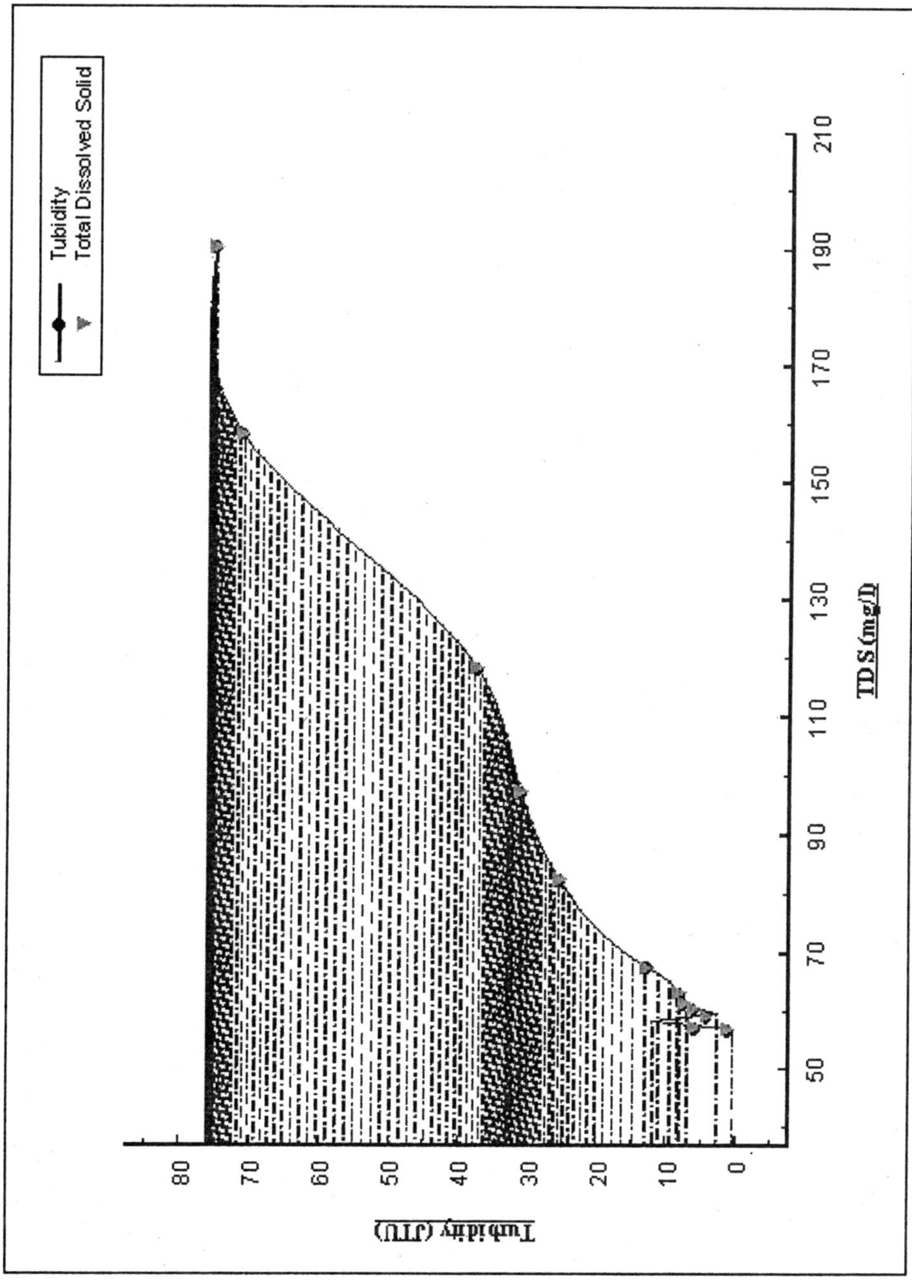

Figure 4.27: Graph Showing Quantitative Variation of Temperature at Sampling Site III during 2006–07

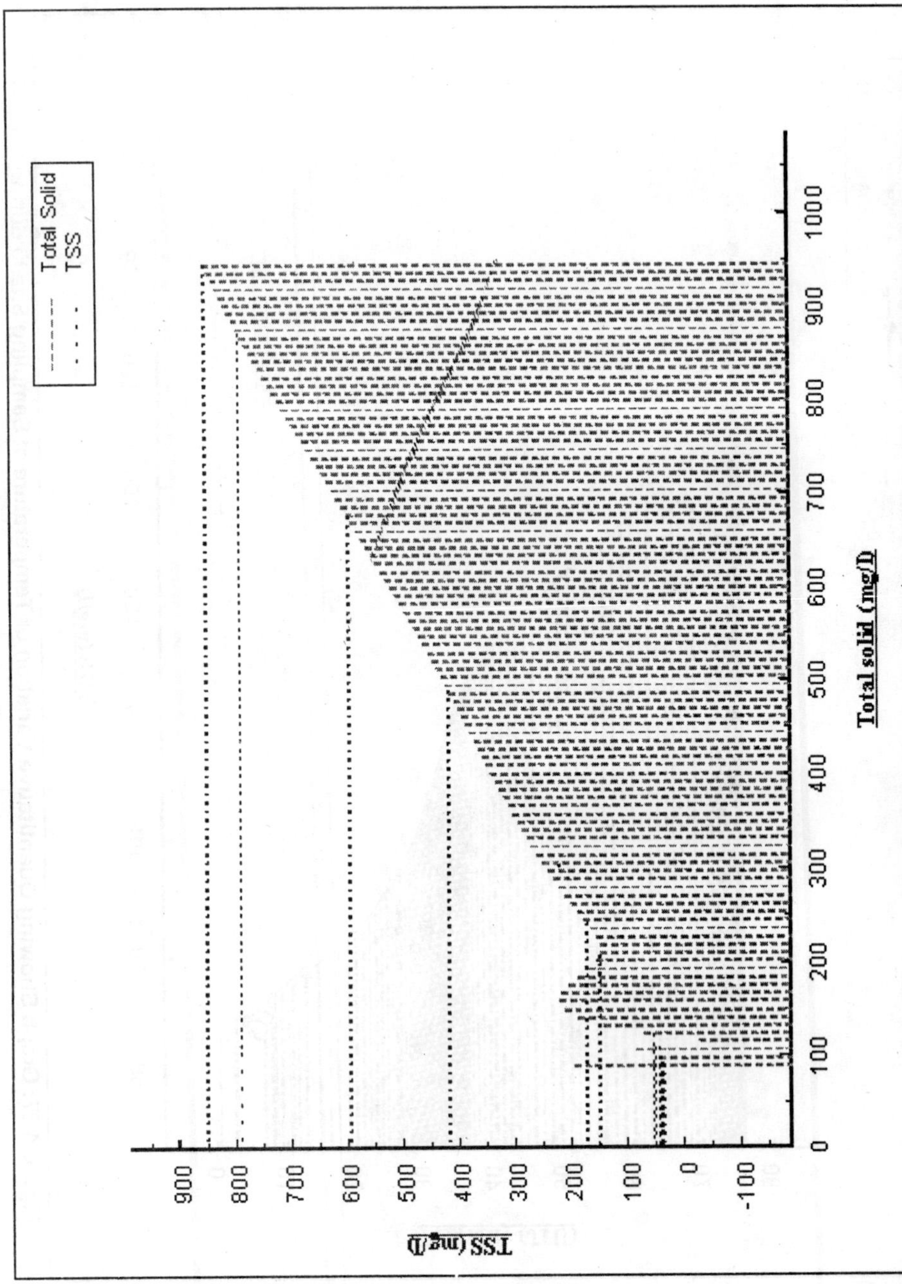

Figure 4.28: Graph Showing Quantitative Variation of TS and TSS at Sampling Site III during 2007–08

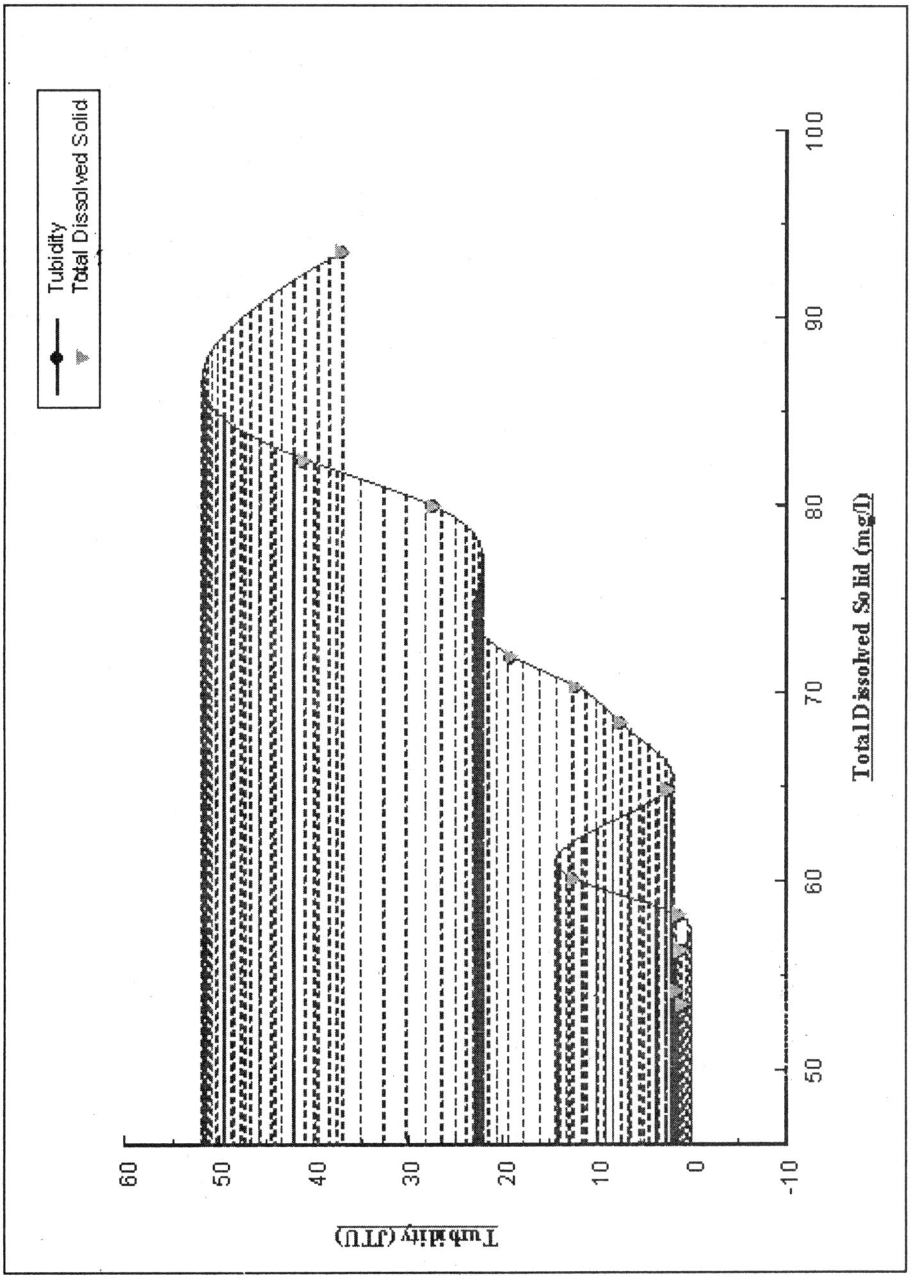

Figure 4.29: Graph Showing Quantitative Variation of Turbidity and TDS at Sampling Site III during 2007–08

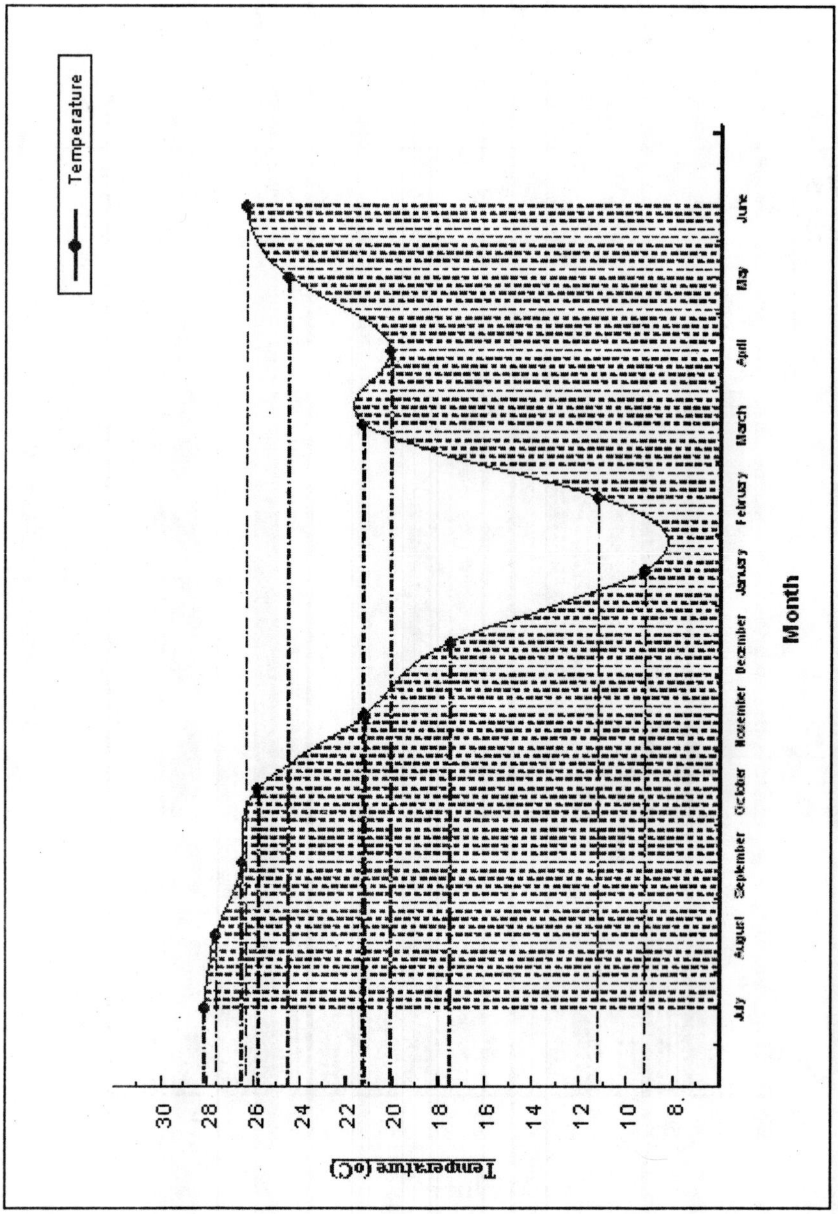

Figure 4.30: Graph Showing Quantitative Variation of Temperature at Sampling Site III during 2007–08

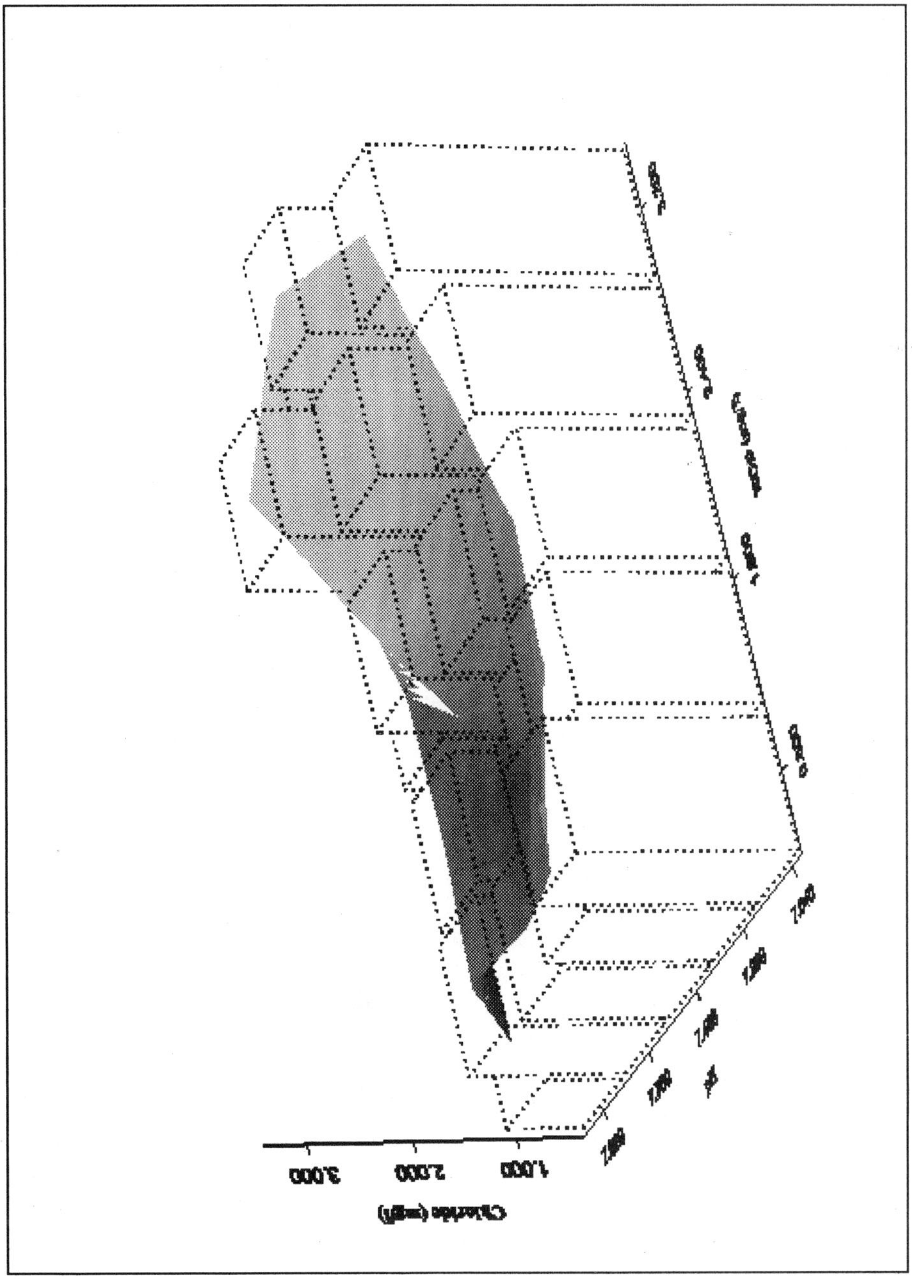

Figure 4.31: Graph Showing Quantitative Variation of Chloride, pH and BOD at Sampling Site III during 2006–07

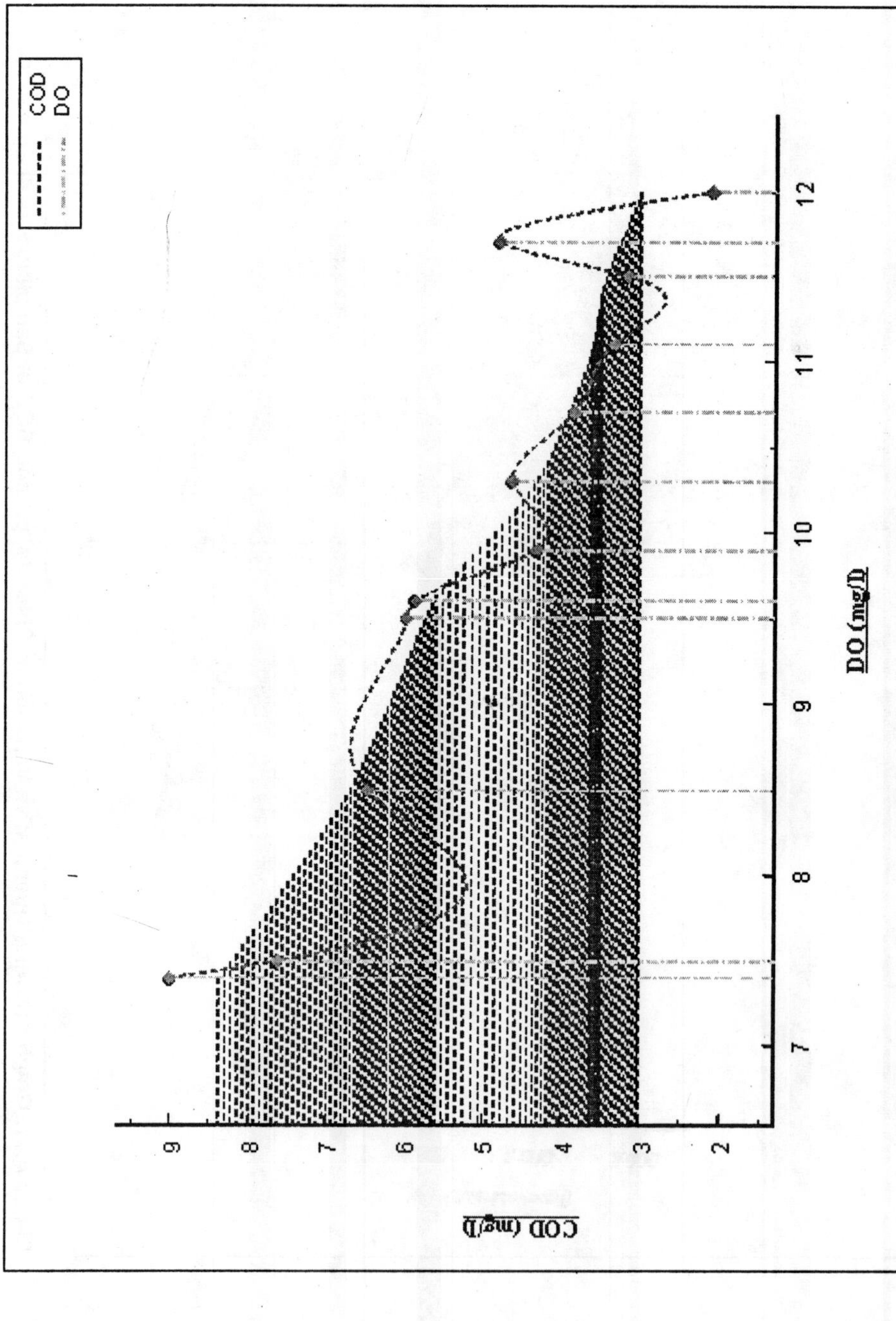

Figure 4.32: Graph Showing Quantitative Variation of COD and DO at Sampling Site III during 2006–07

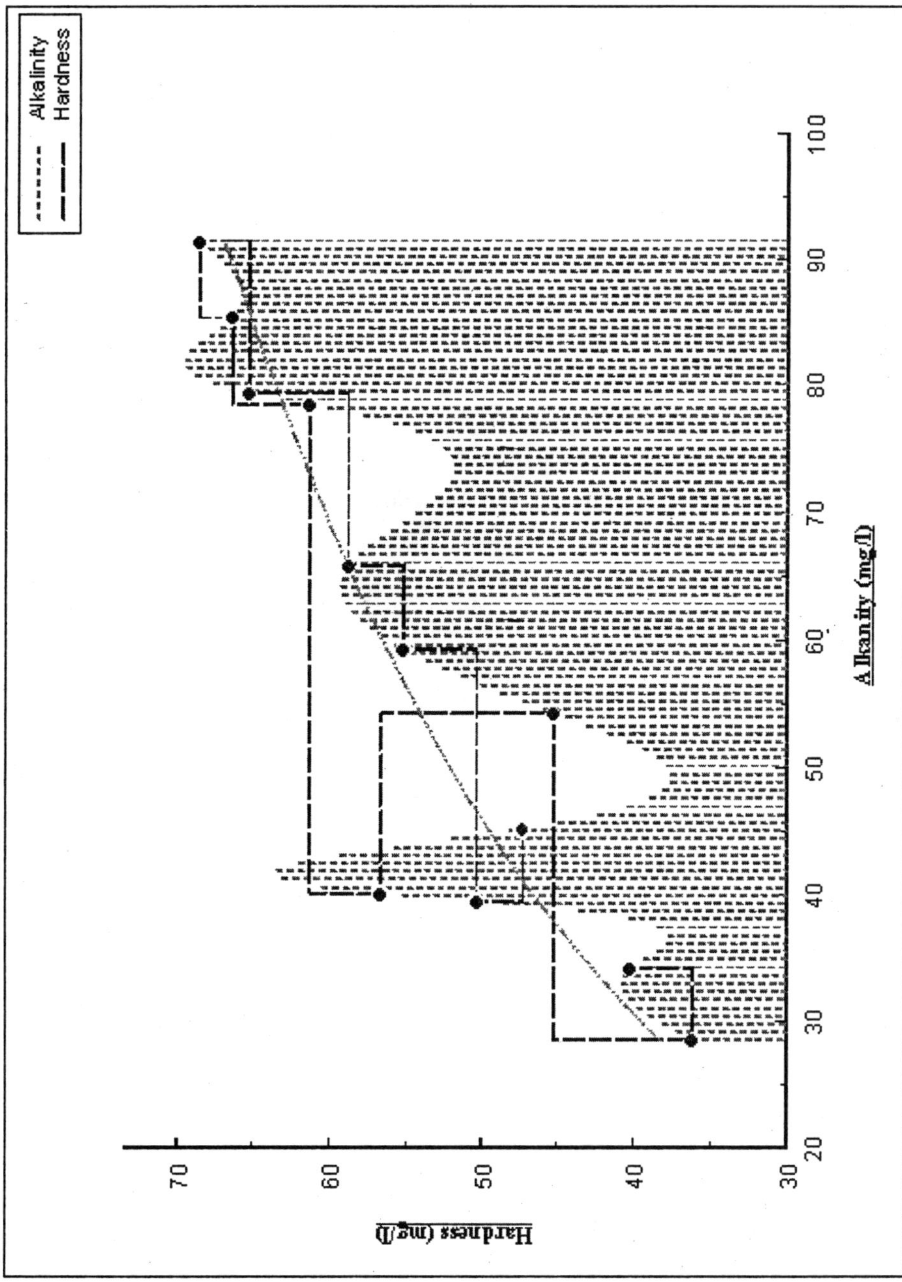

Figure 4.33: Graph Showing Quantitative Variation of Hardness and Alkalinity at Sampling Site III during 2006–07

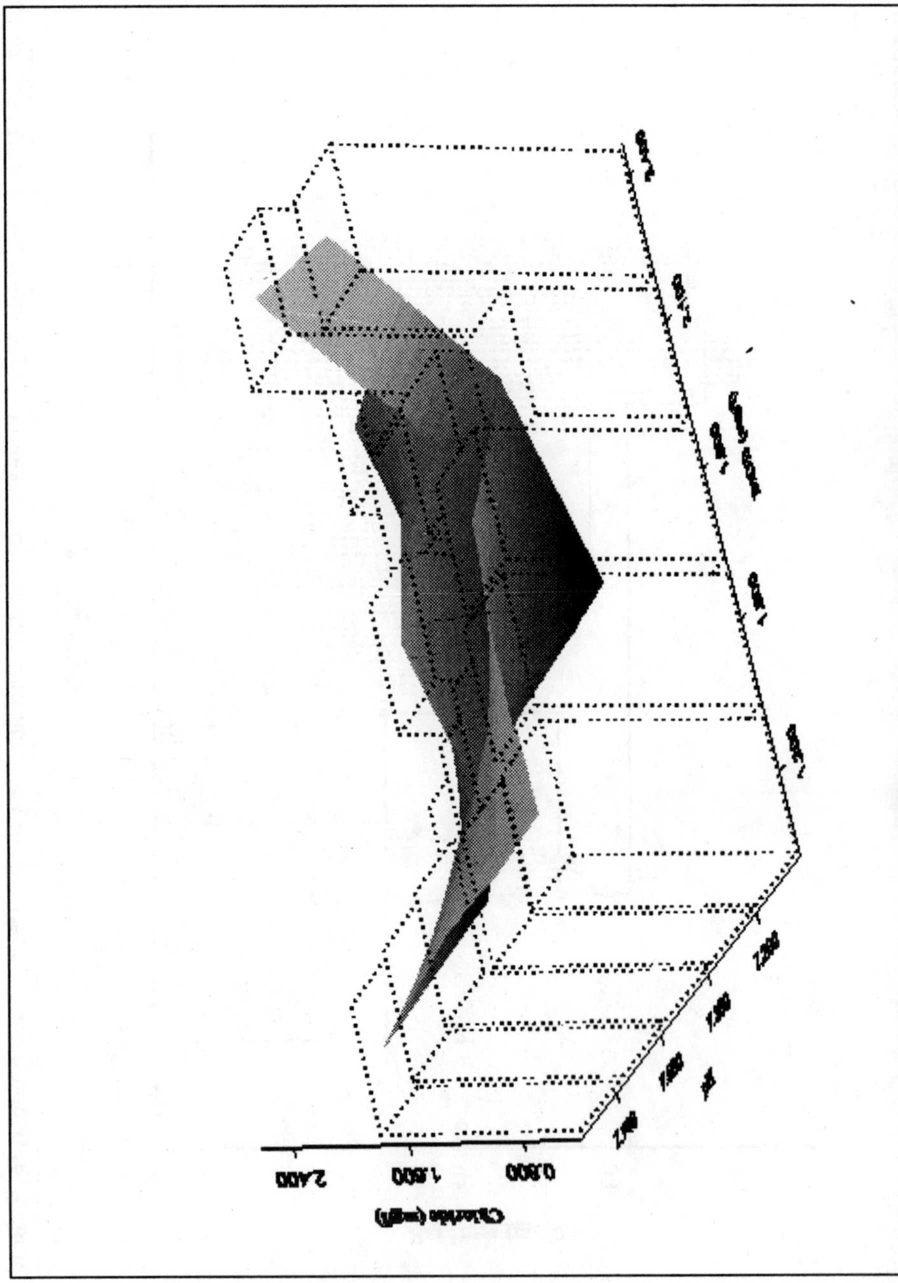

Figure 4.34: Graph Showing Quantitative Variation of Chloride, pH and BOD at Sampling Site III during 2007–08

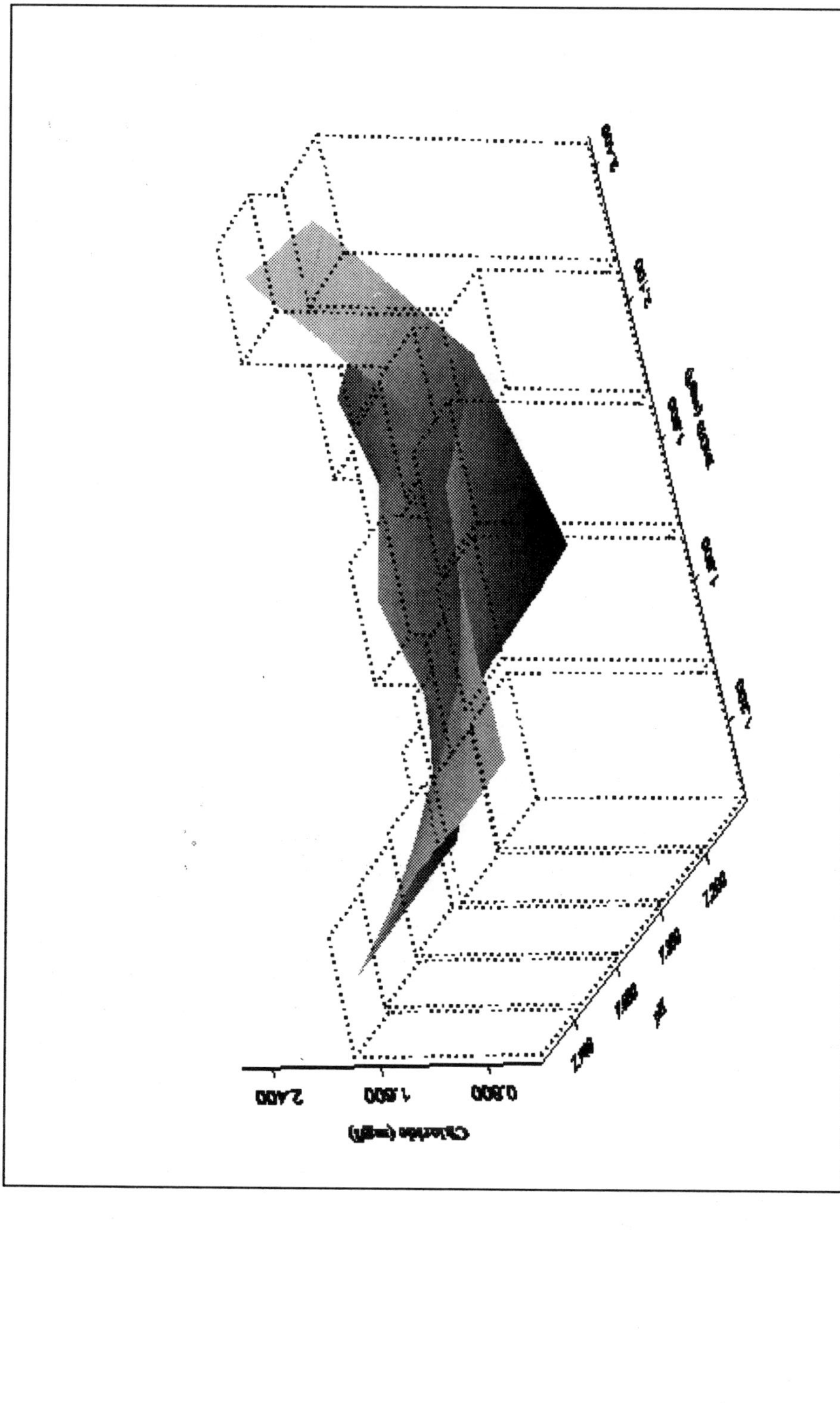

Figure 4.35: Graph Showing Quantitative Variation of COD and DO at Sampling Site III during 2007–08

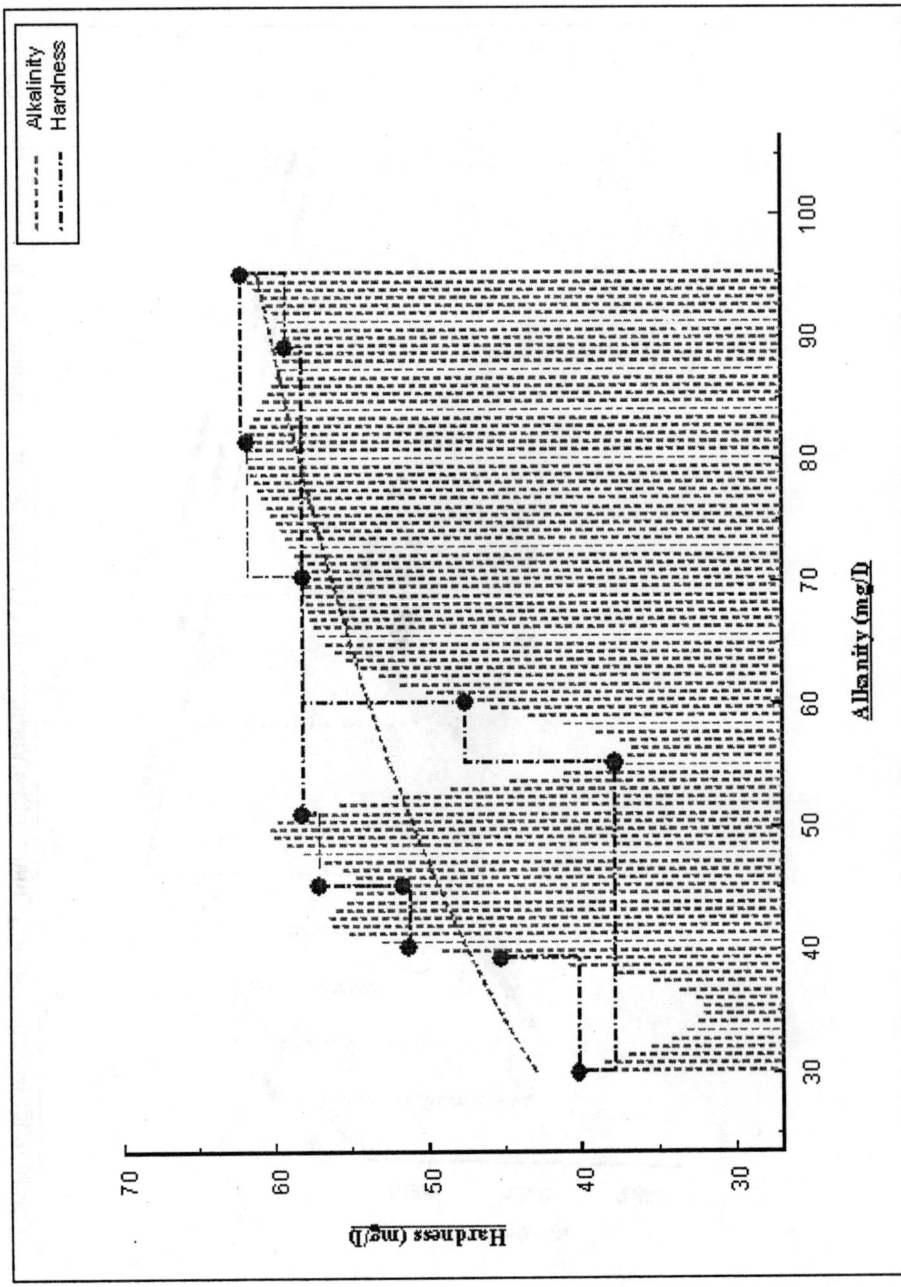

Figure 4.36: Graph Showing Quantitative Variation of Hardness and Alkalinity at Sampling Site III during 2007–08

Chloride (mg/l)

The chloride at this sampling site was recorded minimum1.06 mg/l ±0.05, in the month of March in year 2006-2007 and maximum 2.84 mg/l ±0.29 in the month of July in year 2006-2007. The average value in the year 2006-2007 was recorded 1.86 mg/l ±0.56 and in the year 2007-2008 it was recorded 1.63 mg/l ±0.34.

Sampling Site IV (Koti)

The results of physico-chemical parameters for the sampling site IV are presented in the Tables 4.13 to 4.16 and Figures 4.37 to 4.48.

Water Temperature (°C)

The water temperature at this sampling site was recorded minimum 12.20°C ±0.16, in the month of February in year 2006-2007 and maximum 30.00°C ±0.32 in the month of July in year 2006-2007. The average value in the year 2006-2007 was observed 23.19°C ±6.31 and in the year 2007-2008 it was observed 22.34 °C ±5.20.

Total Solids (mg/l)

The total solids at this sampling site were found minimum 67.00 mg/l ±1.25 in the month of March in 2007-2008 and maximum 1401.50 mg/l ±2.85 in the month of August year 2006-2007. The average value in the year 2006-2007 was found 527.07 mg/l ±520.84 and in the year 2007-2008 it was observed 195.83 mg/l ±156.67.

Total Dissolved Solids (mg/l)

The total dissolved solids at this sampling site were recorded minimum 46.28 mg/l ±1.11 in the month of February in 2007-2008 and maximum 188.30 mg/l ±2.42 in the month of July in 2006-2007. The average value in the year 2006-2007 was observed 92.99 mg/l ±40.36 and in the year 2007-2008 it was observed 67.17 mg/l ±16.85.

Total Suspended Solids (mg/l)

The total suspended solids at this sampling site were observed minimum 19.25 mg/l ±1.00 in the month of March in 2007-2008 and maximum 1213.20 mg/l ±1.52 in the month of July in year 2006-2007. The average value in the year 2006-2007 was noted 434.08 mg/l ±485.84 and in the year 2007-2008 it was noted 128.66 mg/l ±141.18.

Velocity (m/s)

The velocity of surface water at this sampling site was recorded nil throughout the year during the period of study.

Turbidity (JTU)

The turbidity at this sampling site was found minimum 0.00 JTU ±0.00 in the month of January, May and June in 2007-2008 and maximum 39.00 JTU ±0.78 in the month of July in 2006-2007. The average value in the year 2006-2007 was found 16.90 JTU ±14.50 and in the year 2007-2008 it was observed 9.66 JTU ±12.88.

Dissolved Oxygen (mg/l)

The dissolved oxygen at this sampling site was observed minimum 8.80 mg/l ±0.19 in the month of October in 2006-2007 and maximum 12.70 mg/l ±0.22 in the

Table 4.13: Monthly Variation in Physical Parameters of Tehri Reservoir at Sampling Station IV (Koti) during 2006–2007

Month	*Temp. (°C)*	*Total Solids (mg/l)*	*TDS (mg/l)*	*TSS (mg/l)*	*Velocity (m/s)*	*Turbidity (JTU)*
July	30.00±0.32	1401.50±2.85	188.30±2.42	1213.20±1.52	0.00±0.00	39.00±0.78
August	28.50±0.24	1295.20±2.45	141.40±1.25	1153.80±1.30	0.00±0.00	38.50±0.88
September	28.00±0.15	1087.30±2.68	103.50±1.16	983.80±1.26	0.00±0.00	31.00±0.79
October	27.80±0.46	989.35±2.01	99.30±1.16	890.05±1.29	0.00±0.00	30.00±0.72
November	22.50±0.41	587.87±1.45	98.40±1.24	489.47±1.46	0.00±0.00	18.50±0.71
December	16.30±0.27	280.55±1.25	91.40±1.35	189.15±1.35	0.00±0.00	16.00±0.35
January	13.10±0.51	181.50±1.22	89.60±1.42	91.90±1.27	0.00±0.00	13.90±0.28
February	12.20±0.16	162.12±1.08	86.30±1.04	75.82±1.57	0.00±0.00	8.50±0.11
March	19.80±0.31	67.25±1.23	47.55±1.11	19.70±1.11	0.00±0.00	1.80±0.00
April	26.00±0.52	69.50±0.21	49.31±1.21	20.19±1.21	0.00±0.00	1.90±0.20
May	26.50±0.11	74.21±1.05	51.43±1.05	22.78±1.85	0.00±0.00	2.30±0.05
June	27.60±0.28	128.50±1.35	69.41±1.49	59.09±1.90	0.00±0.00	1.50±0.03
Average± SD	23.19±6.31	527.07±520.84	92.99±40.36	434.08±485.84	0.00±0.00	16.90±14.50

Table 4.14: Monthly Variation in Physical Parameters of Tehri Reservoir at Sampling Station IV (Koti) during 2007–2008

Month	*Temp. (°C)*	*Total Solids (mg/l)*	*TDS (mg/l)*	*TSS (mg/l)*	*Velocity (m/s)*	*Turbidity (JTU)*
July	28.90±1.05	427.25±1.85	91.75±1.11	335.50±2.11	0.00±0.00	35.00±0.85
August	27.50±0.25	479.84±1.29	93.63±1.21	386.21±1.85	0.00±0.00	31.00±0.41
September	27.10±0.34	221.50±1.56	76.38±1.34	145.12±1.48	0.00±0.00	16.00±0.21
October	26.20±0.22	421.73±1.44	85.39±1.21	336.34±1.54	0.00±0.00	21.20±0.20
November	22.30±0.13	192.30±1.23	71.05±1.04	121.25±1.34	0.00±0.00	6.40±0.24
December	18.20±0.15	91.50±1.27	51.96±1.08	39.54±1.22	0.00±0.00	2.60±0.12
January	15.20±0.17	77.78±1.38	55.28±1.22	22.50±1.15	0.00±0.00	0.00±0.00
February	13.90±0.16	84.25±1.11	46.28±1.11	37.97±1.42	0.00±0.00	1.10±0.02
March	19.20±0.28	67.00±1.25	47.75±1.03	19.25±1.00	0.00±0.00	0.50±0.01
April	18.10±0.34	94.00±1.90	54.79±1.22	39.21±1.02	0.00±0.00	2.10±0.02
May	25.10±0.10	101.50±1.23	70.50±1.25	31.00±1.00	0.00±0.00	0.00±0.00
June	26.40±0.11	91.25±1.59	61.25±0.88	30.00±1.00	0.00±0.00	0.00±0.00
Average±SD	22.34±5.20	195.83±156.67	67.17±16.85	128.66±141.18	0.00±0.00	9.66± 12.88

Table 4.15: Monthly Variation in Chemical Parameters of Tehri Reservoir at Sampling Site IV (Koti) during 2006–2007

Month	*DO (mg/l)*	*COD (mg/l)*	*BOD (mg/l)*	*Alkalinity (mg/l)*	*pH*	*Total Hardness (mg/l)*	*Chloride (mg/l)*
July	10.00±0.15	4.75± 0.10	1.73±0.03	41.20±1.45	7.04±0.09	45.07±0.12	2.70±0.06
August	9.60±0.20	5.36±0.05	2.00±0.09	27.50±1.28	7.62±0.03	41.10±1.45	2.56±0.05
September	9.10±0.16	5.76±0.14	2.24±0.08	38.50±0.06	7.13±0.08	47.31±1.28	2.56±0.10
October	8.80±0.19	6.51±0.19	2.31±0.10	57.50±0.15	7.32±0.12	55.21±1.33	1.84±0.16
November	10.80±0.05	3.21±0.08	1.26±0.15	62.00±1.42	7.60±0.15	65.31±0.56	1.10±0.13
December	10.10±0.24	4.15±0.06	1.74±0.13	74.00±1.25	7.41±0.11	66.15±0.25	1.20±0.11
January	12.70±0.22	1.37± 0.07	0.46±0.12	91.00±0.64	7.89±0.09	67.68±0.34	1.77±0.05
February	10.90±0.18	3.42±0.11	1.32±0.04	78.20±1.13	7.35±0.03	61.15±0.27	1.60±0.03
March	10.70±0.11	4.27±0.17	1.46±0.05	55.74±0.48	7.76±0.15	59.13±0.47	1.50±0.12
April	9.80±0.16	4.74±0.14	1.83±.0.15	55.40±0.25	7.69±0.02	60.36±0.35	1.84±0.20
May	9.70±0.10	4.35±0.06	1.89±0.11	50.50±0.14	7.25±0.07	55.50±1.58	1.41±0.16
June	9.50±0.08	6.16±0.09	2.10±0.02	51.20±0.16	7.24±0.10	57.98±1.65	2.12±0.08
Average±SD	10.14±1.04	4.50±1.42	1.70±0.51	56.90±17.74	7.44±0.27	56.83±8.50	1.85±0.54

Table 4.16: Monthly Variation in Chemical Parameters of Tehri Reservoir at Sampling Site IV (Koti) during 2007–2008

Month	*DO (mg/l)*	*COD (mg/l)*	*BOD (mg/l)*	*Alkalinity (mg/l)*	*pH*	*Total Hardness (mg/l)*	*Chloride (mg/l)*
July	9.10±0.25	6.57±0.20	2.23±0.06	31.20±0.52	7.16±0.02	51.78±2.15	2.10±0.03
August	9.90±0.28	4.61±0.13	1.83±0.03	37.20±0.43	7.14±0.09	52.72±1.54	1.71±0.05
September	10.10±0.12	4.52±0.25	1.77±0.15	27.20±0.48	7.30±0.10	46.23±1.35	1.12±0.10
October	9.70±0.10	5.81± 0.17	2.13±0.21	65.32±0.12	7.20±0.15	55.21±0.52	1.71±0.18
November	10.50±0.24	3.94±0.13	1.46±0.15	69.30±0.24	7.19±0.09	61.00±0.15	1.10±0.10
December	10.40±0.15	3.55±0.10	1.50±0.10	79.80±0.17	7.70±0.04	62.13±1.64	1.10±0.05
January	10.20±0.08	4.45±0.05	1.66±0.06	91.40±1.52	7.36±0.13	64.10±0.28	0.90±0.03
February	11.40±0.13	2.33±0.09	1.00±0.11	75.40±1.34	7.40±0.07	63.85±0.46	1.45±0.08
March	10.90±0.05	3.52± 0.10	1.20±0.06	48.40±1.49	7.46±0.11	56.10±1.27	1.70±0.10
April	10.50±0.06	4.20±0.19	1.48±0.18	59.20±0.65	7.30±0.08	56.29±0.35	1.80±0.15
May	10.10±0.13	4.31± 013	1.73±0.13	47.50±0.26	7.40±0.10	56.43±0.24	2.10±016
June	9.20±0.18	6.47± 0.18	2.17±0.04	43.50±0.18	7.50±0.13	55.23±1.28	2.75±0.13
Average±SD	10.17±0.65	4.52±1.24	1.68±0.38	56.29±20.35	7.34±0.16	57.76±5.30	1.63±0.53

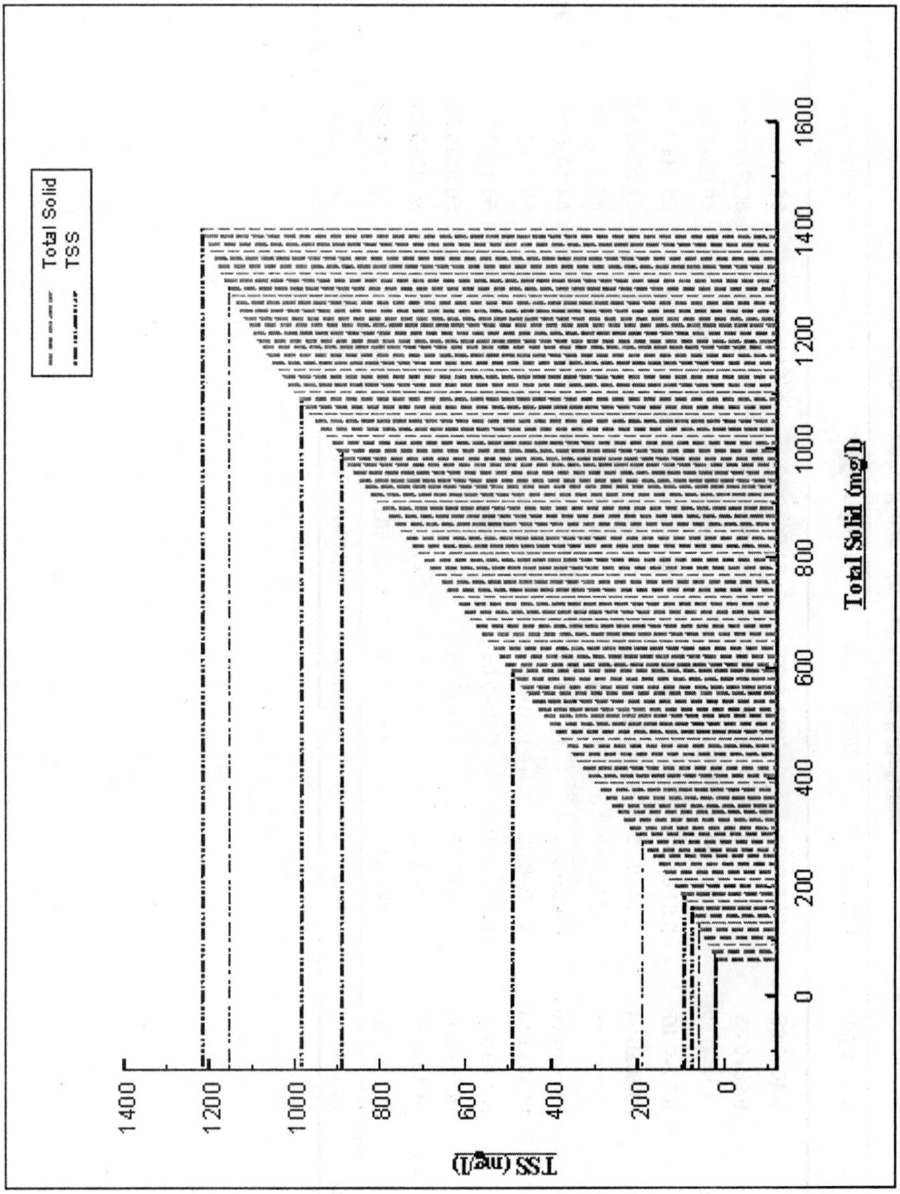

Figure 4.37: Graph Showing Quantitative Variation of TS and TSS at Sampling Site IV during 2006–07

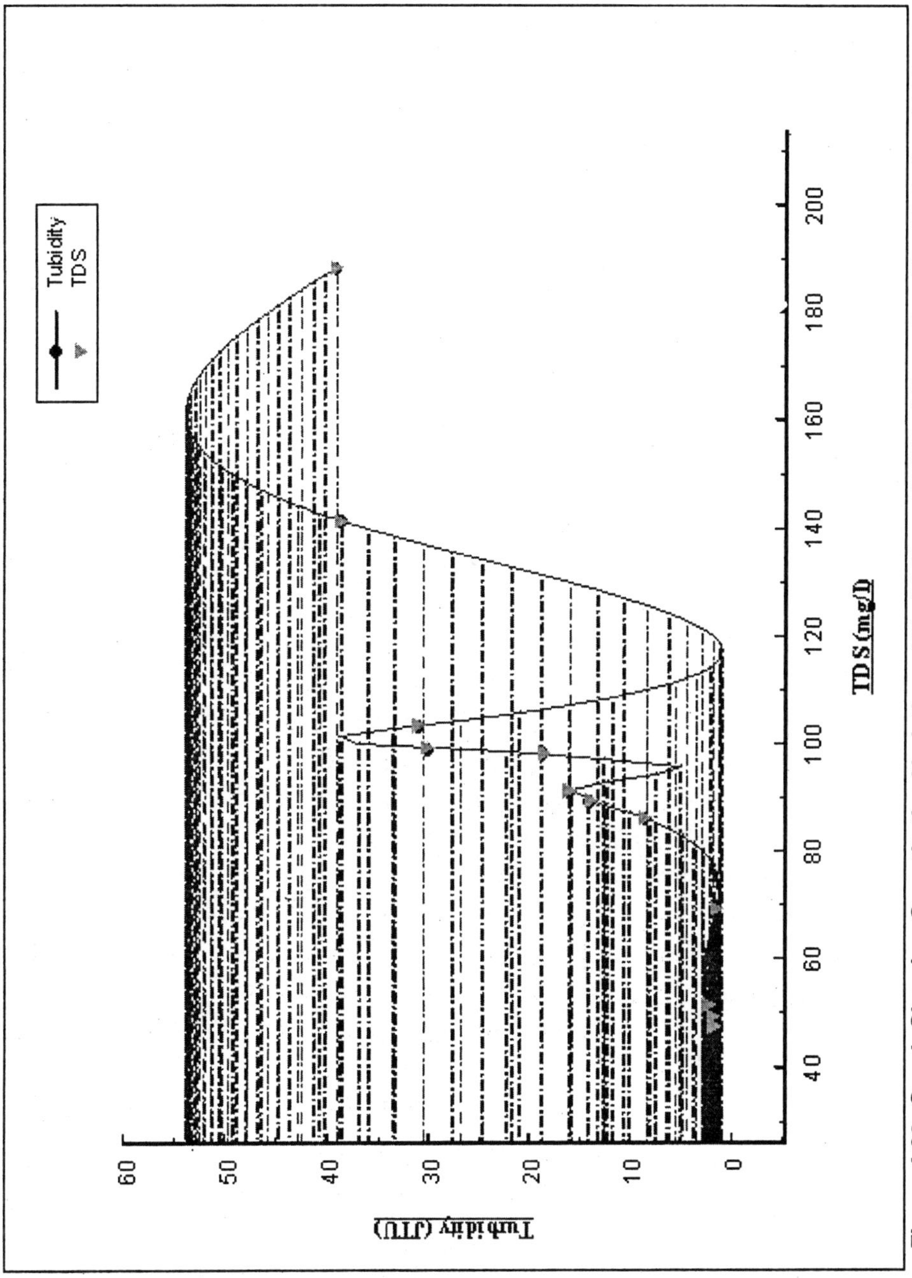

Figure 4.38: Graph Showing Quantitative Variation of Turbidity and TDS at Sampling Site IV during 2006–07

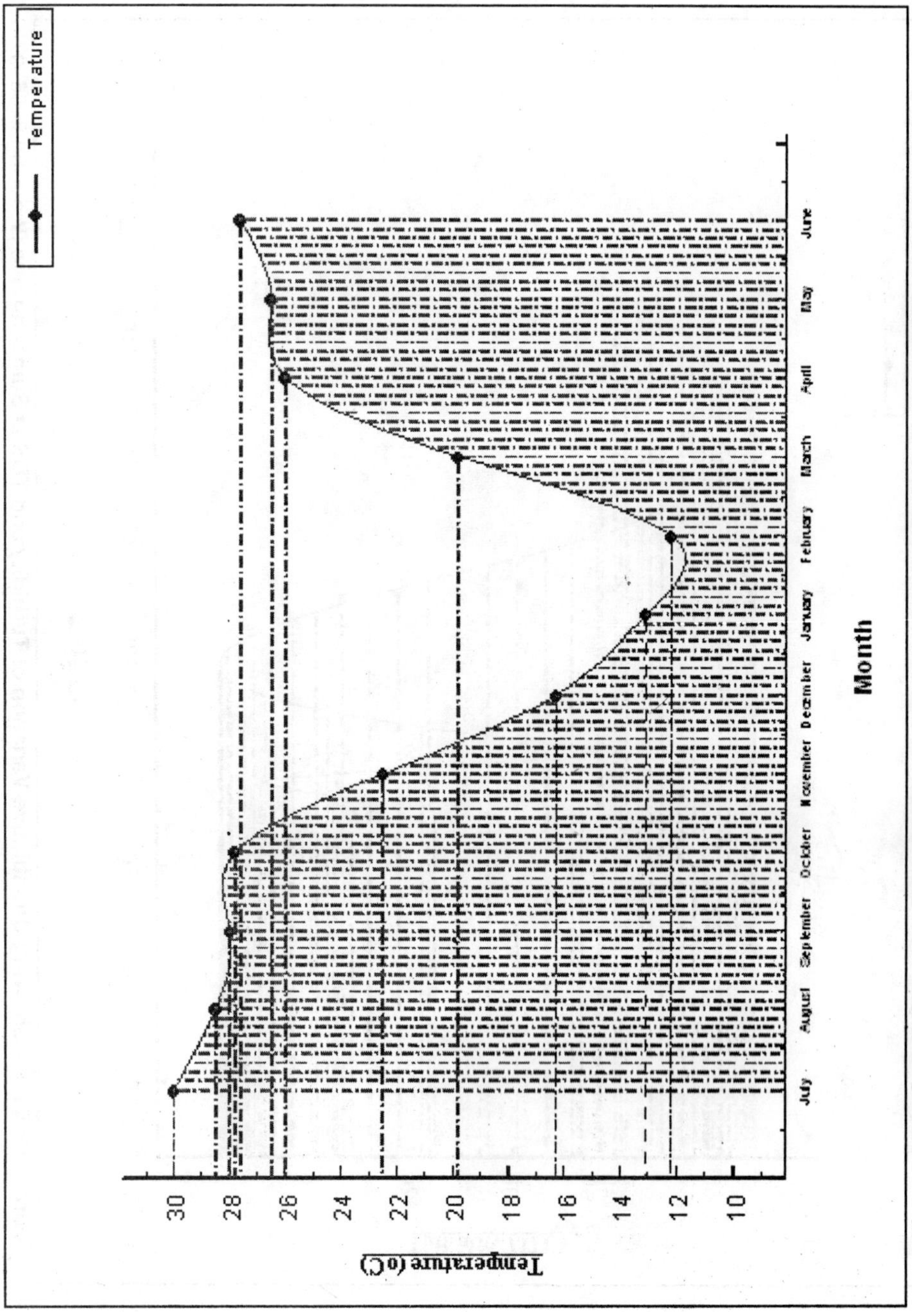

Figure 4.39: Graph Showing Quantitative Variation of Temperature at Sampling Site IV during 2006–07

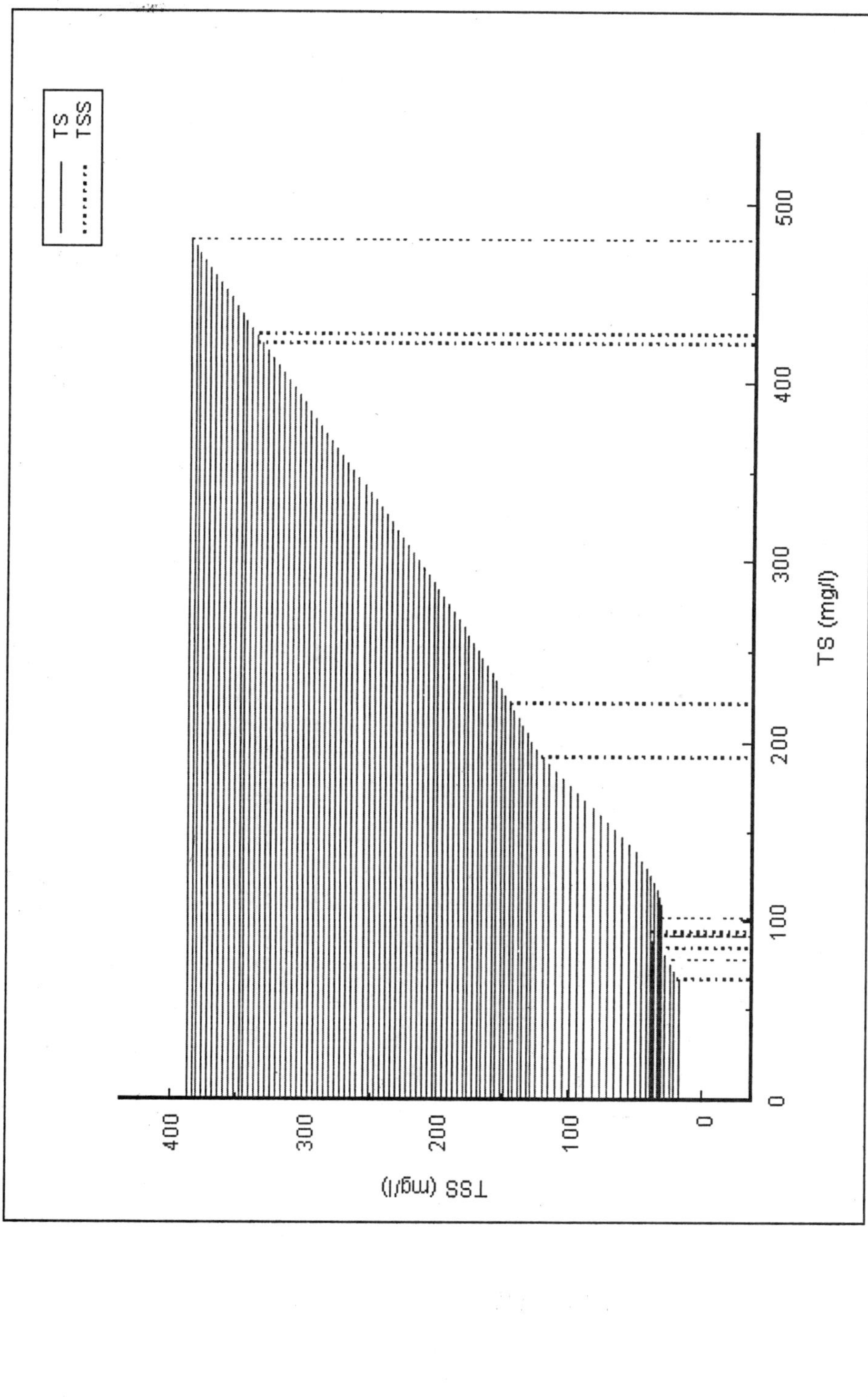

Figure 4.40: Graph Showing Quantitative Variation of TS and TSS at Sampling Site IV during 2007–08

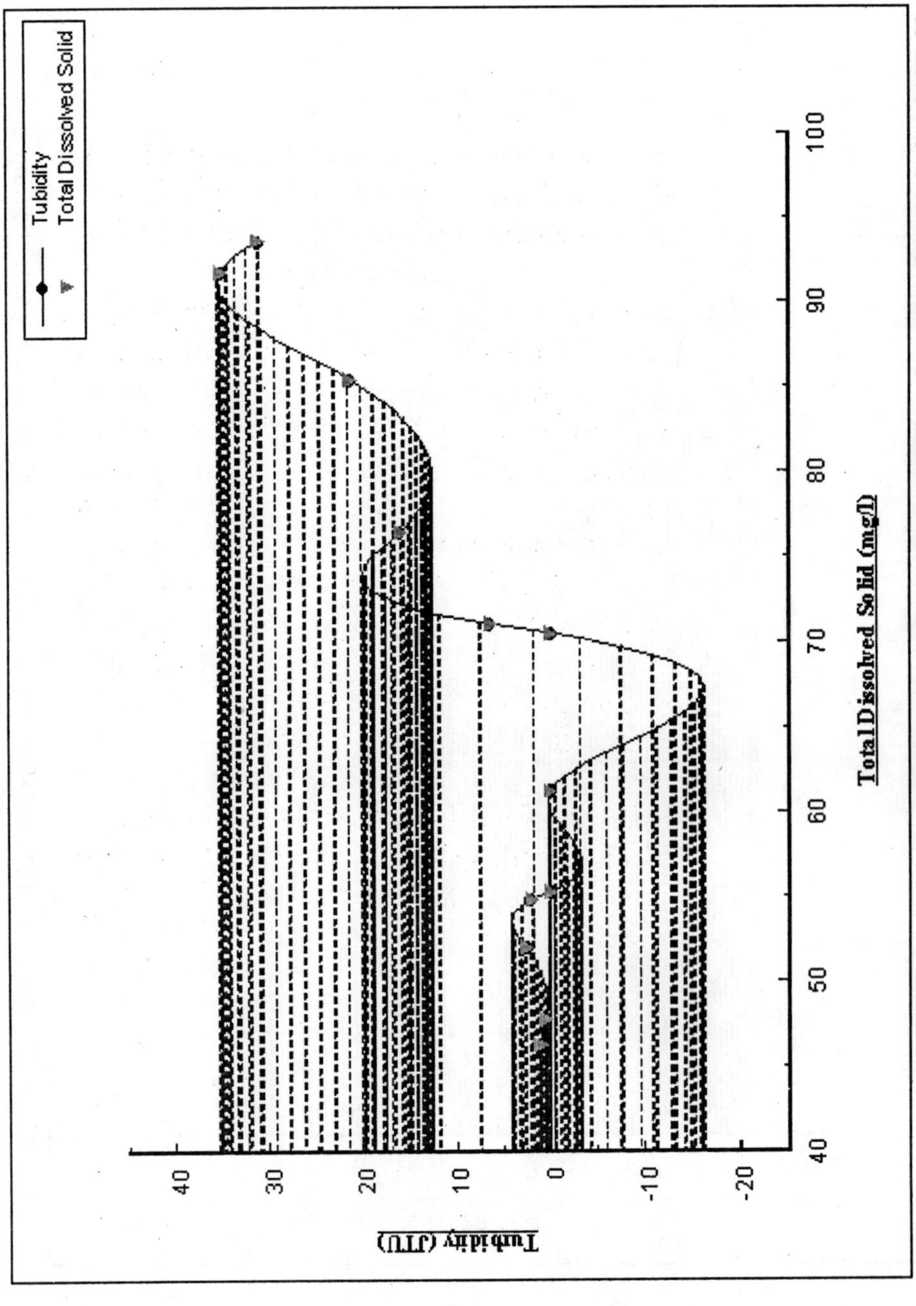

Figure 4.41: Graph Showing Quantitative Variation of Turbidity and TDS at Sampling Site IV during 2007–08

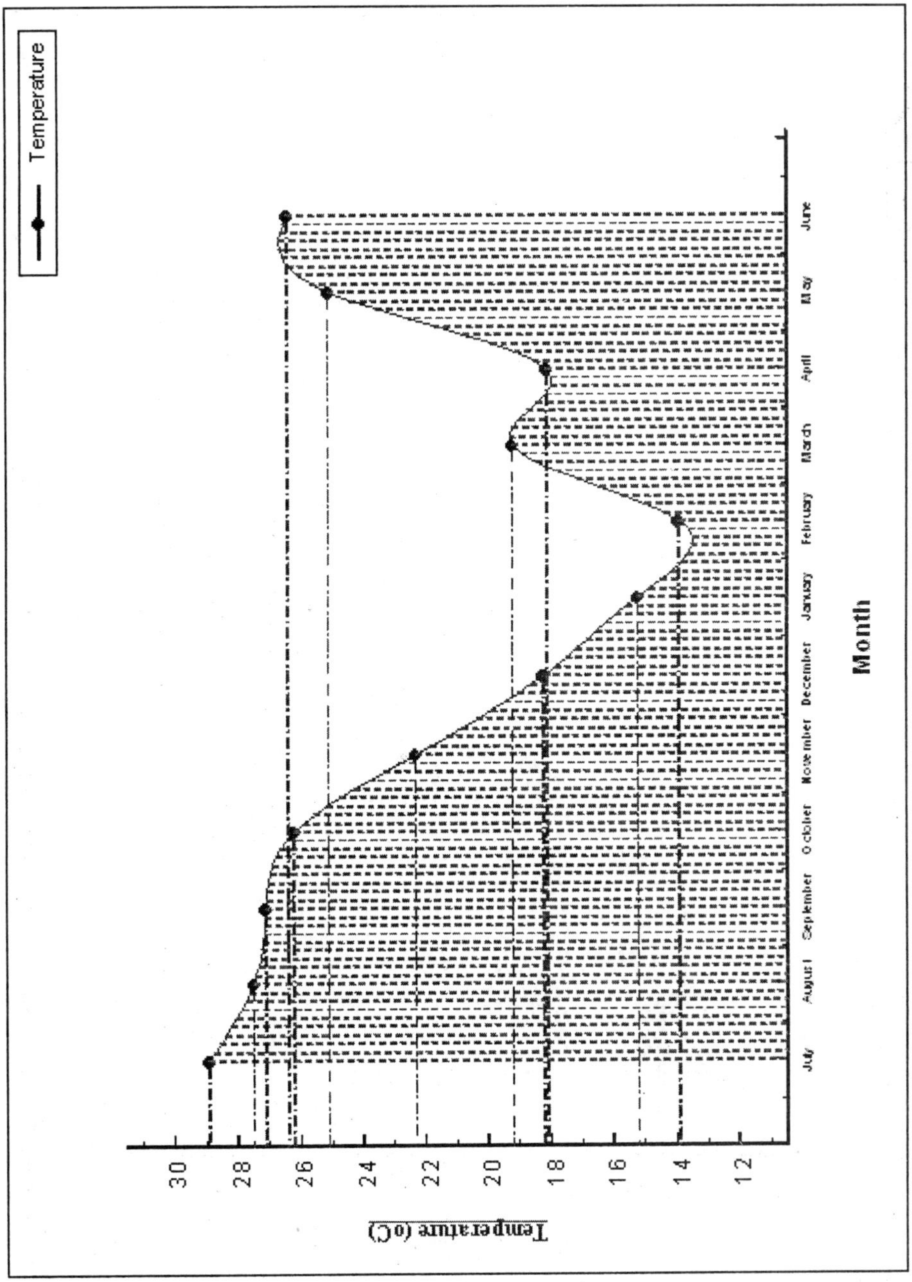

Figure 4.42: Graph Showing Quantitative Variation of Temperature at Sampling Site IV during 2007–08

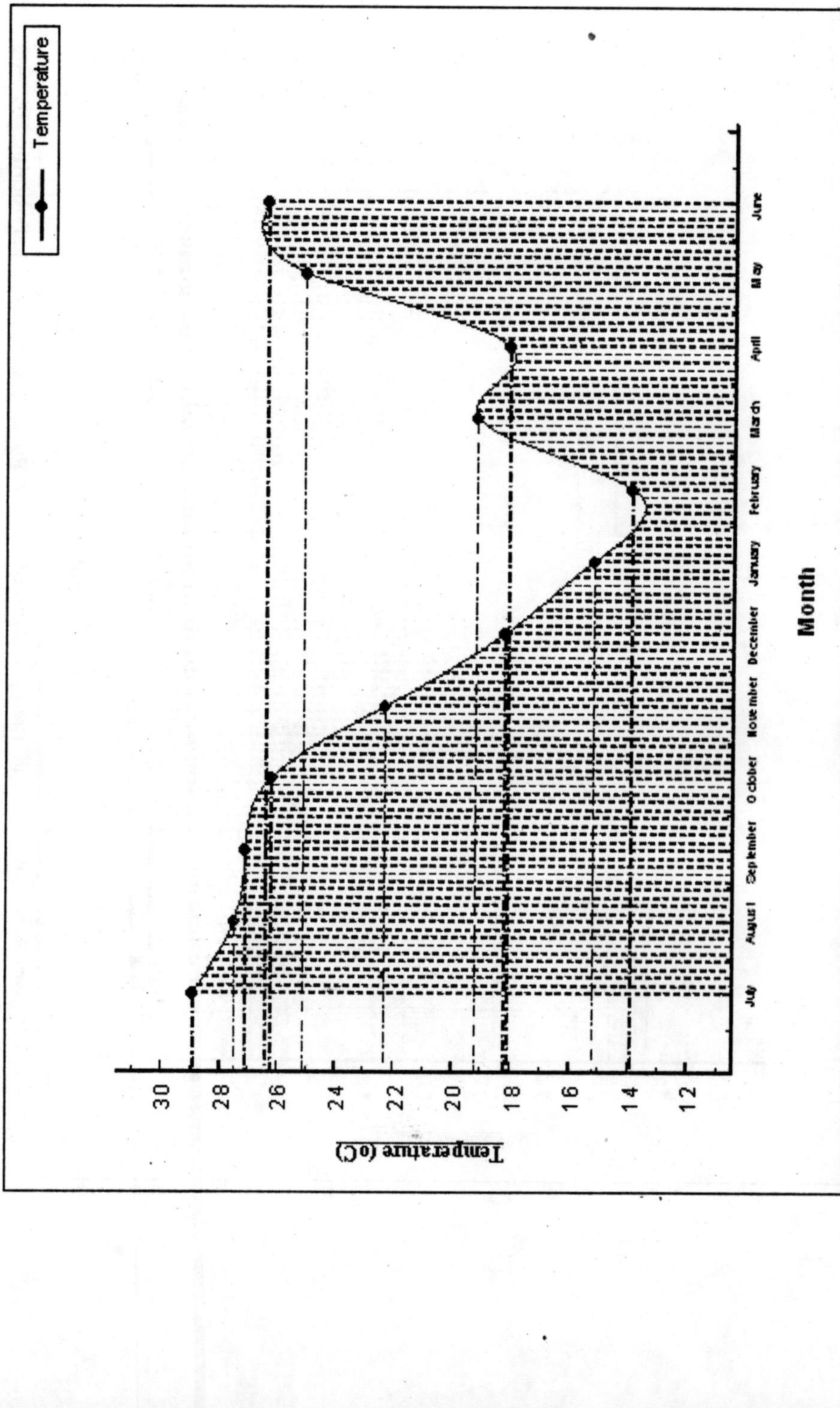

Figure 4.43: Graph Showing Quantitative Variation of Chloride, pH and BOD at Sampling Site IV during 2006–07

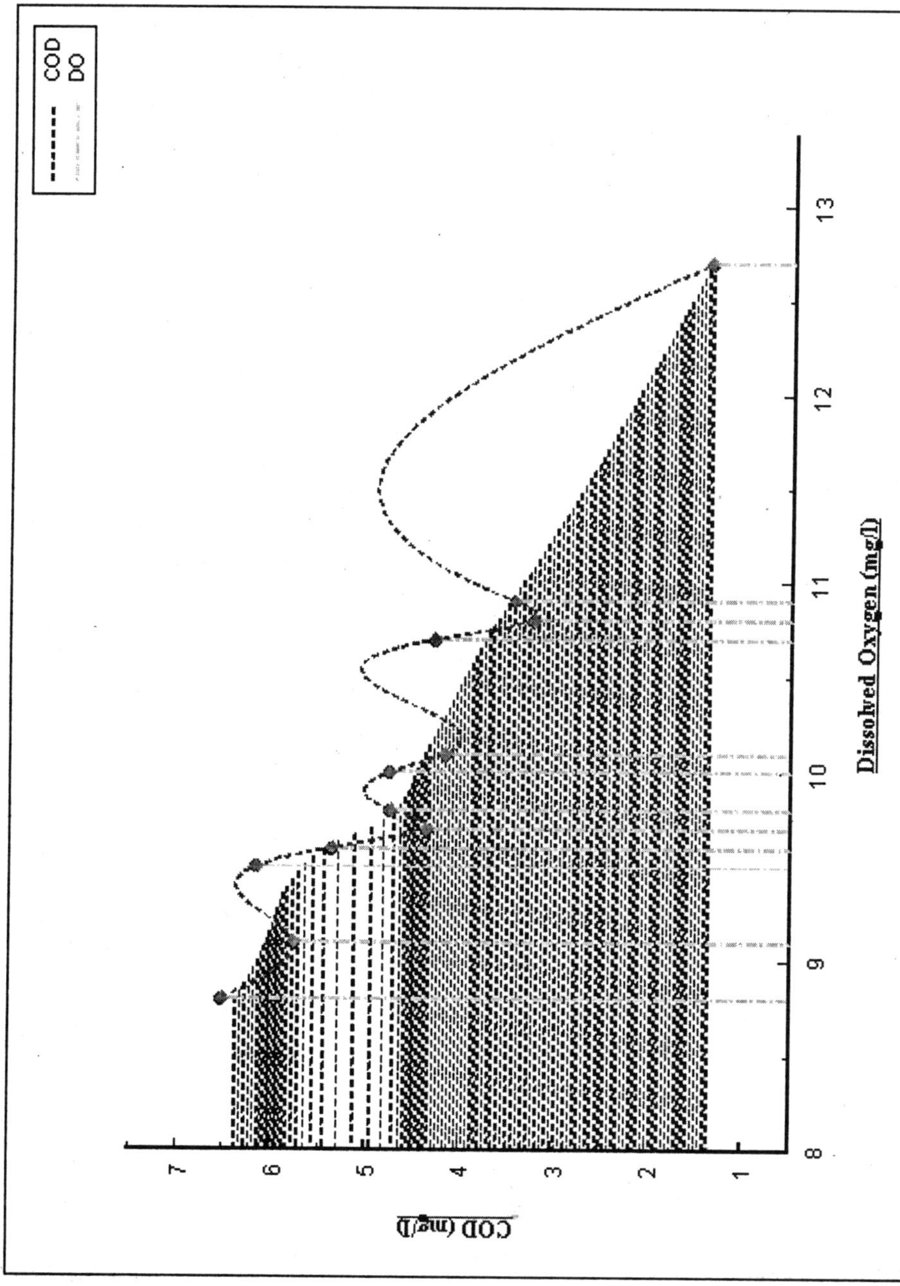

Figure 4.44: Graph Showing Quantitative Variation of COD and DO at Sampling Site IV during 2006–07

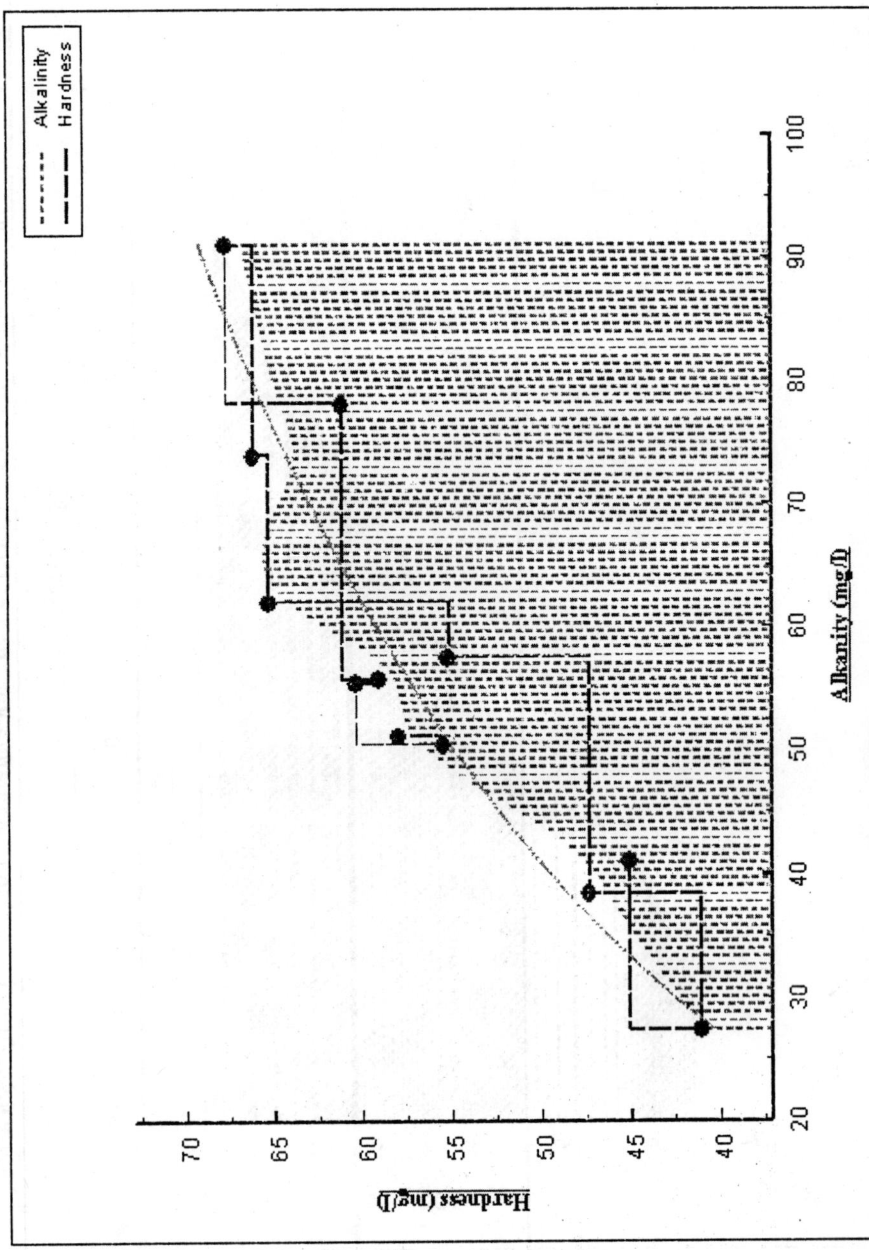

Figure 4.45: Graph Showing Quantitative Variation of Hardness and Alkalinity at Sampling Site IV during 2006–07

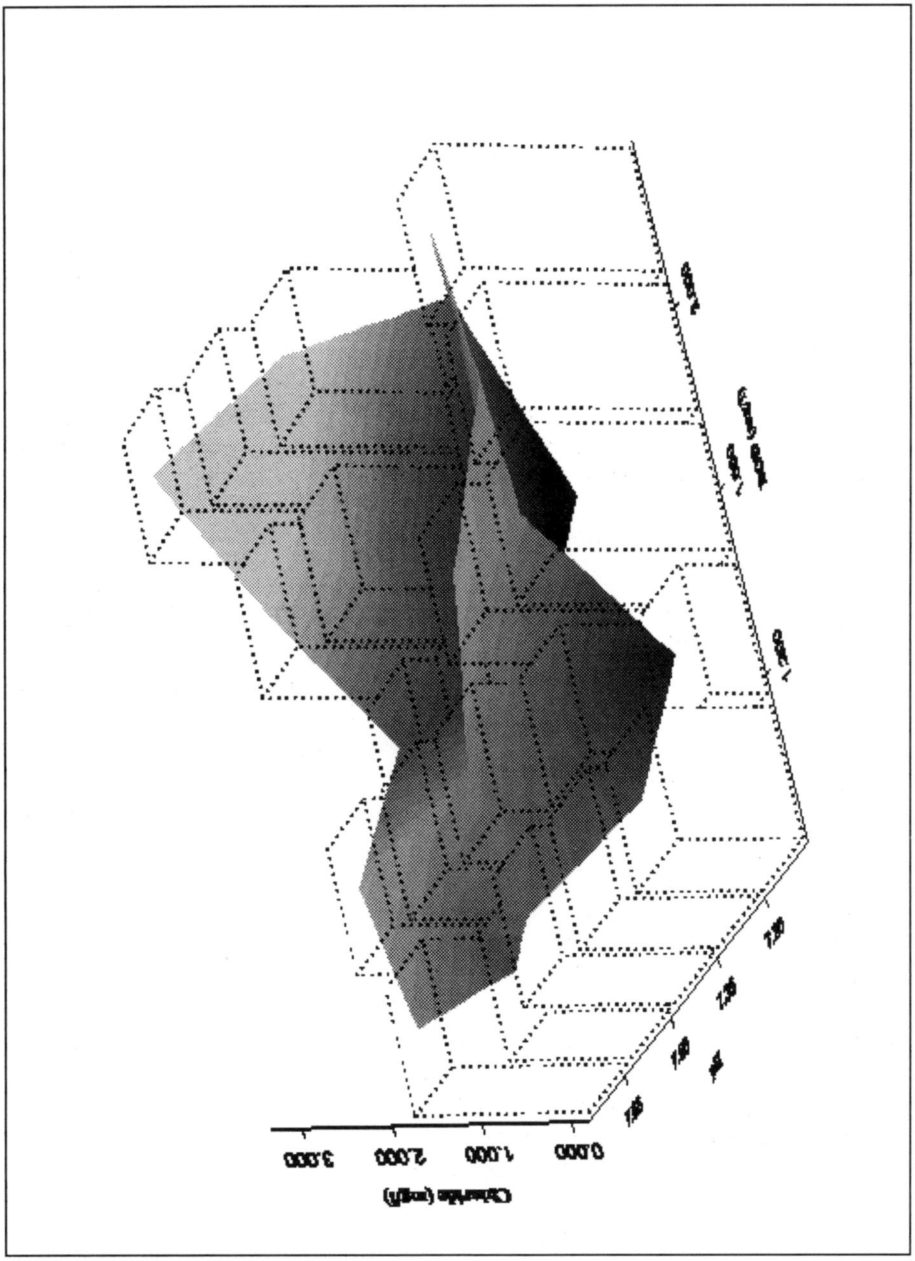

Figure 4.46: Graph Showing Quantitative Variation of Chloride, pH and BOD at Sampling Site IV during 2007–08

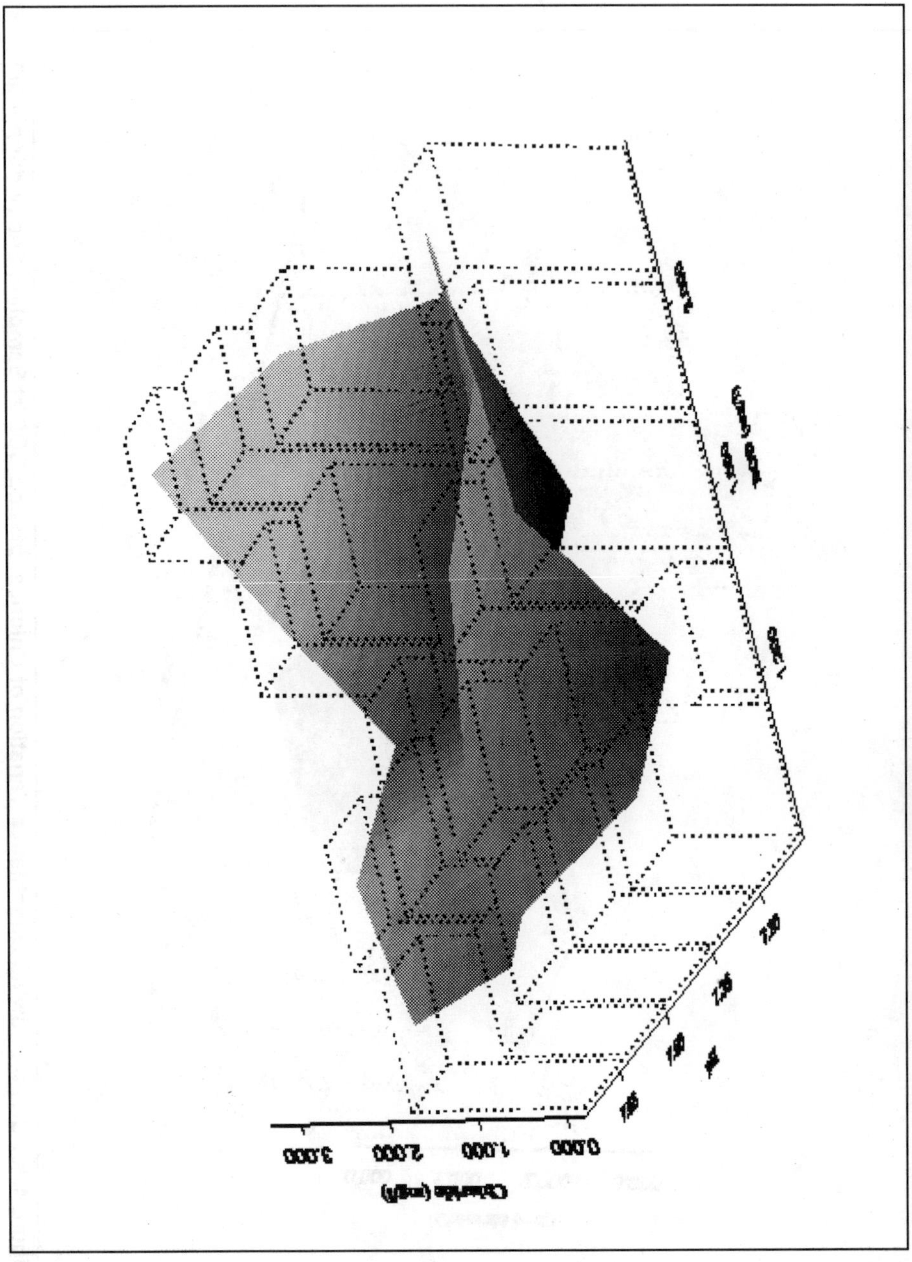

Figure 4.47: Graph Showing Quantitative Variation of COD and DO at Sampling Site IV during 2007–08

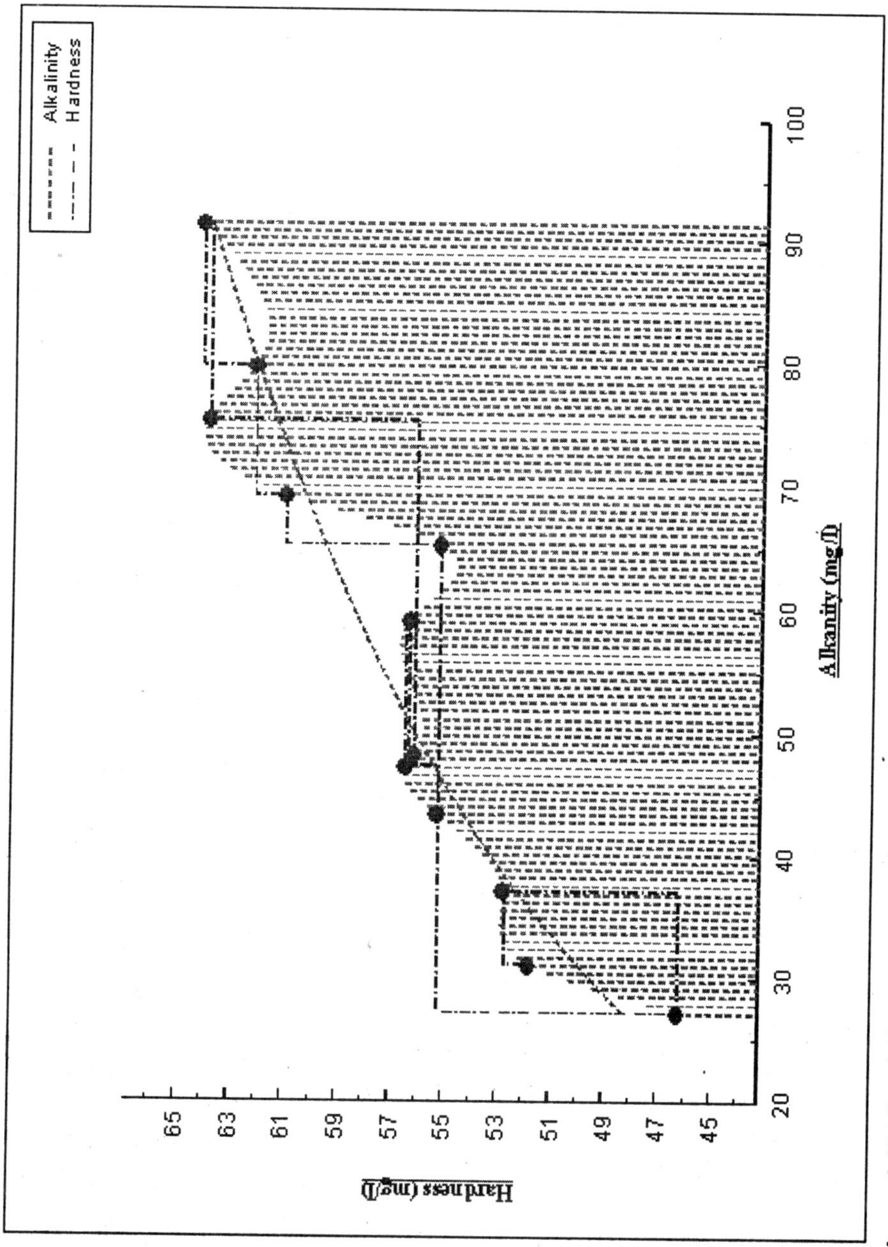

Figure 4.48: Graph Showing Quantitative Variation of Hardness and Alkalinity at Sampling Site IV during 2007–08

month of January in 2006-2007. The average value in the year 2006-2007 was observed 10.14 mg/l ±1.04 and in the year 2007-2008 it was found 10.17 mg/l ±0.65.

Chemical Oxygen Demand (mg/l)

The COD at this sampling site was recorded minimum 1.37 mg/l ±0.07 in the month of January in 2006-2007 and maximum 6.57 mg/l ±0.20 in the month of July 2007-2008. The average value in the year 2006-2007 was observed 4.50 mg/l ±1.42 and in the year 2007-2008 it was noted 4.52 mg/l ±1.24.

Biochemical Oxygen Demand (mg/l)

The BOD at this sampling site was observed minimum 0.46 mg/l ±0.12, in the month of January in 2006-2007 and maximum 2.24 mg/l ±0.08 in the month of September in year 2006-2007. The average value in the year 2006-2007 was observed 1.70 mg/l ±0.51 and in the year 2007-2008 it was found 1.68 mg/l ±0.38.

Alkalinity (mg/l)

The alkalinity at this sampling site was recorded minimum 27.20 mg/l ±0.48 in the month of September in 2007-2008 and maximum 91.40 mg/l ±1.52 in the month of January 2007-2008. The average value in the year 2006-2007 was found 56.90 mg/l ±17.74 and in the year 2007-2008 it was observed 56.29 mg/l ±20.35.

pH

pH at this sampling site was observed minimum 7.04 ±0.09 in the month of July in 2006-2007 and maximum 7.89 ±0.09 in the month of January in 2006-2007. The average value in the year 2006-2007 was noted 7.44 ±0.27 and in the year 2007-2008 it was recorded 7.34 ±0.16.

Total Hardness (mg/l)

The Total hardness at this sampling site was found minimum 41.10 mg/l ±1.45 in the month of August in 2006-2007 and maximum 67.68 mg/l ±0.34 in the month of January in 2006-2007. The average value in the year 2006-2007 was observed 56.83 mg/l ±8.50 and in the year 2007-2008 it was observed 57.76 mg/l ±5.30.

Chloride (mg/l)

The chloride at this sampling site was observed minimum 0.90 mg/l ±0.03 in the month of January in year 2007-2008 and maximum 2.75 mg/l ±0.13 in the month of July in year 2007-2008. The average value in the year 2006-2007 was recorded 1.85 mg/l ±0.54 and in the year 2007-2008 it was recorded 1.63 mg/l ±0.53.

Sampling Site V (Dam Out Near Chopra Vvillage)

The results of physico-chemical parameters for the sampling site V are presented in the Tables 4.17 to 4.20 and Figures 4.49 to 4.60.

Water Temperature (°C)

The water temperature at this sampling site was found minimum 8.50°C ±0.29 in the month of January in 2007-2008 and maximum 20.10 °C ±0.22 in the month of May in 2007-2008. The average value in the year 2006-2007 was recorded 14.67°C ± 3.12 and in the year 2007-2008 it was recorded 16.28°C ±3.14.

Table 4.17: Monthly Variation in Physical Parameters of Bhagirathi River at Sampling Station V (Dam Outlet near Chopra village) during 2006–2007

Month	*Temp. (°C)*	*Total Solids (mg/l)*	*TDS (mg/l)*	*TSS (mg/l)*	*Velocity (m/s)*	*Turbidity (JTU)*
July	17.50±0.48	1531.1±2.15	216.30±1.05	1314.80±1.85	1.70±0.08	135.00±1.12
August	18.00±0.75	2179.34±2.38	321.34±2.20	1858.00±1.88	1.90±0.02	180.00±1.24
September	17.00±0.76	931.31±2.01	108.40±1.46	822.91±1.51	1.65±0.03	55.10±1.01
October	16.80±0.89	1080.12±2.10	112.20±1.76	967.92±1.24	0.80±0.05	58.20±1.11
November	15.30±0.99	701.00±1.89	101.30±1.58	599.70±1.63	0.60±0.01	17.00±1.02
December	12.20±0.54	512.10±1.87	97.50±1.11	414.60±1.52	0.60±0.06	12.20±1.10
January	10.90±0.38	251.20±1.79	89.30±1.24	161.90±1.49	0.50±0.04	5.30±0.85
February	8.90±0.64	159.57±1.59	78.40±1.21	81.17±1.11	0.45±0.07	3.50±0.25
March	11.90±0.29	110.20±1.94	75.30±1.24	34.90±1.25	0.51±0.02	1.50±0.44
April	13.20±0.21	99.00±1.22	73.40±1.21	25.60±1.10	0.53±0.01	0.00±0.00
May	16.20±0.11	525.44±1.58	99.30±1.25	426.14±1.54	0.56±0.02	12.50±0.85
June	18.10±0.22	805.00±1.25	106.50±2.03	698.50±1.72	0.65±0.01	18.50±0.88
Average±SD	14.67±3.12	740.45±627.90	123.27±72.73	617.18±560.82	0.87±0.54	41.57± 58.23

Table 4.18: Monthly Variation in Physical Parameters of Bhagirathi River at Sampling Station V (Dam Outlet near Chopra village) during 2007–2008

Month	*Temp. (°C)*	*Total Solids (mg/l)*	*TDS (mg/l)*	*TSS (mg/l)*	*Velocity (m/s)*	*Turbidity (JTU)*
July	17.60±0.58	1641.50±2.52	230.50±1.29	1411.00±2.16	1.65±0.21	170.50±1.54
August	18.00±0.54	204.50±2.16	90.50±1.15	114.00±2.13	0.25±0.12	25.20±1.12
September	18.20±0.68	1232.39±2.31	131.50±1.21	1100.89±2.11	1.50±0.13	96.50±1.15
October	17.50±0.42	750.29±1.85	107.72±1.12	642.57±1.18	1.25±0.01	66.30±1.11
November	17.30±0.75	651.24±1.94	103.91±1.16	547.33±1.57	0.90±0.01	24.10±1.01
December	13.40±0.38	397.50±1.78	99.50±1.18	298.00±1.46	1.50±0.02	15.10±0.89
January	8.50±0.29	101.00±1.65	81.30±1.14	19.70±1.21	1.90±0.03	5.50±0.87
February	12.50±0.46	95.07±1.28	78.40±1.11	16.67±1.11	0.60±0.01	1.20±0.21
March	17.10±0.16	204.34±1.11	85.40±1.08	118.94±1.23	0.53±0.04	3.30±0.00
April	15.90±0.24	87.57±1.22	71.20±1.09	16.37±1.52	0.50±0.03	0.80±0.01
May	20.10±0.22	397.39±1.31	82.14±0.85	315.25±1.16	0.65±0.03	16.30±1.02
June	19.30±0.13	791.00±1.89	111.50±0.55	679.50±1.15	0.80±0.04	27.50±0.16
Average±SD	16.28±3.14	546.15±471.97	106.13±40.87	440.02±434.93	1.00±0.51	37.69±48.51

Table 4.19: Monthly Variation in Chemical Parameters of River Bhagirathi at Sampling Site V (Dam Outlet near Chopra village) during 2006–07

Month	*DO (mg/l)*	*COD (mg/l)*	*BOD (mg/l)*	*Alkalinity (mg/l)*	*pH*	*Total Hardness (mg/l)*	*Chloride (mg/l)*
July	7.56±0.15	8.68±0.24	3.11±0.05	29.50±0.53	7.09±0.09	50.47±2.13	3.20±0.12
August	10.10±0.13	4.24 ±0.12	1.65±0.14	31.20±0.89	7.17±0.05	54.98±1.52	2.42±0.10
September	10.10±0.05	4.10±0.19	1.72±0.02	39.40±0.24	7.20±0.11	54.12±1.40	2.13±0.06
October	9.20±0.06	6.26±0.08	2.16±0.06	40.50±0.35	7.10±0.14	55.93±0.92	2.13±0.08
November	10.10±0.10	4.22±0.03	1.68±0.15	73.50±1.24	7.15±0.15	59.31±0.54	0.99±0.03
December	10.80±0.17	3.12±0.04	1.28±0.13	72.50±1.14	7.57±0.03	65.21±1.32	1.19±0.00
January	8.40±0.16	6.23±0.10	2.48±0.05	88.50±1.24	7.46±0.09	66.12±0.38	1.98±0.13
February	9.30±0.13	4.65±0.05	2.00±0.03	86.30±1.3	7.30±0.15	59.60±0.45	1.77±0.15
March	10.10±0.10	4.37±0.19	1.60±0.14	62.30±1.40	7.69±0.13	57.53±1.35	2.20±0.02
April	9.30±0.05	5.98±0.13	2.14±0.12	61.20±0.52	7.50±0.05	55.12±0.47	2.70±0.03
May	9.00±0.03	5.10±0.18	2.27±0.07	54.20±0.36	7.10±0.14	56.42±0.31	2.90±0.08
June	9.30±0.00	6.68± 0.13	2.24±0.16	42.30±0.56	7.10±.0.11	56.25±1.25	3.10±0.06
Average±SD	9.44±0.88	5.30±1.52	2.03±0.49	56.78±20.53	7.29.±0.21	57.58±4.47	2.23±0.70

Table 4.20: Monthly Variation in Chemical Parameters of River Bhagirathi at Sampling Site V (Dam Outlet near Chopra village) during 2007–08

Month	*DO (mg/l)*	*COD (mg/l)*	*BOD (mg/l)*	*Alkalinity (mg/l)*	*pH*	*Total Hardness (mg/l)*	*Chloride (mg/l)*
July	8.70±0.24	6.56±0.05	2.34±0.06	35.50±0.58	7.12±0.32	49.20±1.54	3.20±0.18
August	9.80±0.13	4.66± 0.24	1.81±0.05	32.70±0.24	7.20±0.15	55.15±1.35	3.56±0.29
September	9.70±0.16	4.79±0.15	1.95±0.03	33.10±0.35	7.40±0.28	47.79±1.48	2.50±0.16
October	9.50±0.19	5.72± 0.13	2.06±0.08	38.90±0.18	7.25±0.16	53.60±0.35	1.10±0.06
November	9.80±0.21	5.26±0.23	1.90±0.07	71.20±1.25	7.60±0.23	57.00±0.34	1.05±0.08
December	10.70±0.26	3.23±0.24	1.25±0.15	85.30±1.47	7.95±0.15	58.49±0.25	2.00±0.35
January	10.80±0.14	3.11±0.11	1.18±0.12	91.30±1.65	7.40±0.08	61.50±0..36	2.56±0.21
February	10.10±0.21	4.28±0.16	1.76±0.05	61.20±1.27	7.40±0.16	59.10±0.98	2.50±0.10
March	9.10±0.11	6.22±0.13	2.17±.0.16	57.30±0.58	7.20±0.07	54.87±1.24	2.10±0.15
April	8.50±0.09	6.37±0.18	2.40±.0.15	49.50±0.24	7.15±0.11	55.30±1.58	2.50±0.13
May	10.20±0.08	3.51±0.12	1.56±0.02	47.50±0.64	7.30±0.16	51.00±0.65	2.60±0.15
June	9.00±0.03	6.65±0.06	2.26±0.00	49.50±0.12	7.20±0.02	51.40±0.24	2.95±0.09
Average±SD	9.66±0.74	5.03±1.30	1.89±0.40	54.42±19.70	7.35±0.23	54.53±4.15	2.39±0.75

Total Solids (mg/l)

The total solids at this sampling site were recorded minimum 87.57 mg/l ±1.22 in the month of April in 2007-2008 and maximum 2179.34 mg/l ±2.38 in the month of August in 2006-2007. The average value in the year 2006-2007 was found 740.45 mg/l ±627.90 and in the year 2007-2008 it was observed 546.25 mg/l ±471.97.

Total Dissolved Solids (mg/l)

The total dissolved solids at this sampling site were observed minimum 71.20 mg/l ±1.09 in the month of April in 2007-2008 and maximum 321.34 mg/l ±2.20 in the month of August in 2006-2007. The average value in the year 2006-2007 was noted 123.27 mg/l ± 72.73 and in the year 2007-2008 it was found 106.13 mg/l ±40.87.

Total Suspended Solids (mg/l)

The total suspended solids at this sampling site were recorded minimum 16.37 mg/l ±1.52 in the month of April in 2007-2008 and maximum 1858.0 mg/l ±1.88 in the month of August in 2006-2007. The average value in the year 2006-2007 was observed 617.18 mg/l ±560.82 and in the year 2007-2008 it was observed 440.02 mg/l ±434.93.

Velocity (m/s)

The velocity at this sampling site was observed minimum 0.25 m/s ±0.12 in the month of August in 2007-2008 and maximum 1.90 m/s ±0.02 in the months of August 2006-2007 and in the months of January 2007-2008. The average value in the year 2006-2007 was noted 0.87 m/s ±0.54 and in the year 2007-2008 it was recorded 1.00 m/s ±0.51.

Turbidity (JTU)

The turbidity at this sampling site was found minimum 0.00 JTU ±0.00 in the month of April in 2006-2007 and maximum 180.00 JTU ±1.24 in the month of August in 2006-2007. The average value in the year 2006-2007 was recorded 41.57 JTU ±58.23 and in the year 2007-2008 it was observed 37.69 JTU ±48.51.

Dissolved Oxygen (mg/l)

The DO at this sampling site was recorded minimum 7.56 mg/l ±0.15 in the month of July in 2006-2007 and maximum 10.80 mg/l ±0.17 in the months of December 2006-2007 and January 2007-2008. The average value in the year 2006-2007 was found 9.44 mg/l ±0.88 and in the year 2007-2008 it was noted 9.66 mg/l ±0.74.

Chemical Oxygen Demand (mg/l)

The COD at this sampling site was found maximum 8.68 mg/l ±0.24 in the month of July in 2006-2007 and minimum 3.11 mg/l ±0.11 in the month of January in 2007-2008. The average value in the year 2006-2007 was observed 5.30 mg/l ±1.52 and in the year 2007-2008 it was recorded 5.03 mg/l ±1.30.

Biochemical Oxygen Demand (mg/l)

The BOD at this sampling site was recorded minimum 1.18 mg/l ±0.12 in the month of January in year 2007-2008 and maximum 3.11 mg/l ±0.05 in the month of

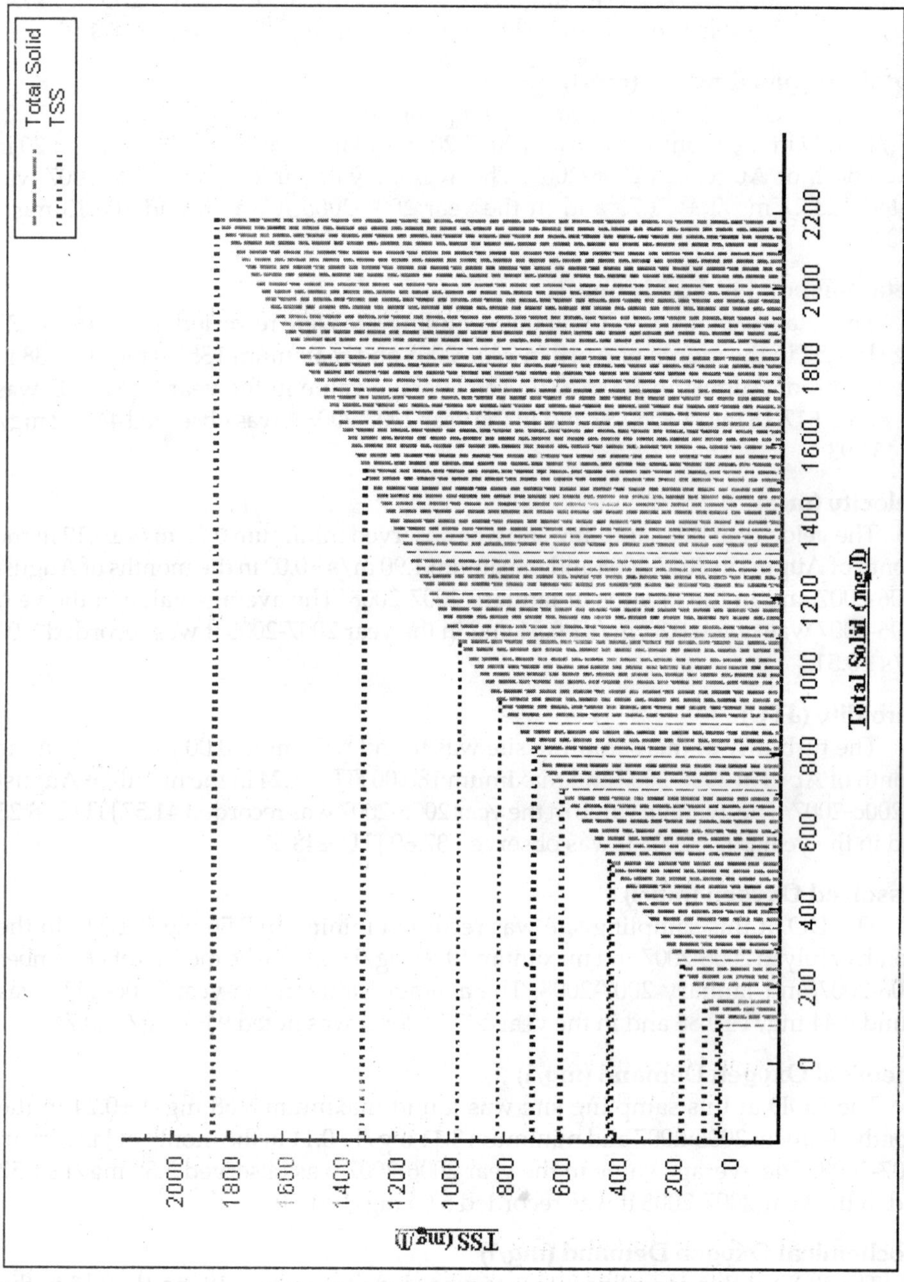

Figure 4.49: Graph Showing Quantitative Variation of TS and TSS at Sampling Site V during 2006–07

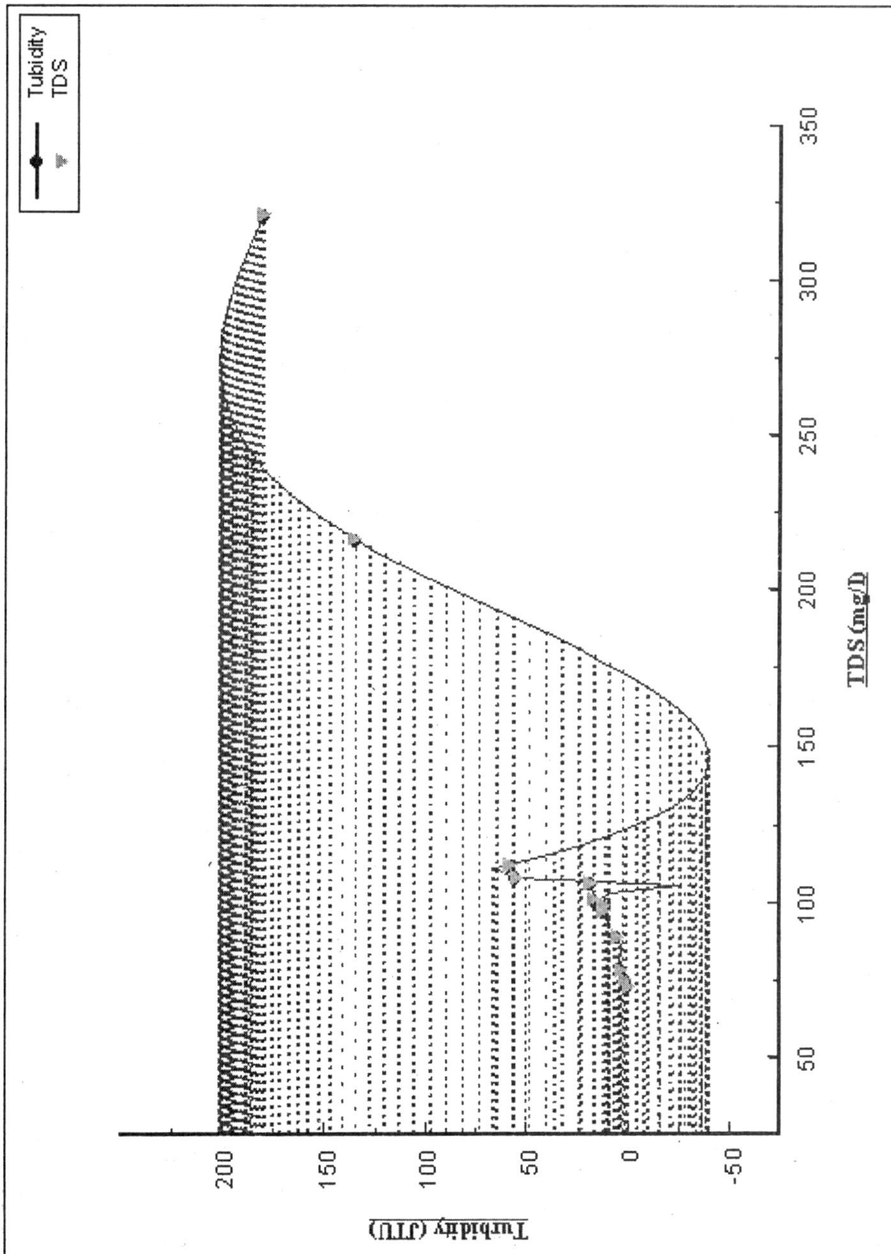

Figure 4.50: Graph Showing Quantitative Variation of Turbidity and TDS at Sampling Site V during 2006–07

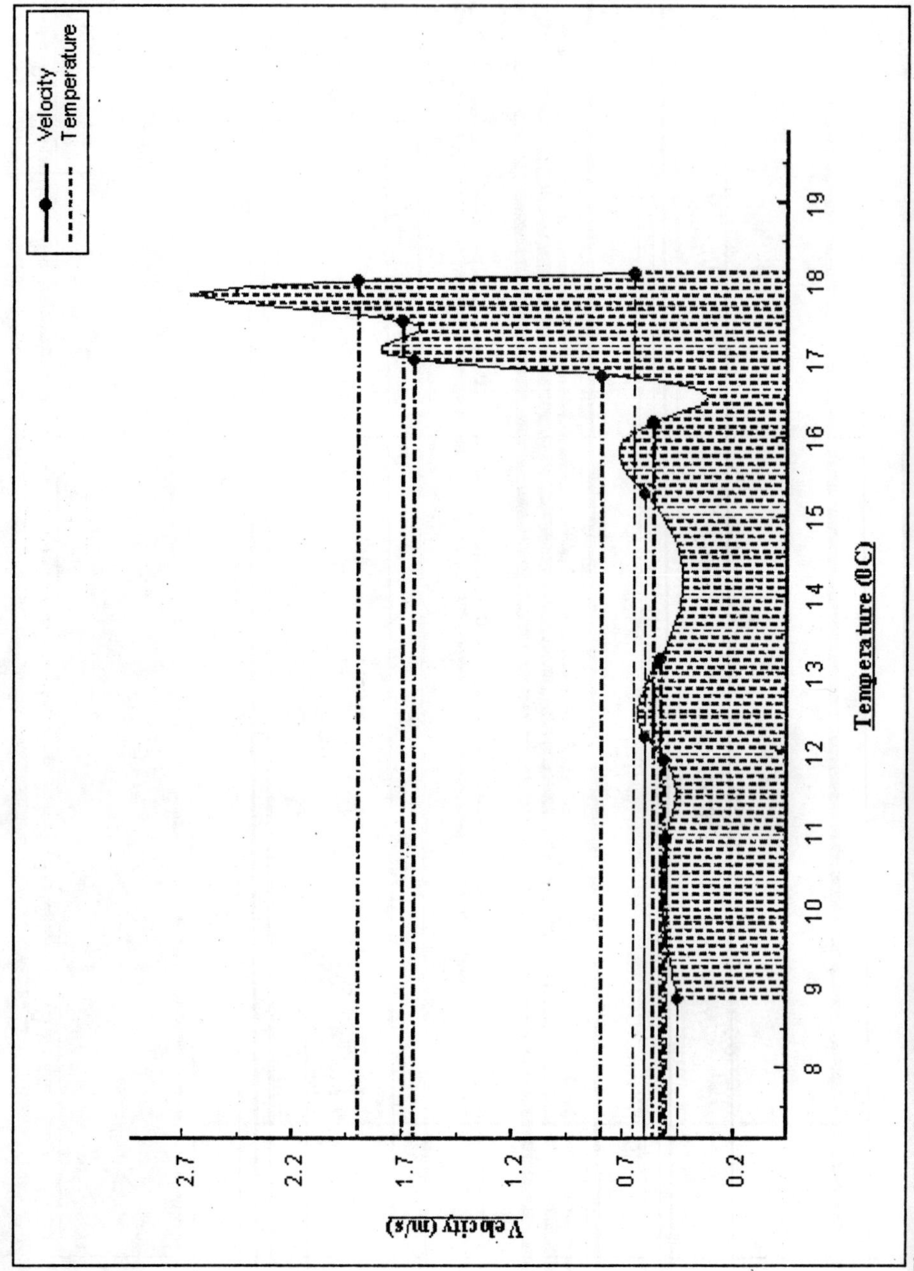

Figure 4.51: Graph Showing Quantitative Variation of Velocity and Temperature at Sampling Site V during 2006–07

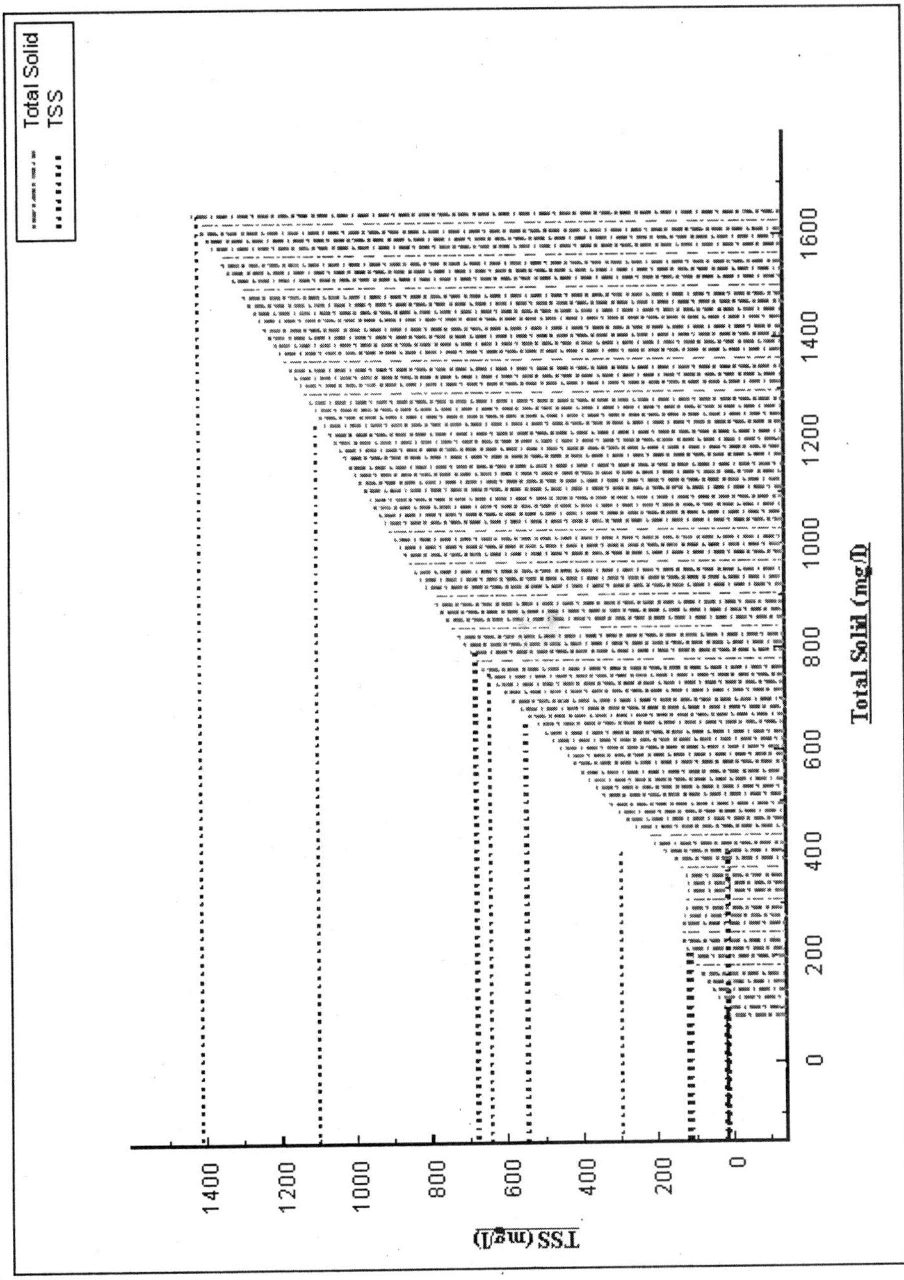

Figure 4.52: Graph Showing Quantitative Variation of TS and TSS at Sampling Site V during 2007–08

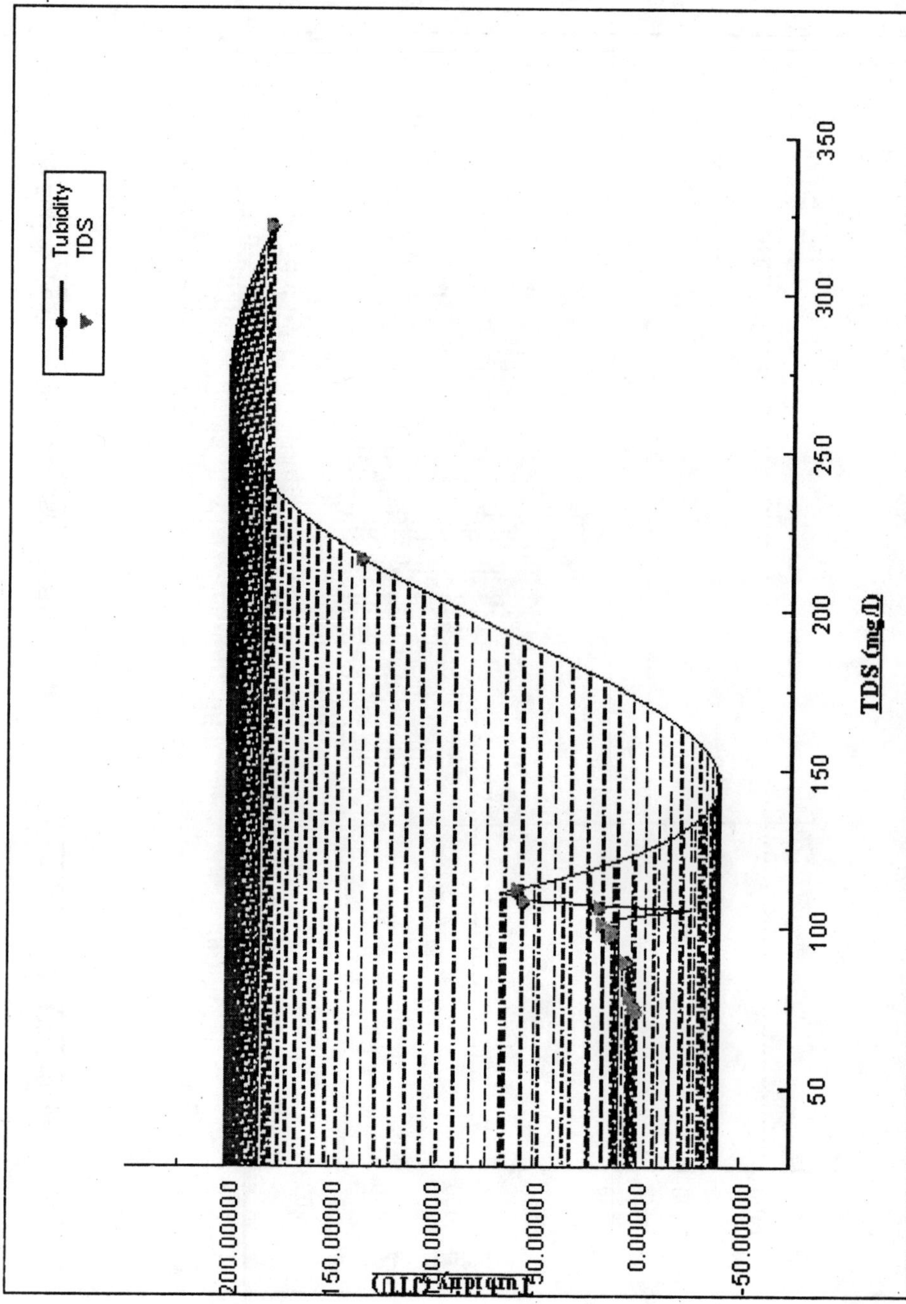

Figure 4.53: Graph Showing Quantitative Variation of Turbidity and TDS at Sampling Site V during 2007–08

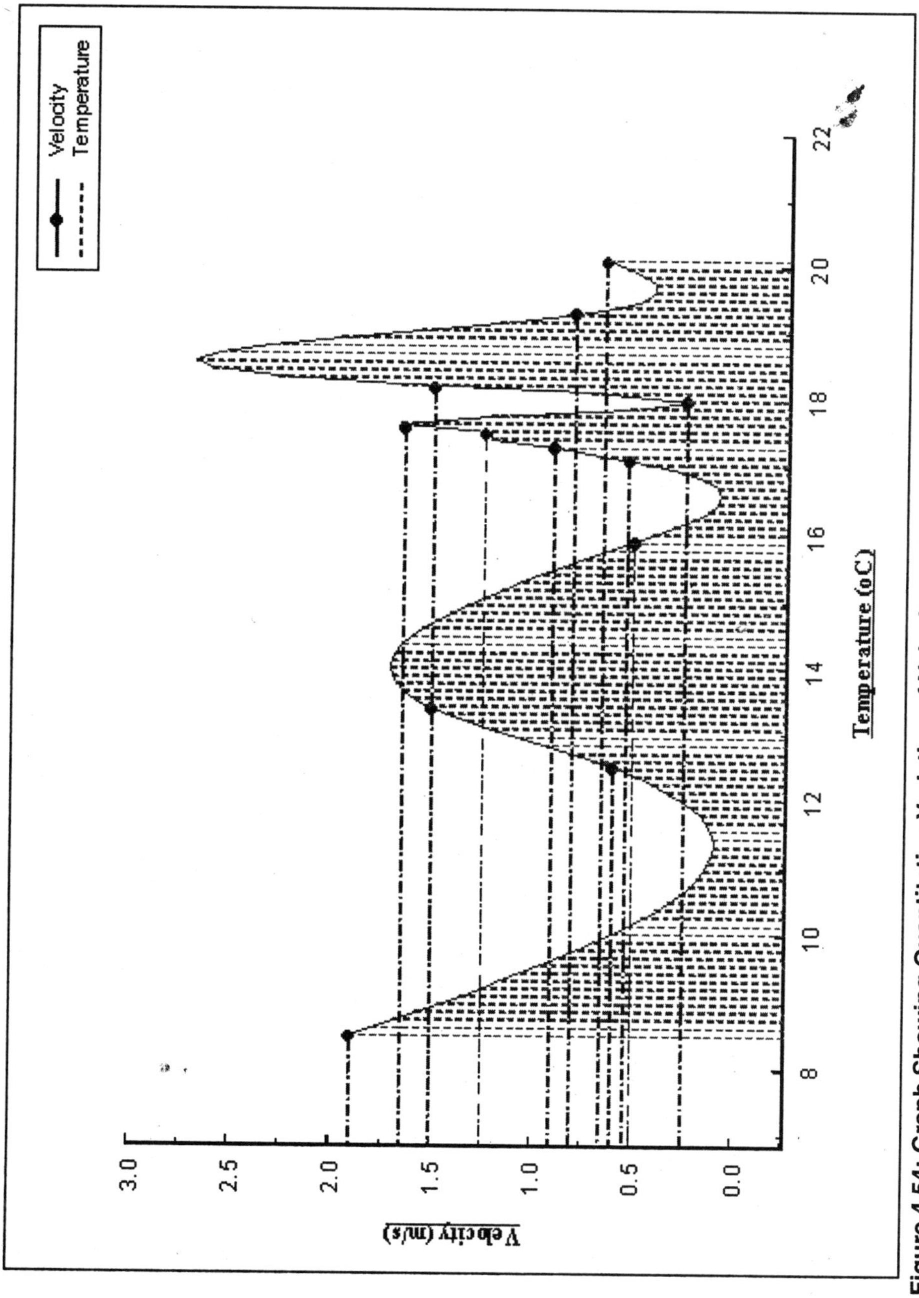

Figure 4.54: Graph Showing Quantitative Variation of Velocity and Temperature at Sampling Site V during 2007–08

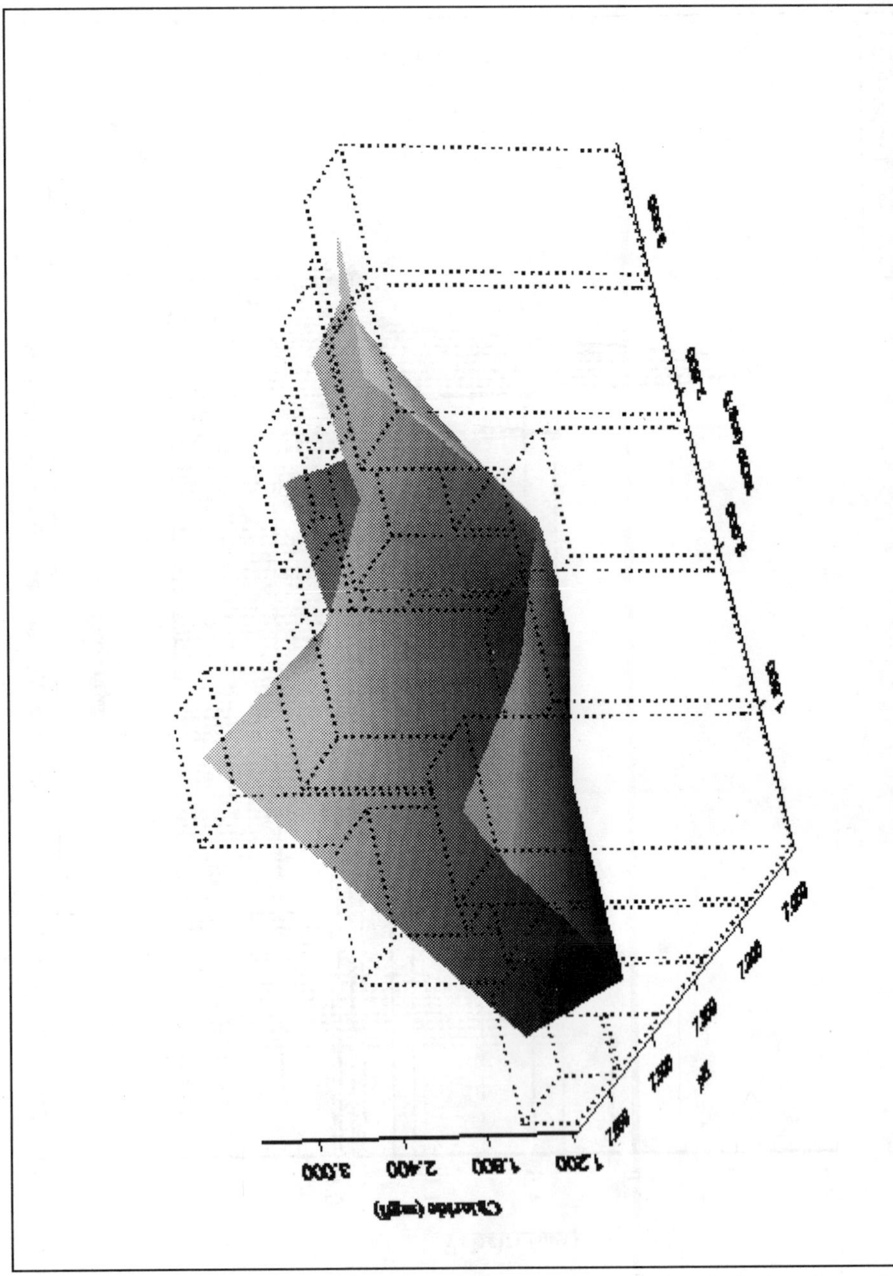

Figure 4.55: Graph Showing Quantitative Variation of Chloride, pH and BOD at Sampling Site V during 2006–07

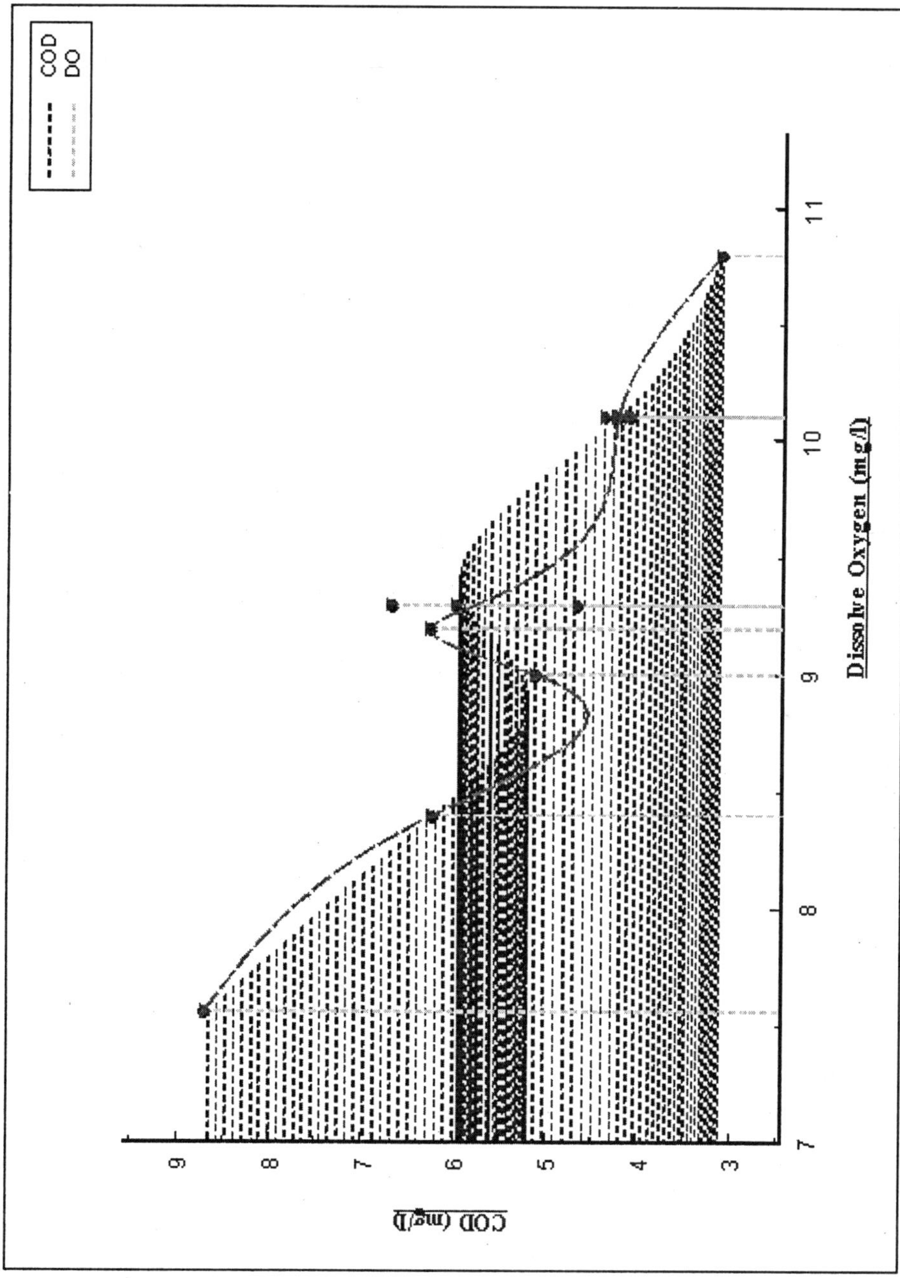

Figure 4.56: Graph Showing Quantitative Variation of COD and DO at Sampling Site V during 2006–07

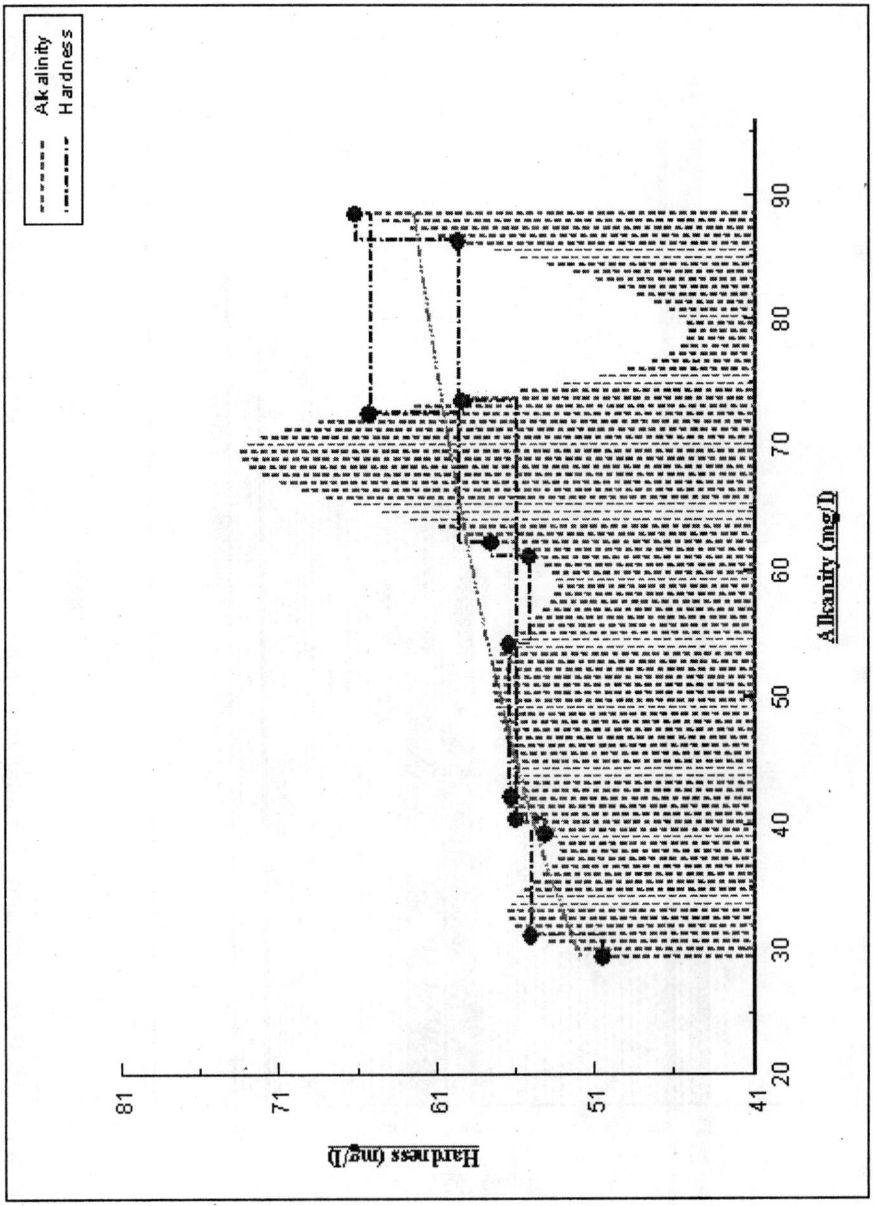

Figure 4.57: Graph Showing Quantitative Variation of Hardness and Alkalinity at Sampling Site V during 2006–07

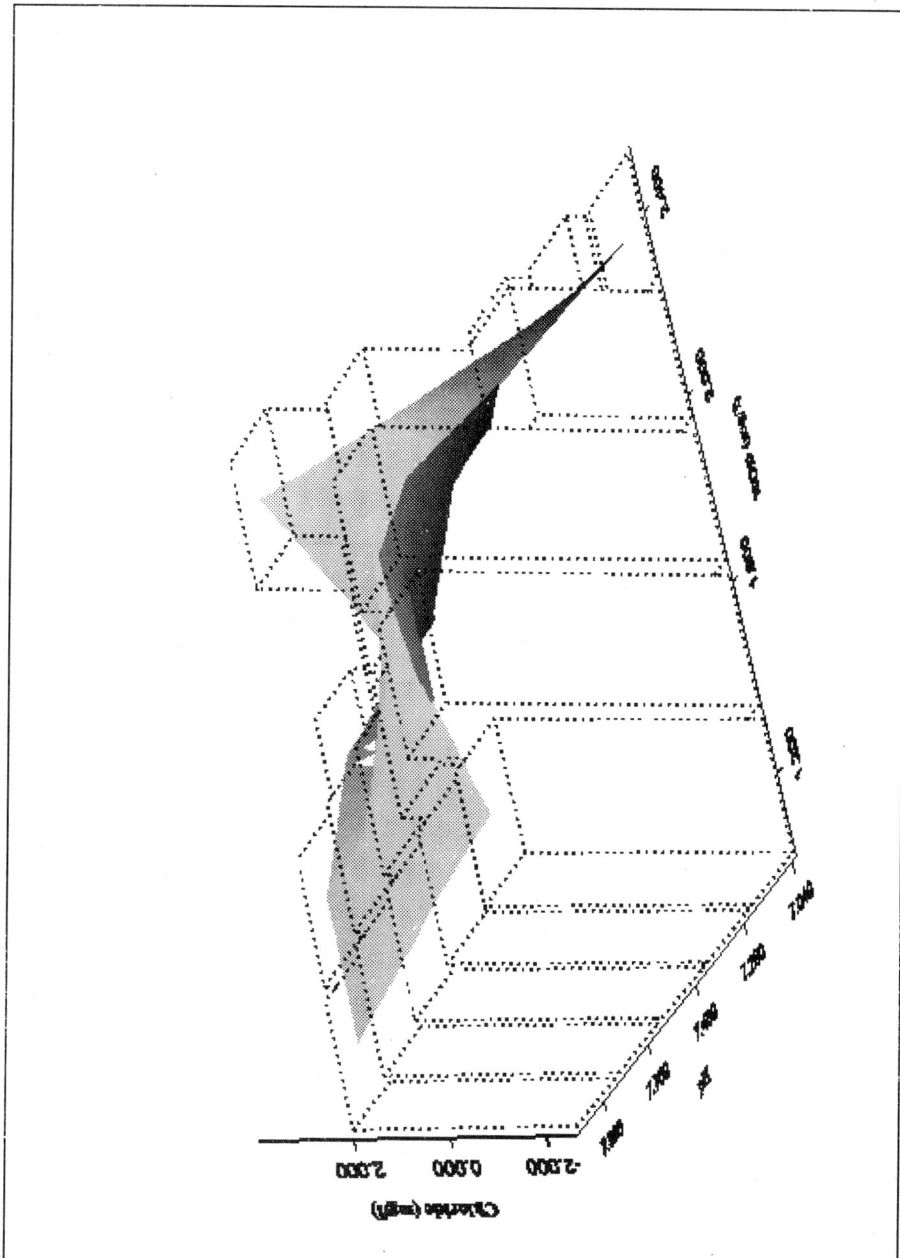

Figure 4.58: Graph Showing Quantitative Variation of Chloride, pH and BOD at Sampling Site V during 2007–08

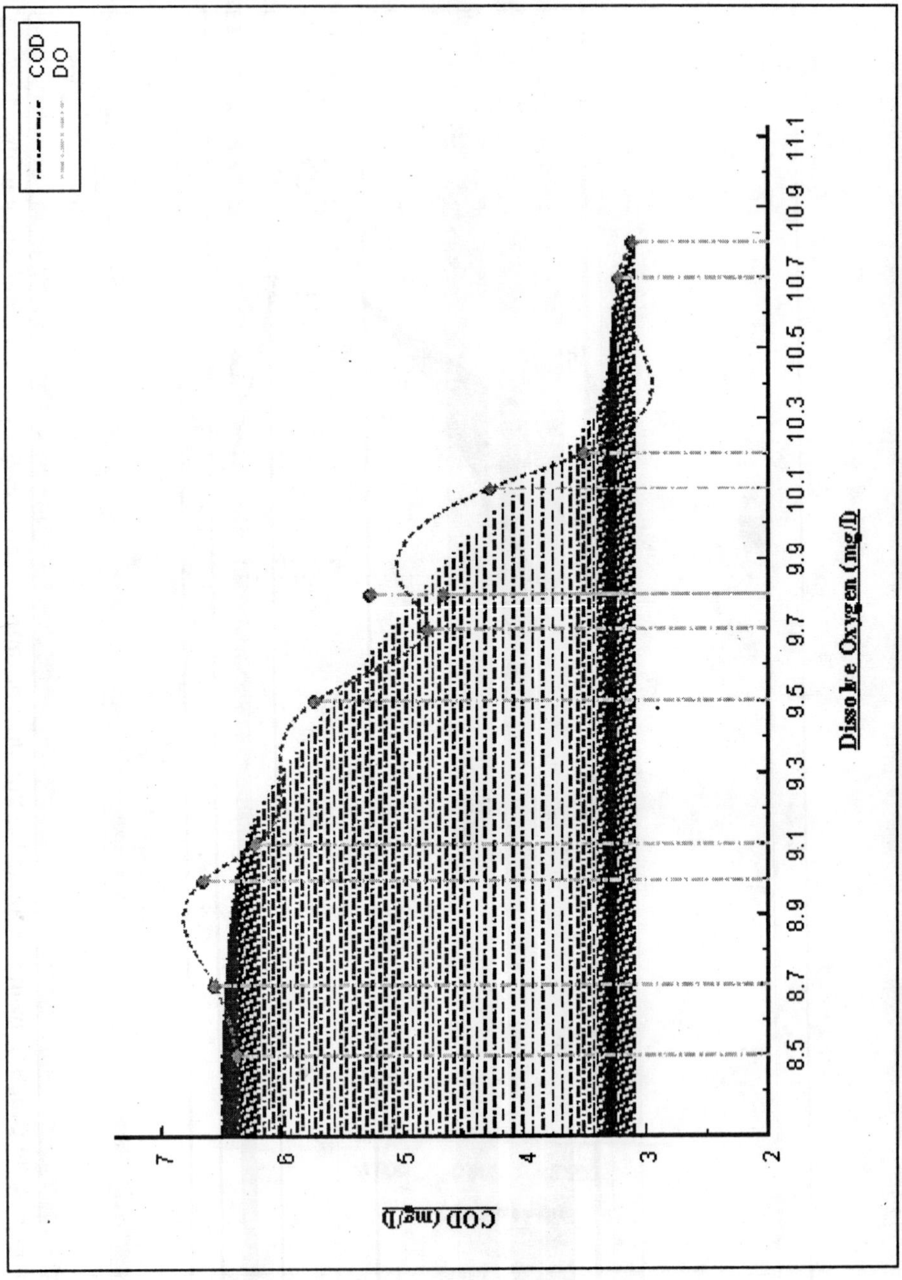

Figure 4.59: Graph Showing Quantitative Variation of COD and DO at Sampling Site V during 2007–08

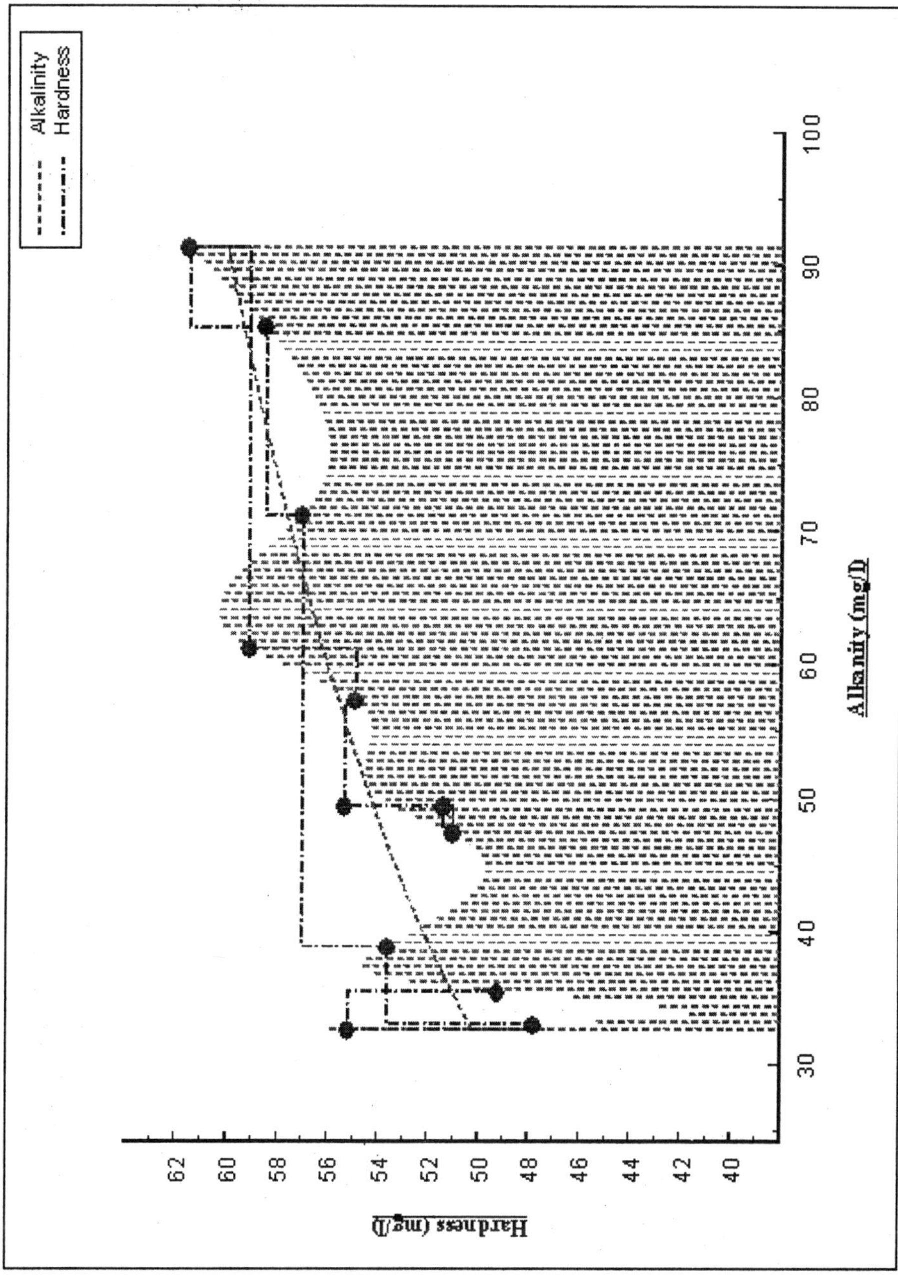

Figure 4.60: Graph Showing Quantitative Variation of Hardness and Alkalinity at Sampling Site V during 2007–08

July in 2006-2007. The average value of BOD in the year 2006-2007 was noted 2.03 mg/l ±0.49 and in the year 2007-2008 it was obtained 1.89 mg/l ±0.40.

Alkalinity (mg/l)

The alkalinity at this sampling site was found minimum 29.5 mg/l ±0.53 in the month of July in 2006-2007 and maximum 91.3 mg/l ±1.65 in the month of January in 2007-2008. The average value in the year 2006-2007 was observed 56.78 mg/l ±20.53 and in the year 2007-2008 it was noted 54.42 mg/l ±19.70.

pH

The pH at this sampling site was observed minimum 7.09 ±0.09 in the months of July in 2006-2007 and maximum 7.95 ±0.15 in the months of December in 2007-2008. The average value in the year 2006-2007 was obtained 7.29 ± 0.21 and in the year 2007-2008 it was recorded 7.35 ±0.23.

Total Hardness (mg/l)

The Total hardness at this sampling site was recorded minimum 47.79 mg/l ±1.48 in the month of September in 2007-2008 and maximum 66.12 mg/l ±0.38 in the month of January in 2006-2007. The average value in the year 2006-2007 was observed 57.58 mg/l ±4.74 and in the year 2007-2008 it was found 54.53 mg/l ±4.15.

Chloride (mg/l)

The chloride at this sampling site was found minimum 0.99 mg/l ±0.03 in the month of November in 2006-2007 and maximum 3.56 mg/l ±0.29 in the month of August in 2007-2008. The average value in the year 2006-2007 was noted 2.23 mg/l ±0.70 and in the year 2007-2008 it was observed 2.39 mg/l ±0.75.

Biological Parameters

A study on the biological parameters of the river Bhagirathi and Tehri dam reservoir includes the primary producers like the phytoplankton, primary consumers like zooplankton and tertiary consumers like fishes. Station wise details of the observations of biological parameters are mentioned in tables 4.21 to 4.46.

Plankton

During course of study three broad categories of phytoplankton *i.e.* Diatom, Green algae, and blue green algae were found while among zooplankton Protozoa, Rotifera, Cladocerans and Copepods were observed, which are further divided in to genus from Protozoa –Amoeba and Arcella, Rotifera –Keratella, Notholca and Philodima, Cladocera-Bosmina and Daphnia and Copepods-Cyclops.

Sampling Site I (Dharanshu)

The result of phytoplankton and zooplankton for this site are given in Tables 4.21 to 4.24 and 4.41 and Figures 4.61 to 4.64.

Plankton

Quantitative analysis of plankton at this sampling site revealed that the total plankton number per liter of water was observed maximum 5466.00 unit/l in the month of February in the year 2006-2007 and minimum was recorded 569.83 unit/l in the month of July in year 2006-2007.

Phytoplankton

Phytoplankton were noted lowest of 471.83 unit/l in the month of July in year 2006-2007 and highest of 5170.00 unit/l in the month of February in year 2006-2007. The average numbers of cells were observed 2241.26 unit/l ±1542.89 in year 2006-2007and recorded 2352.67 unit/l ±1516.03 in year 2007-2008.

In phytoplankton Diatoms were dominated and class Blue Green Algae were found least in both the years of study. At this sampling station Diatoms were observed maximum in the month of February of year 2006-2007 (4705.00 unit/l), it is 91 per cent of total phytoplankton and the minimum in the month of July of the year 2006-2007 (442.21unit/l) which is 93.72 per cent of total phytoplankton. The average count of cells in 2006-2007 was observed 1990 unit/l ±1422.47 and in 2007-2008 it was recorded 2069.64 unit/l ±1374.51.

Green algae were found to be highest in the month of January year 2007-08 (481.25 unit/l), it is 13.26 per cent of total phytoplankton and lowest (25.12 unit/l), which is 5.33 per cent of total phytoplankton, in the month of July year 2006-2007. The average count of green algae in 2006-2007 was observed 205.81unit/l ±134.88 and in 2007-2008 was noted 228.90 unit/l ±140.19.

Blue green algae showed optimum peak during the month of April in the year 2007-08 (122.40 unit/l) it is 2.88 per cent of total phytoplankton and least peak during the month of August in the year 2006-2007 (3.20 unit/l), which is 0.46 per cent of total phytoplankton. In 2006-2007 the average value of Blue green algae was noted 45.48 unit/l ±28.55 while in 2007-2008 it was observed 54.14 unit/l ±36.69.

Zooplankton

Among zooplankton protozoa (*Amoeba* and *Arcella*), Rotifera (*Keratella, Notholca* and *Philodima*), Cladocera (*Bosmina* and *Daphnia*) and Copepods (*Cyclops*) were observed. Zooplankton were observed minimum in the month of August in 2006-2007 (96 unit /l) and maximum 378 unit/l in the month of May in 2007-08. The average value of zooplankton for year 2006-2007 was observed 199.11 unit/l ±60.97 and for year 2007-2008 it was noted 279.67 unit/l ±98.30.

Protozoa

Protozoa were observed maximum 95.00 unit/l in the month of January 2006-2007 and minimum 10.00 unit/l in the month of August 2006-2007. The annual mean recorded for Protozoa in the first year 2006-2007 was 41.50 unit/l ±25.16 and in year 2007-2008 it was observed 53.75 unit/l ±24.45.

Rotifera

The rotifera were recorded maximum 211.00 unit/l in the month of May 2007-08 and the minimum 37.00 unit/l in the month of July 2006-2007. The annual mean of Rotifera recorded in the first year 2006-2007 was 76.33 unit/l ±33.93 and for second year 2007-2008 it was noted 110.42 unit/l ±56.96.

Table 4.21: No. and Per cent of Different Groups among the Phytoplankton at Sampling Site I (Dharanshu) during 2006–2007 and 2007–2008

Month	Total Phytoplankton/ Liter of Water		Diatom/Liter of Water		Green Algae/ Liter of Water		Blue Green Algae/ Liter of Water		Percentage of Different Groups					
									Diatoms		Green Algae		Blue Green Algae	
	2006-07	2007-08	2006-07	2007-08	2006-07	2007-08	2006-07	2007-08	2006-07	2007-08	2006-07	2007-08	2006-07	2007-08
July	471.83	616.08	442.21	521.00	25.13	85.23	4.50	9.85	93.72	84.56	5.33	13.84	0.95	1.60
August	709.74	672.36	655.31	587.12	51.23	76.62	3.20	8.62	92.33	87.32	7.21	11.39	0.46	1.28
September	940.27	1119.11	813.24	958.34	101.23	125.57	25.80	35.20	86.50	85.62	10.76	11.22	2.74	3.16
October	1266.06	1154.85	1024.00	973.35	180.56	154.30	61.50	27.20	80.88	84.29	14.26	13.36	4.86	2.35
November	1356.58	1408.21	1098.00	1159.00	221.28	201.21	37.30	48.00	80.93	82.30	16.32	14.28	2.75	3.42
December	2099.43	2926.00	1681.12	2503.30	351.23	340.30	67.20	82.40	80.07	85.55	16.73	11.63	3.20	2.82
January	3299.59	3629.70	2825.35	3050.00	398.24	481.25	76.00	98.45	85.62	84.02	12.06	13.26	2.30	2.72
February	5170.00	4220.60	4705.00	3702.15	409.00	457.15	56.00	61.30	91.00	87.72	7.92	10..83	1.08	1.45
March	4284.69	4507.75	3911.30	4125.25	310.39	320.00	63.30	62.50	91.28	91.52	7.24	7.10	1.48	1.38
April	3738.55	4255.70	3510.00	3918.00	174.24	215.30	54.30	122.40	93.88	92.06	4.66	5.06	1.46	2.88
May	2379.58	2696.35	2111.15	2425.00	182.23	192.85	86.20	78.50	88.71	89.94	7.65	7.15	3.62	2.91
June	1178.82	1025.35	1103.32	913.15	65.00	97.00	10.50	15.20	93.59	89.05	5.52	9.47	0.89	1.48
Average± SD	2241.26± 1542.89	2352.67± 1516.03	1990± 1422.47	2069.64± 1374.51	205.81± 134.88	228.90± 140.19	45.48± 28.55	54.14± 36.69	88.21± 5.30	87.00± 3.13	9.63± 4.29	10.71± 3.11	2.16± 1.27	2.29± 0.79

Table 4.22: Quantitative Analysis for Zooplankton in Different Months of River Bhagirathi at Sampling Site I (Dharanshu) during 2006–2007

Month	*Protozoa*		*Rotifera*			*Cladocera*		*Copepoda*	*Total (Unit/l)*
	Amoeba	*Arcella*	*Keratella*	*Notholca*	*Philodima*	*Bosmina*	*Daphnia*	*Cyclops*	
July	+	+	+	+	+	+	+	+	98
August	+	+	+	+	+	+	+	+	96
September	+	+	+	+	+	+	+	+	145
October	+	+	+	+	+	+	+	+	202
November	+	+	+	+	-	+	+	+	167
December	+	+	+	+	+	+	+	+	259
January	+	+	+	+	+	+	+	+	224
February	+	+	+	+	+	+	+	+	296
March	+	+	+	+	+	+	+	+	224
April	+	+	+	+	+	-	+	+	155
May	+	+	+	+	+	+	+	+	154
June	+	+	+	+	+	+	+	+	120

Table 4.23: Quantitative Analysis for Zooplankton in Different Months of River Bhagirathi at Sampling Site I (Dharanshu) during 2007–2008

Month	*Protozoa*		*Rotifera*			*Cladocera*		*Copepoda*	*Total (Unit/l)*
	Amoeba	*Arcella*	*Keratella*	*Notholca*	*Philodima*	*Bosmina*	*Daphnia*	*Cyclops*	
July	+	-	+	+	+	+	+	+	100
August	-	+	+	+	+	+	+	+	110
September	+	+	+	+	+	+	+	-	163
October	+	+	+	+	-	+	+	+	178
November	+	+	+	+	+	-	+	+	207
December	+	+	+	+	+	+	+	+	302
January	+	+	+	+	+	+	+	+	320
February	+	+	+	+	+	+	+	+	359
March	+	+	+	+	+	+	+	+	270
April	+	+	+	+	+	+	+	+	361
May	+	+	+	+	+	+	+	+	378
June	+	+	+	+	+	+	+	+	142

Table 4.24: No. of Different Genera among Zooplankton at Sampling Site I (Dharanshu) during 2006–07 and 2007–08

Months	*Protozoa*		*Rotifera*		*Cladocera*		*Copepoda*		*Total (Unit/liter)*	
	2006-07	*2007-08*	*2006-07*	*2007-08*	*2006-07*	*2007-08*	*2006-07*	*2007-08*	*2006-07*	*2007-08*
July	13.00	16.00	37.00	47.00	31.00	25.00	17.00	12.00	98.00	100.00
August	10.00	20.00	46.00	54.00	23.00	21.00	17.00	15.00	96.00	110.00
September	16.00	32.00	62.00	99.00	28.00	32.00	39.00	00.00	145.00	163.00
October	24.00	30.00	61.00	77.00	72.00	50.00	45.00	21.00	202.00	178.00
November	55.00	50.00	54.00	95.00	54.00	25.00	04.00	37.00	167.00	207.00
December	73.00	79.00	93.00	106.00	72.00	78.00	21.00	39.00	259.00	302.00
January	95.00	73.00	91.00	85.00	18.00	115.00	20.00	47.00	224.00	320.00
February	49.00	60.00	144.00	184.00	84.00	90.00	19.00	25.00	296.00	359.00
March	49.00	66.00	132.00	125.00	31.00	59.00	12.00	20.00	224.00	270.00
April	35.00	75.00	85.00	193.00	28.00	62.00	07.00	31.00	155.00	361.00
May	39.00	90.00	64.00	211.00	17.00	58.00	25.00	19.00	145.00	378.00
June	40.00	54.00	47.00	49.00	25.00	29.00	08.00	10.00	120.00	142.00
Average± SD	41.50± 25.16	53.75± 24.45	76.33± 33.93	110.42± 56.96	40.25± 23.66	53.67± 29.53	19.50± 12.29	23.00± 13.51	199.11± 60.97	279.67± 98.30

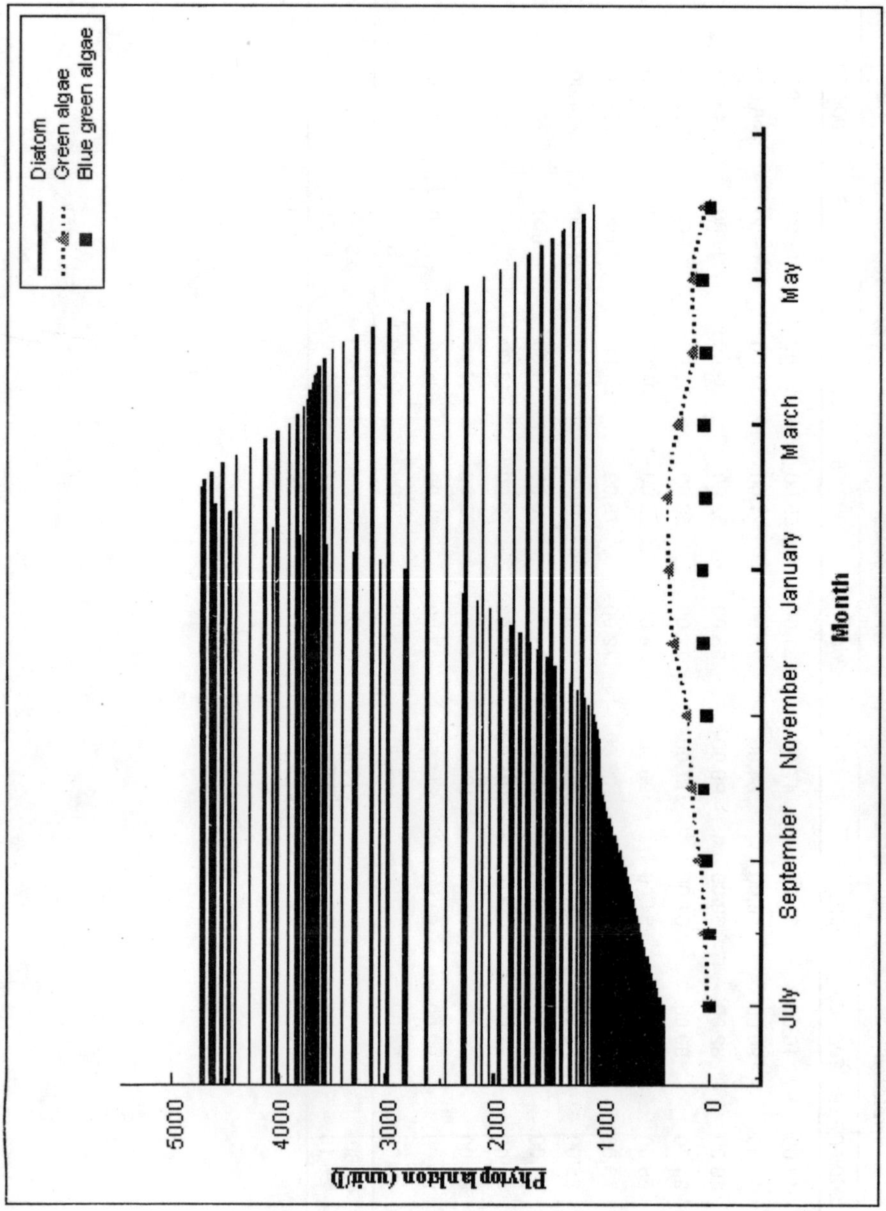

Figure 4.61: Graph Showing Quantitative Variation of Diatom, Green Algae and Blue Green Algae at Sampling Site I during 2006–07

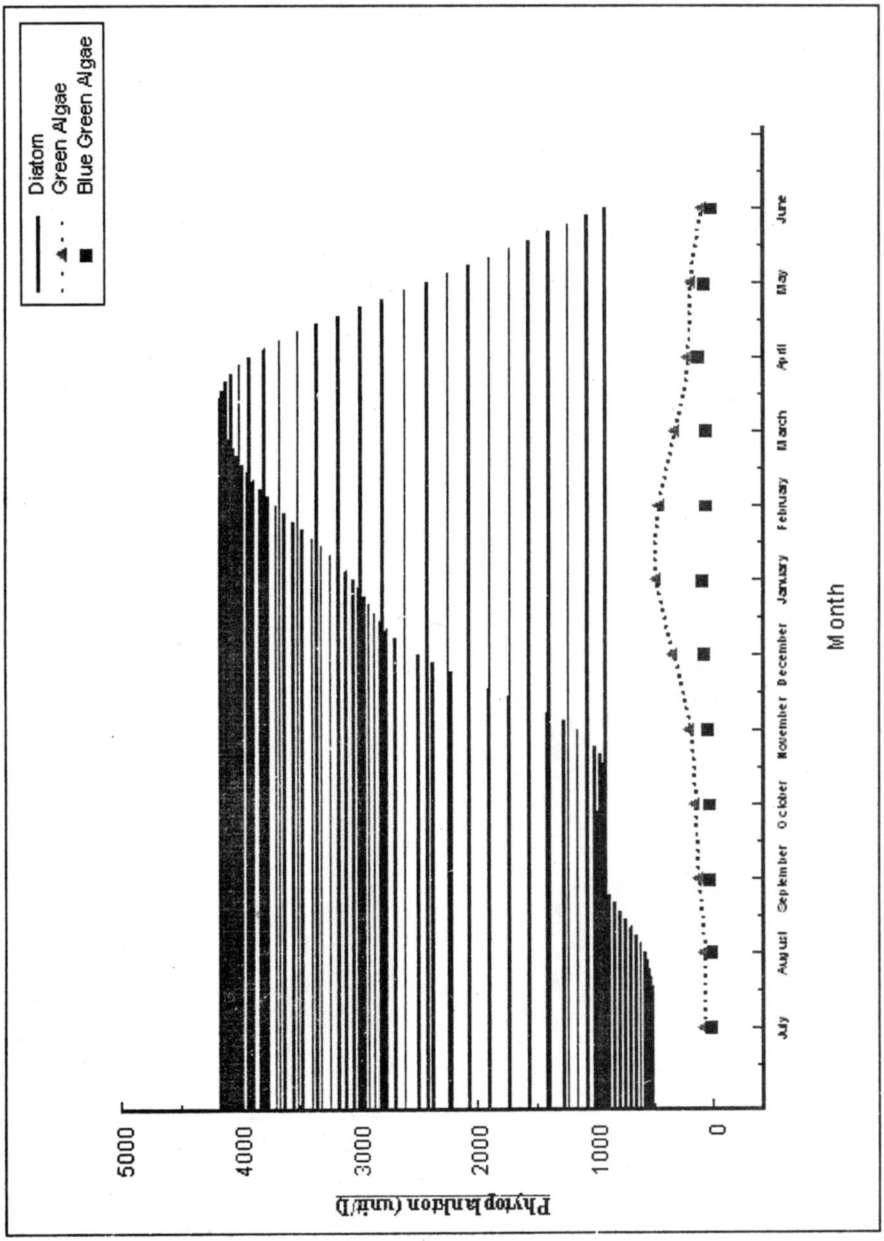

Figure 4.62: Graph Showing Quantitative Variation of Diatom, Green Algae and Blue Green Algae at Sampling Site I during 2007–08

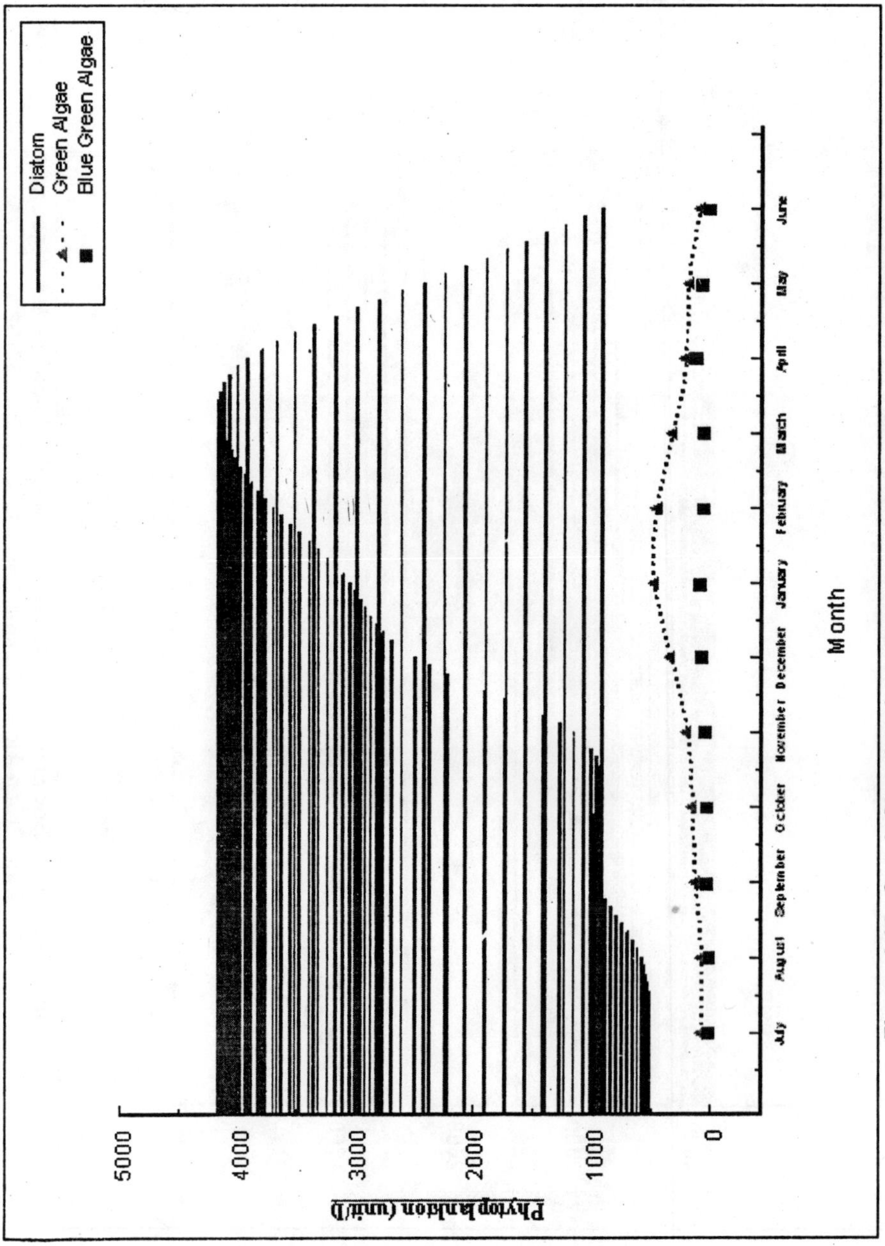

Figure 4.63: Graph Showing Quantitative Variation of Copepoda, Cladocera, Rotifera and Protozoa at Sampling Site I during 2006–07

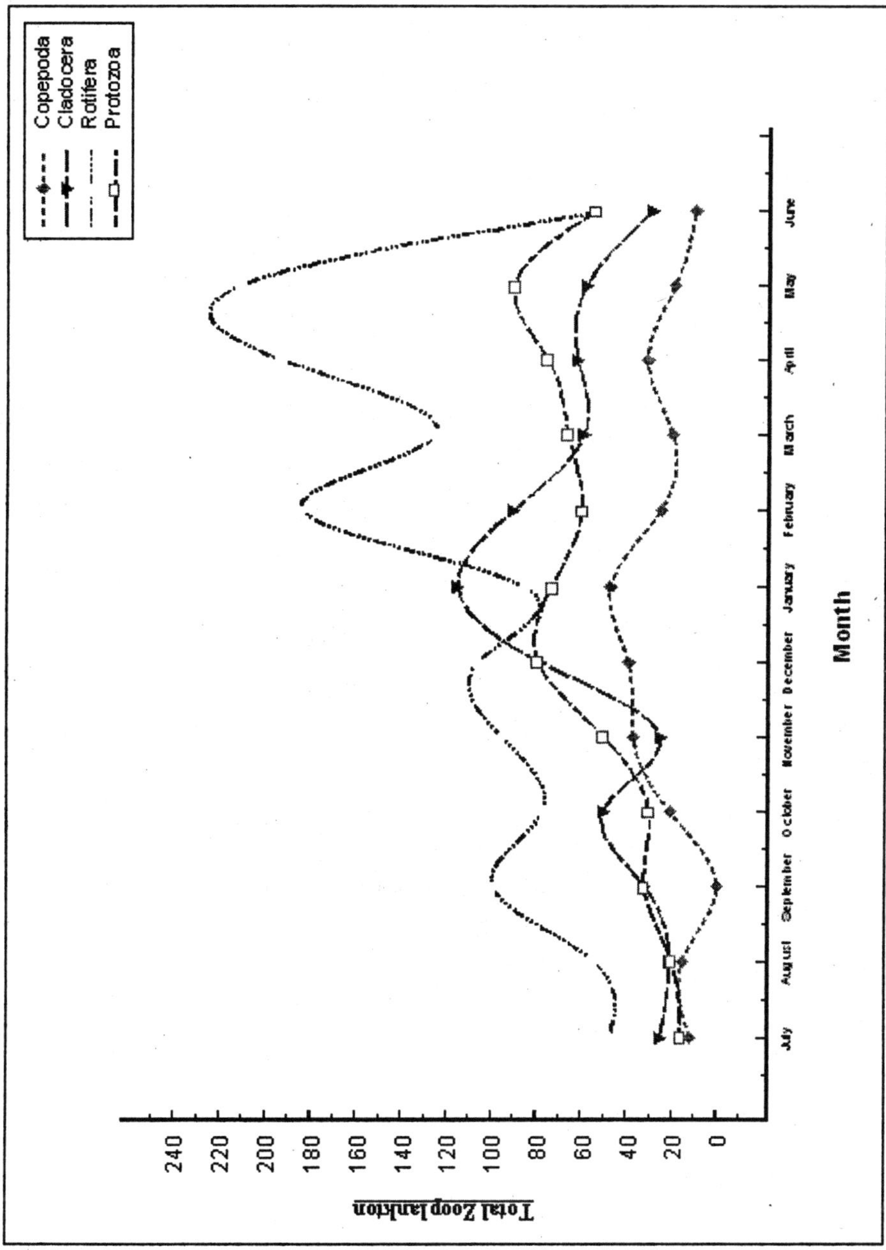

Figure 4.64: Graph Showing Quantitative Variation of Copepoda, Cladocera, Rotifera and Protozoa at Sampling Site I during 2007–08

Cladocera

The Cladocera were found maximum 115.00 unit/l in the month of January 2007-08 and minimum 17.00 unit/l in the month of May 2006-2007. The annual mean recorded for Cladocera in the first year 2006-2007 was 40.25 unit/l ±23.66 and for second Year 2007-2008 it was observed 53.67unit/l ±29.53.

Copepods

Cyclops were observed maximum 47.00 unit/l in the month of January in 2007-2008 and minimum 00.00 unit/l was found in September month of 2007-2008. The annual mean recorded for the first year 2006-2007 was 19.50 unit/l ±12.29 and for second year 2007-2008 it was noted 23.00 unit/l ±13.51.

Sampling Site II (Chinyalisaur)

The result of phytoplankton and zooplankton for this site are given in Tables 4.25 to 4.28 and 4.42 and Figures 4.65 to 4.68.

Plankton

Quantitative analysis of plankton at sampling site revealed that the total plankton number per liter of water was maximum 5663.08 unit/l in the month of February in the year 2006-2007 and minimum was observed 961.58 unit/l in the month of July in year 2007.

Phytoplankton

Phytoplankton were observed lowest of 799.58 unit/l in the month of July in year 2007-2008 and highest of 5439.08 unit/l in the month of February in year 2006-2007. The average number of cells was observed 2668.31 unit/l ±1562.82 in year 2006-2007 and 2722.67 unit/l ±1483.70 in year 2007-2008.

In phytoplankton Diatoms were dominated and class Blue Green Algae were found least in both the years of study. At this sampling station Diatoms were observed maximum in the month of February of year 2006-2007 (4908.21 unit/l), it is 90.24 per cent of total phytoplankton and the minimum in the month of July of the year 2007-08 (724.12 unit/l), which is 90.56 per cent of total phytoplankton. The average count of cells in 2006-2007 was observed 2293.54 unit/l ±1521.12 and in 2007-2008 it was noted 2403.09 unit/l ±1327.36. Green algae were found to be highest in the month of February in 2007-08 (488.37 unit/l), it is 10.37 per cent of total phytoplankton and lowest in the month of July in 2006-2007 (39.00 unit/l), which is 4.39 per cent of total phytoplankton. The average count of green algae in the year 2006-2007 was noted 240.38 unit/l ±145.70 and in 2007-2008 it was observed 254.17 unit/l ±150.74. Blue green algae showed optimum peak during the month of January in the 2007-08 (101.94 unit/l), it is 2.37 per cent of total phytoplankton and least peak during the month of August in the 2006-2007 (8.45 unit/l) which is 0.84 per cent of total phytoplankton. In 2006-2007 the average value of Blue green algae was recorded 51.06 unit/l ±30.39 while in 2007-2008 it was observed 65.41 unit/l ±35.97.

Zooplankton

Among zooplankton protozoa (*Amoeba* and *Arcella*), Rotifera (*Keratella, Notholca* and *Philodima*), Cladocera (*Bosmina* and *Daphnia*) and Copepods (*Cyclops*) were

Table 4.25: No. and Per cent of Different Group among the Phytoplankton at Sampling Site II (Chinyalisaur) during 2006–2007 and 2007–2008

Month	Total Phytoplankton/ Liter of Water		Diatom/Liter of Water		Green Algae/ Liter of Water		Blue Green Algae/ Liter of Water		Percentage of Different Groups					
									Diatoms		Green Algae		Blue Green Algae	
	2006-07	2007-08	2006-07	2007-08	2006-07	2007-08	2006-07	2007-08	2006-07	2007-08	2006-07	2007-08	2006-07	2007-08
July	887.89	799.58	839.39	724.12	39.00	65.25	9.50	10.21	94.55	90.56	4.39	8.16	1.06	1.28
August	1001.74	1188.82	911.34	1087.34	81.95	89.35	8.45	12.13	90.98	91.46	8.18	7.52	0.84	1.02
September	1558.48	1883.84	1415.95	1719.20	121.25	127.29	21.28	37.35	90.85	91.26	7.78	6.76	1.37	1.98
October	1465.15	1803.14	1224.30	1529.54	199.35	211.21	41.50	62.39	83.56	84.83	13.60	11.71	2.84	3.46
November	1969.47	2117.05	1698.72	1832.41	227.28	225.25	43.47	59.39	86.25	86.56	11.55	10.64	2.20	2.80
December	2578.25	2643.42	2118.35	2125.84	381.47	419.28	78.43	98.30	82.16	80.42	14.79	15.86	3.05	3.72
January	3722.49	4301.90	3211.28	3728.57	425.21	471.39	86.00	101.94	86.27	86.67	11.42	10.96	2.31	2.37
February	5439.08	4705.96	4908.21	4125.24	451.75	488.37	79.12	92.35	90.24	87.67	8.31	10.37	1.45	1.96
March	4668.73	4670.98	4198.21	4229.45	398.34	354.21	72.18	87.32	89.92	90.55	8.53	7.58	1.55	1.87
April	4267.99	4487.04	3921.48	4088.29	284.21	298.54	62.30	100.21	91.88	91.12	6.66	6.65	1.46	2.23
May	3232.80	3025.08	2957.00	2725.50	187.29	201.50	88.51	98.08	91.48	90.12	5.79	6.66	2.73	3.24
June	1227.64	1045.17	1118.20	921.54	87.50	98.35	21.94	25.28	91.09	88.17	7.12	9.42	1.79	2.41
Average± SD	2668.31± 1562.82	2722.67± 1483.70	2293.54± 1521.12	2403.09± 1327.36	240.38± 145.70	254.17± 150.74	51.06± 30.39	65.41± 35.97	89.10± 3.76	88.28± 3.31	9.01± 3.17	9.36± 2.73	1.89± 0.72	2.36± 0.83

Table 4.26: Quantitative Analysis for Zooplankton in Different Months of River Bhagirathi at Sampling Site II (Chinyalisaur) during 2006–2007

Month	*Protozoa*		*Rotifera*			*Cladocera*		*Copepoda*	*Total (Unit/l)*
	Amoeba	*Arcella*	*Keratella*	*Notholca*	*Philodima*	*Bosmina*	*Daphnia*	*Cyclops*	
July	+	+	+	+	+	+	+	+	141
August	–	+	+	+	+	+	+	+	133
September	+	+	+	+	+	+	+	+	185
October	+	+	+	+	+	+	+	+	147
November	+	+	+	+	+	+	+	+	193
December	+	+	+	+	+	+	+	+	191
January	+	+	+	+	+	+	+	+	264
February	+	+	+	=	+	+	+	+	224
March	+	+	+	+	+	+	+	+	233
April	+	+	+	+	–	+	+	+	200
May	+	+	+	+	+	+	+	+	217
June	+	+	+	+	+	+	+	+	160

Table 4.27: Quantitative Analysis for Zooplankton in Different Months of River Bhagirathi at Sampling Site II (Chinyalisaur) during 2007–2008

Month	*Protozoa*		*Rotifera*			*Cladocera*		*Copepoda*	*Total (Unit/l)*
	Amoeba	*Arcella*	*Keratella*	*Notholca*	*Philodima*	*Bosmina*	*Daphnia*	*Cyclops*	
July	+	+	+	+	+	+	+	+	162
August	+	+	+	+	+	+	+	+	165
September	+	-	+	+	+	+	+	+	170
October	+	+	+	+	+	+	+	+	140
November	+	+	+	+	-	+	+	+	143
December	+	+	+	+	+	+	+	+	155
January	+	+	+	+	+	+	+	+	209
February	+	+	+	+	+	+	+	+	208
March	+	+	+	+	+	+	+	+	287
April	+	+	+	+	+	+	+	+	188
May	+	+	+	+	+	+	+	+	212
June	+	+	+	+	+	+	+	+	182

Table 4.28: No. of Different Genera among Zooplankton in Sampling Station II (Chinyalisaur) during 2006–07 and 2007–08

Months	*Protozoa*		*Rotifera*		*Cladocera*		*Copepoda*		*Total (Unit/liter)*	
	2006-07	*2007-08*	*2006-07*	*2007-08*	*2006-07*	*2007-08*	*2006-07*	*2007-08*	*2006-07*	*2007-08*
July	46.00	45.00	47.00	54.00	41.00	46.00	07.00	17.00	141.00	162.00
August	10.00	30.00	54.00	68.00	60.00	49.00	09.00	18.00	133.00	165.00
September	44.00	11.00	89.00	94.00	47.00	50.00	05.00	15.00	185.00	170.00
October	30.00	50.00	81.00	60.00	36.00	27.00	00.00	03.00	147.00	140.00
November	58.00	62.00	91.00	42.00	19.00	24.00	25.00	15.00	193.00	143.00
December	58.00	60.00	78.00	46.00	38.00	27.00	17.00	22.00	191.00	155.00
January	36.00	63.00	101.00	72.00	96.00	56.00	31.00	18.00	264.00	209.00
February	44.00	46.00	89.00	109.00	50.00	32.00	41.00	21.00	224.00	208.00
March	67.00	88.00	83.00	88.00	66.00	59.00	17.00	52.00	233.00	287.00
April	40.00	36.00	96.00	86.00	35.00	39.00	29.00	27.00	200.00	188.00
May	96.00	65.00	77.00	66.00	34.00	46.00	10.00	35.00	217.00	212.00
June	52.00	36.00	42.00	82.00	51.00	45.00	15.00	19.00	160.00	182.00
Average ±SD	48.42± 21.09	49.33± 20.07	77.33± 19.37	72.25± 20.23	47.75± 19.74	41.67± 11.76	17.17± 12.17	21.83± 12.13	203.22± 38.45	191.56± 38.84

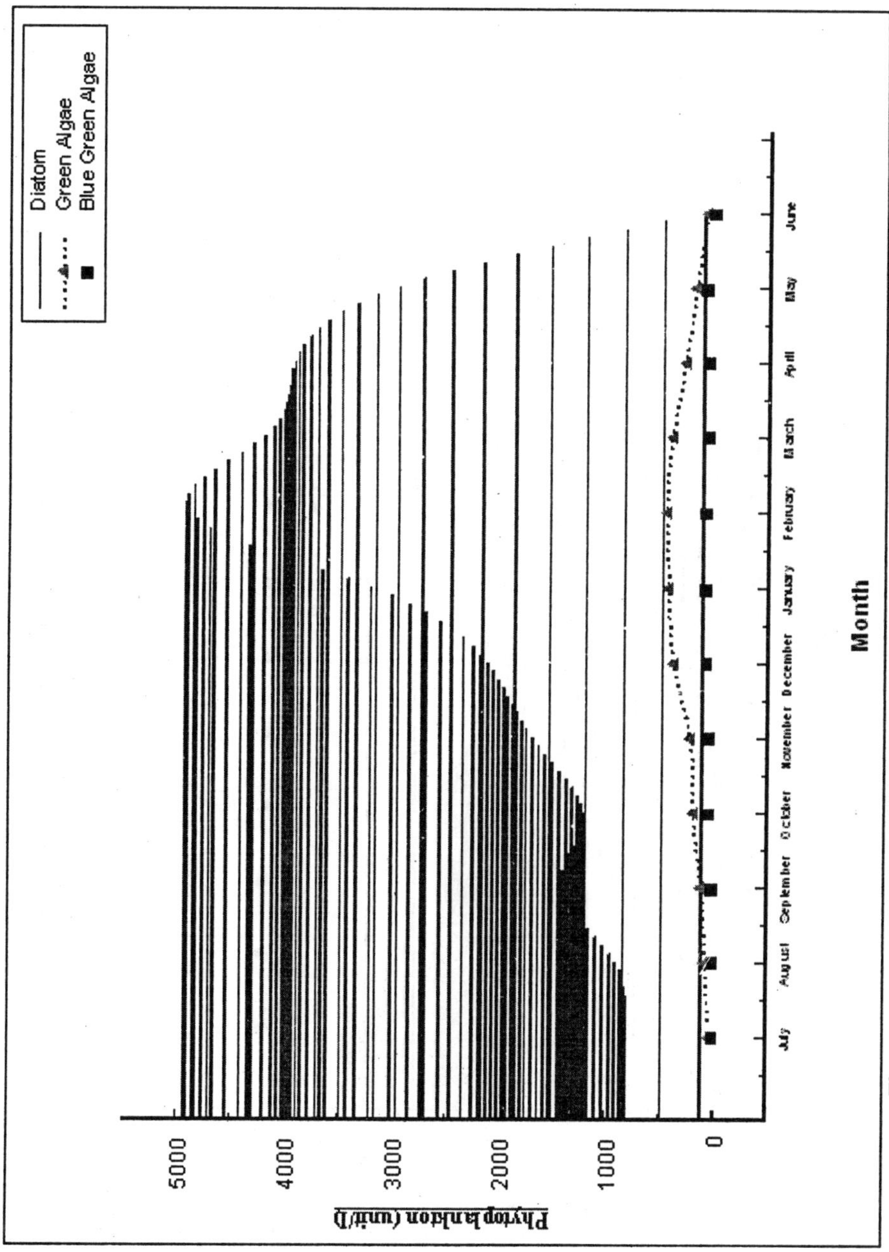

Figure 4.65: Graph Showing Quantitative Variation of Diatom, Green Algae and Blue Green Algae at Sampling Site II during 2006–07

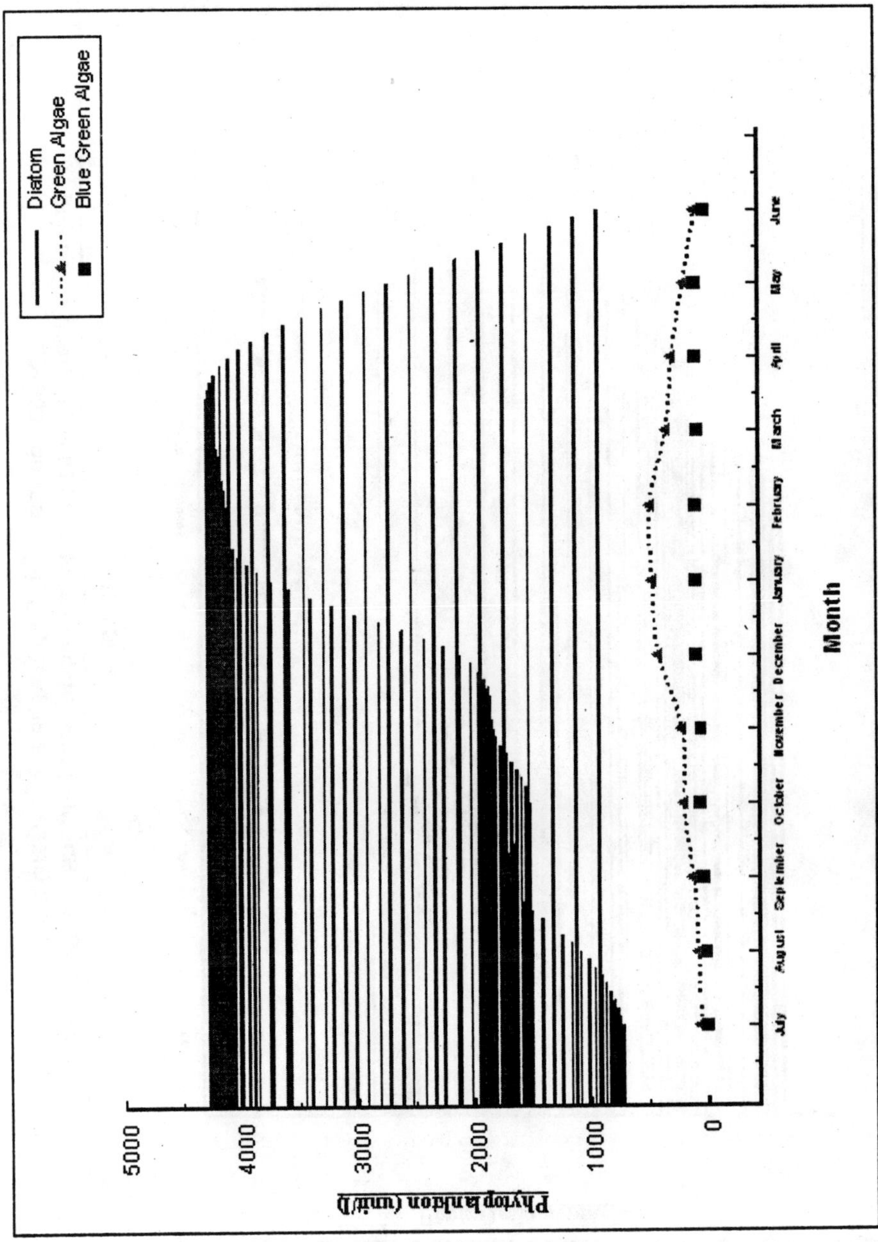

Figure 4.66: Graph Showing Quantitative Variation of Diatom, Green Algae and Blue Green Algae at Sampling Site II during 2007–08

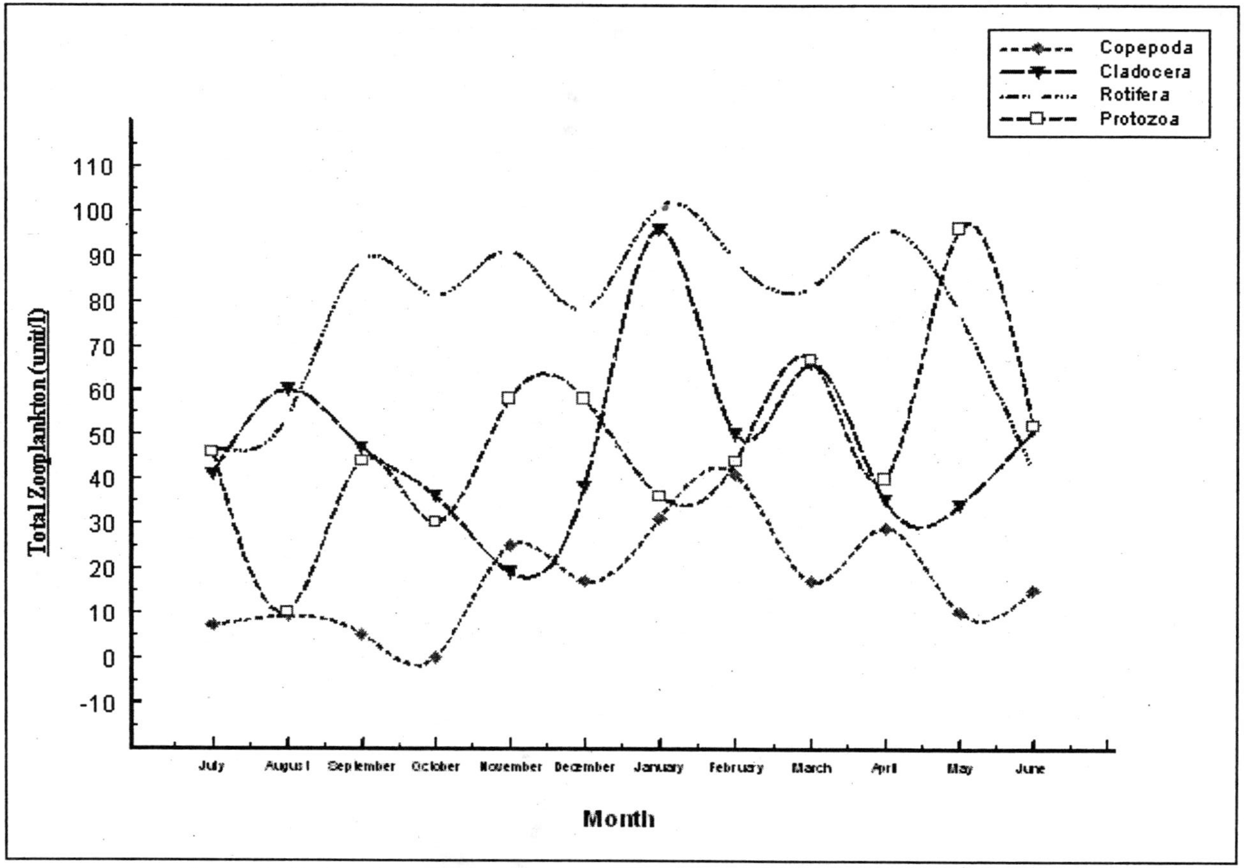

Figure 4.67: Graph Showing Quantitative Variation of Copepoda, Cladocera, Rotifera and Protozoa at Sampling Site II during 2006–07

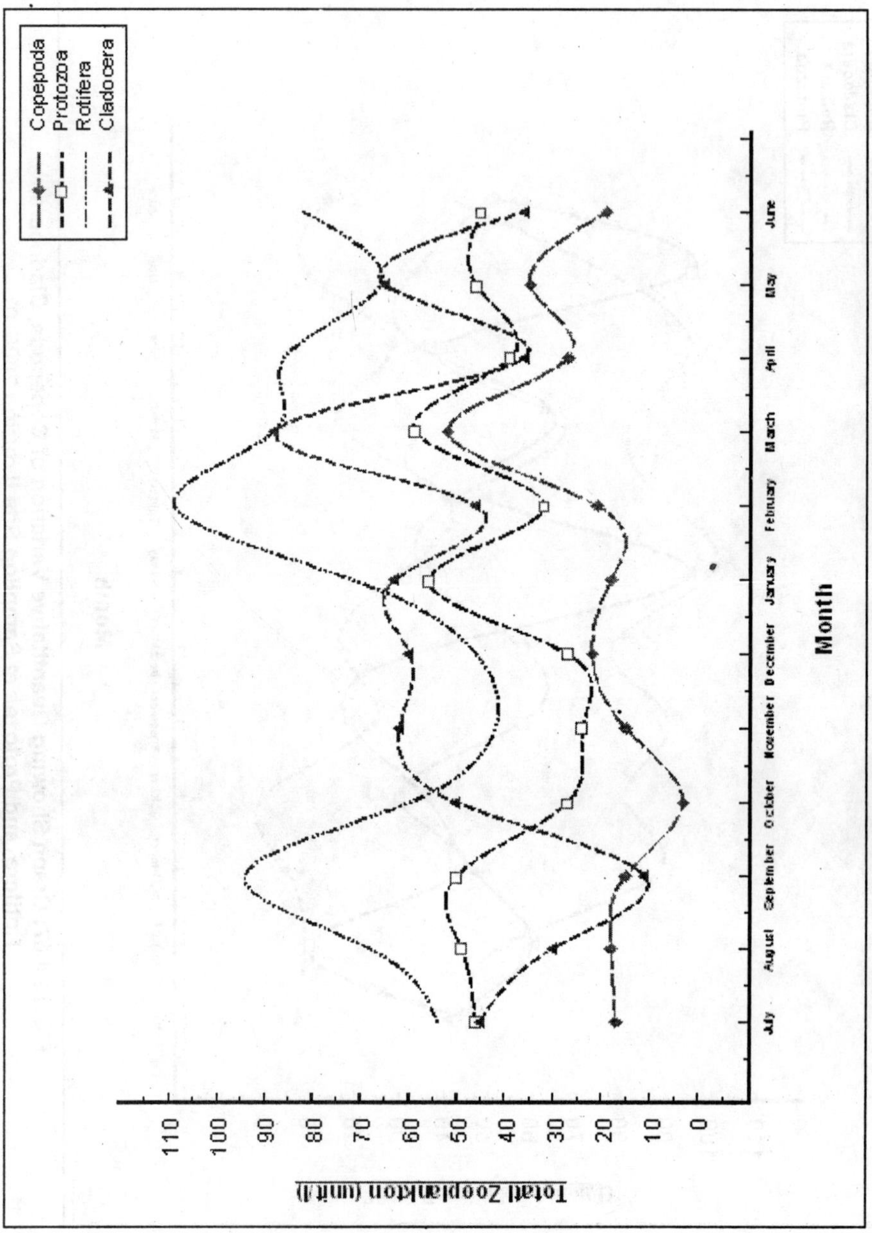

Figure 4.68: Graph Showing Quantitative Variation of Copepoda, Cladocera, Rotifera and Protozoa at Sampling Site II during 2007–08

observed. Zooplankton were noted minimum in the month of August year 2006-2007 (133 unit/l) and found maximum 264 unit/l in the month of January 2006-2007.

The average value of zooplankton for year 2006-2007 was noted 203.22 unit/l ±38.45 and for year 2007-2008 it was recorded 191.56 unit/l ±38.84.

Protozoa

Protozoa were observed maximum 96.00 unit/l in the month of May in 2006-2007 and minimum 10.00 unit/l in the month of August in 2006-2007. The annual mean recorded for Protozoa in first year 2006-2007 was 48.42 unit/l ±21.09 and in year 2007-2008 was observed 49.33 unit/l ±20.07.

Rotifera

The Rotifera were found maximum 109.00 unit/l in the month of February 2007-2008 and the minimum 42.00 unit/l in the month of June, 2006-2007 and November 2007-2008. The annual mean recorded in the first year 2006-2007 was observed 77.33 unit/l ±19.37 and for second year 2007-2008 was recorded 72.25 unit/l ±20.23.

Cladocera

The Cladocera were recorded maximum 96.00 unit/l in the month of January 2006-2007 and minimum 19.00 unit/l in the month of November 2006-2007. The annual mean recorded for first year 2006-2007 was noted 47.75 unit/l ±19.74 and for second Year 2007-2008 it was observed 41.67 unit/l ±11.76.

Copepods

Copepods were found maximum 52.00 unit/l in the month of March in 2007-2008 and minimum 00.00 unit/l was found in October month of 2006-2007. The annual mean recorded for the first year 2006-2007 was recorded 17.17 ±12.17 unit/l and for second year 2007-2008 it was observed 21.83 unit/l ±12.13.

Sampling Site III (Bhaldiyana)

The result of phytoplankton and zooplankton for this site are given in Tables 4.29 to 4.32 and 4.43 and Figures 4.69 to 4.72.

Plankton

Quantitative analysis of plankton at sampling site resulted that the total plankton number per liter of water was observed maximum 5024.31 unit/l in the month of February in the year 2006-2007 and minimum was recorded 1275.81 unit/l in the month of June in year 2006-2007.

Phytoplankton

Phytoplankton were noted lowest of 1051.81 unit/l in the month of June in year 2006-2007 and highest of 4740.31unit/l in the month of February in year 2006-2007. The average number of cells was noted 2831.50 unit/l ±1237.03 in year 2006-2007 and observed 2955.91 unit/l ±1105.77 in year 2007-2008.

In phytoplankton Diatoms were dominated and class Blue Green Algae were found least in both the years of study. At this sampling station Diatoms were observed

Table 4.29: No. and Per cent of Different Groups among the Phytoplankton at Sampling Site III (Bhaldiyana) during 2006–2007 and 2007–2008

Month	Total Phytoplankton/ Liter of Water		Diatom/Liter of Water		Green Algae/ Liter of Water		Blue Green Algae/ Liter of Water		Percentage of Different Groups					
									Diatoms		Green Algae		Blue Green Algae	
	2006-07	2007-08	2006-07	2007-08	2006-07	2007-08	2006-07	2007-08	2006-07	2007-08	2006-07	2007-08	2006-07	2007-08
July	1477.05	1457.39	1339.24	1334.64	118.29	101.47	19.52	21.28	90.67	91.58	8.00	6.96	1.33	1.46
August	1860.32	1967.08	1719.35	1823.34	121.47	128.54	19.50	15.20	92.43	92.69	6.53	6.53	1.04	0.77
September	2570.59	2370.99	2333.41	2148.21	211.94	201.39	25.24	21.39	90.77	90.60	8.25	8.49	0.98	0.90
October	2461.57	2256.50	2134.39	1973.45	300.34	251.34	26.84	31.71	86.71	87.46	12.20	11.13	1.09	1.40
November	2749.21	2868.96	2291.94	2519.78	385.00	281.90	72.27	67.29	83.36	87.83	14.00	9.82	2.64	2.34
December	3010.19	3581.83	2547.35	3145.25	425.52	388.20	37.32	48.35	84.62	87.81	14.13	10.83	1.25	1.35
January	4510.05	3692.05	3945.21	3141.35	488.54	463.39	76.30	87.31	87.48	85.08	10.83	12.54	1.69	2.36
February	4740.31	4665.02	4239.45	4125.28	399.32	419.53	101.54	120.21	89.43	88.43	8.43	8.99	2.14	2.57
March	4032.84	4581.33	3625.25	4128.47	331.21	381.47	76.28	71.39	89.99	90.11	8.22	8.32	1.89	1.56
April	3925.10	3973.99	3615.39	3616.25	222.47	278.39	87.24	79.35	92.11	90.99	5.67	7.00	2.22	2.00
May	1588.98	2284.80	1413.34	2125.21	111.34	101.29	64.30	58.30	88.94	93.01	7.02	4.43	4.04	2.55
June	1051.81	1770.97	925.84	1631.34	97.52	113.24	28.45	26.39	88.02	92.11	9.27	6.39	2.71	1.49
Average± SD	2831.50± 1237.03	2955.9± 1105.77	2510.85± 1105.59	2642.71± 964.61	267.75± 138.87	259.18± 131.93	52.90± 29.63	54.01± 32.42	88.70± 2.81	89.82± 2.46	9.38± 2.81	8.45± 2.34	1.92±. 9.31	1.73± 0.62

Table 4.30: Quantitative Analysis for Zooplankton in Different Months of Tehri Reservoir at Sampling Site III (Bhaldiyana) during 2006–2007

Month	*Protozoa*		*Rotifera*			*Cladocera*		*Copepoda*	*Total (Unit/l)*
	Amoeba	*Arcella*	*Keratella*	*Notholca*	*Philodima*	*Bosmina*	*Daphnia*	*Cyclops*	
July	+	+	+	+	+	+	+	+	174
August	+	+	+	+	+	+	+	+	222
September	+	+	+	+	+	+	+	+	214
October	-	+	+	+	+	+	+	+	258
November	+	+	+	+	+	+	+	+	139
December	+	+	+	+	+	+	+	+	223
January	+	+	+	+	+	+	+	+	309
February	+	+	+	+	+	+	+	+	284
March	+	+	+	+	+	+	+	+	258
April	+	+	+	+	+	+	+	+	310
May	+	+	+	+	+	+	+	+	267
June	+	+	+	+	+	+	+	+	224

Table 4.31: Quantitative Analysis for Zooplankton in Different Months of Tehri Reservoir at Sampling Site III (Bhaldiyana) during 2007–2008

Month	*Protozoa*		*Rotifera*			*Cladocera*		*Copepoda*	*Total (Unit/l)*
	Amoeba	*Arcella*	*Keratella*	*Notholca*	*Philodima*	*Bosmina*	*Daphnia*	*Cyclops*	
July	+	+	+	+	+	+	+	+	136
August	+	+	+	+	+	+	+	+	220
September	+	+	+	+	+	+	+	-	139
October	-	+	-	+	+	+	+	+	231
November	+	+	+	+	+	+	+	+	214
December	+	+	+	+	+	+	+	+	251
January	+	+	+	+	+	+	+	+	237
February	+	+	+	+	+	+	+	+	163
March	+	+	+	+	+	+	+	+	215
April	+	+	+	+	+	+	-	+	232
May	+	+	+	+	+	-	+	+	123
June	+	+	+	+	+	+	+	+	148

Table 4.32: No. of Different Genera among Zooplankton in Sampling Station III (Bhaldiyana) during 2006–07 and 2007–08

Months	Protozoa		Rotifera		Cladocera		Copepoda		Total (Unit/liter)	
	2006-07	2007-08	2006-07	2007-08	2006-07	2007-08	2006-07	2007-08	2006-07	2007-08
July	44.00	37.00	88.00	43.00	33.00	35.00	09.00	21.00	174.00	136.00
August	56.00	32.00	78.00	97.00	64.00	59.00	24.00	32.00	222.00	220.00
September	43.00	27.00	80.00	56.00	88.00	56.00	03.00	00.00	214.00	139.00
October	15.00	71.00	133.00	87.00	59.00	28.00	51.00	45.00	258.00	231.00
November	21.00	61.00	45.00	93.00	46.00	29.00	27.00	31.00	139.00	214.00
December	52.00	78.00	66.00	103.00	58.00	49.00	47.00	21.00	223.00	251.00
January	95.00	53.00	73.00	119.00	96.00	48.00	45.00	17.00	309.00	237.00
February	61.00	48.00	69.00	86.00	83.00	11.00	71.00	18.00	284.00	163.00
March	59.00	73.00	101.00	69.00	67.00	62.00	31.00	11.00	258.00	215.00
April	93.00	56.00	97.00	112.00	100.00	39.00	20.00	25.00	310.00	232.00
May	125.00	20.00	60.00	55.00	68.00	37.00	14.00	11.00	267.00	123.00
June	56.00	47.00	74.00	51.00	85.00	32.00	09.00	18.00	224.00	148.00
Average± SD	60.00± 31.23	50.25± 18.72	80.33± 22.63	80.92± 25.48	70.58± 20.34	40.42± 14.92	29.25± 20.54	20.83± 11.63	252.44± 49.22	201.56± 44.63

maximum in the month of February of year 2006-2007 (4239.45 unit/l), it is 89.43 per cent of total phytoplankton and the minimum in the month of June of the year 2006-2007 (925.84 unit/l), which is 88.02 per cent of total phytoplankton. The average count of cells in 2006-2007 was 2510.85 unit/l ±1105.59 and in 2007-2008 was 2642.71 unit/l ±964.61. Green algae were found to be highest in the month of January year 2007-08 (488.54 unit/l), it is 10.83 per cent of total phytoplankton and lowest in the month of June year 2006-2007 (97.52 unit/l) which is 9.27 per cent of total phytoplankton. The average count of green algae in 2006-2007 was 267.74 unit/l ±138.87and in 2007-2008 was observed 259.18 unit/l ±131.93.

Blue green algae showed optimum peak during the month of February in the year 2007-08 (120.21unit/l), it is 2.57 per cent of total phytoplankton and least peak during the month of August in the year 2007-08 (15.20 unit/l) which is 0.77 per cent of total phytoplankton. In 2006-2007 the average value of Blue green algae was observed 52.90 unit/l ±29.63, while in 2007-2008 it was recorded 54.01 unit/l ±32.42.

Zooplankton

Among zooplankton protozoa (*Amoeba* and *Arcella*), Rotifera (*Keratella, Notholca* and *Philodima*), Cladocera (*Bosmina* and *Daphnia*) and Copepods (*Cyclops*) were observed. Zooplankton were observed minimum in the month of May year 2007-08 (123.00 unit /l) and recorded maximum 310.00 unit/l in the month of April of the year 2006-2007. The average value of zooplankton for year 2006-2007 was recorded 252.44 unit/l ±49.22 and for year 2007-2008 it was noted 201.56 unit/l ±44.63.

Protozoa

Protozoa were noted maximum 125 unit/l in the month of May in year 2006-2007 and minimum 15 unit/l in the month of October 2006-2007. The annual mean recorded for first year 2006-2007 was observed 60.00 unit/l ±31.23 and in year 2007-2008 it was found 50.25 unit/l ±18.72.

Rotifera

The rotifera were found maximum 133.00 unit/l in the month of October 2006-2007 and observed minimum 43.00 unit/l in the month of July 2007-08. The annual mean recorded in the first year 2006-2007 was recorded 80.33 unit/l ±22.63 and for second year 2007-2008 was observed 80.92 unit/l ±25.48.

Cladocera

The Cladocera were observed maximum 100.00 unit/l in the month of April 2006-2007 and minimum 11.00 unit/l in the month of February 2007-08. The annual mean recorded for first year 2006-2007 was recorded 70.58 unit/l ±20.34 and for second Year 2007-2008 it was noted 40.42 unit/l ±14.92.

Copepods

Copepods were observed maximum 71.00 unit/l in the month of February 2006-2007 and minimum 00.00 unit/l were found in September month of 2007-08. The annual mean recorded for the first year 2006-2007 was observed 29.25 unit/l ±20.54 and for second year 2007-2008 it was noted 20.83 unit/l ±11.63.

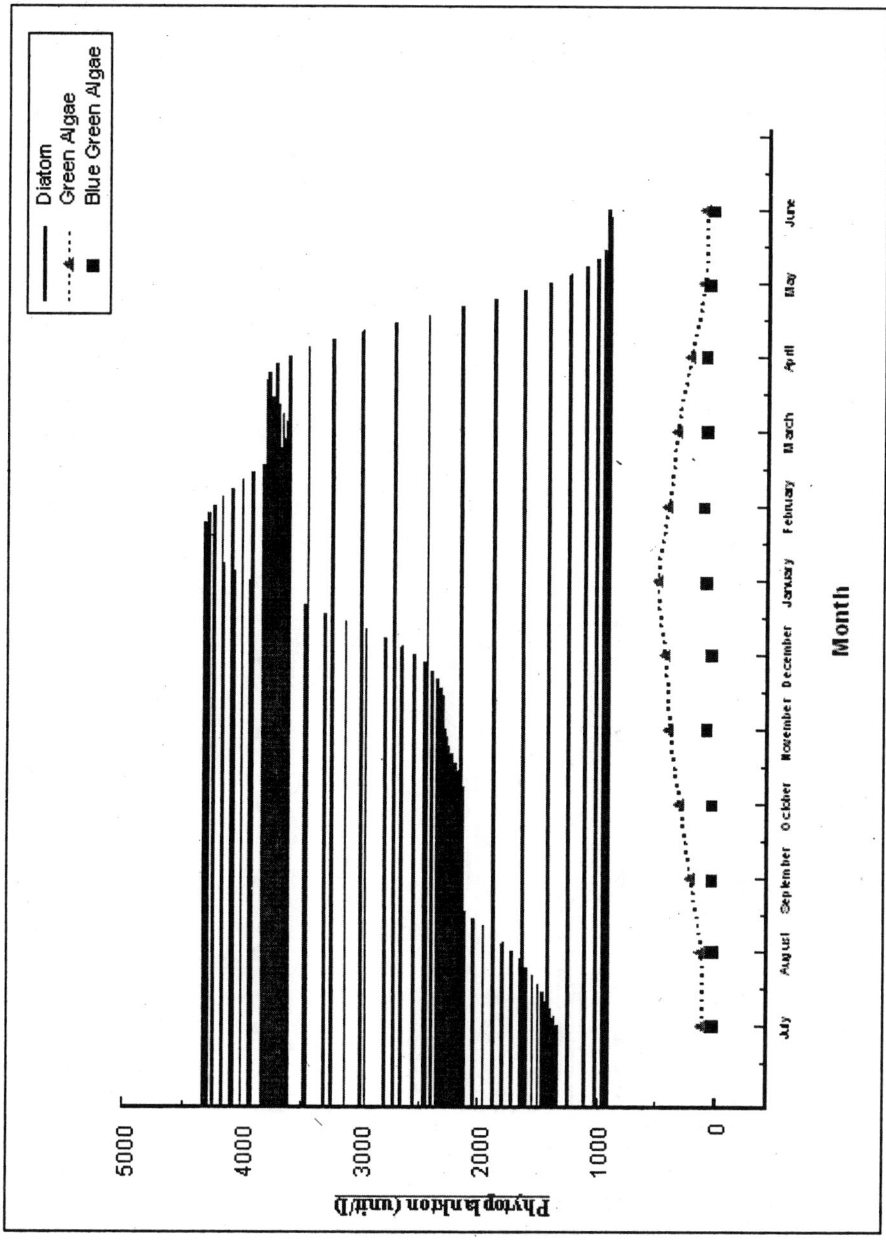

Figure 4.69: Graph Showing Quantitative Variation of Diatom, Green Algae and Blue Green Algae at Sampling Site III during 2006–07

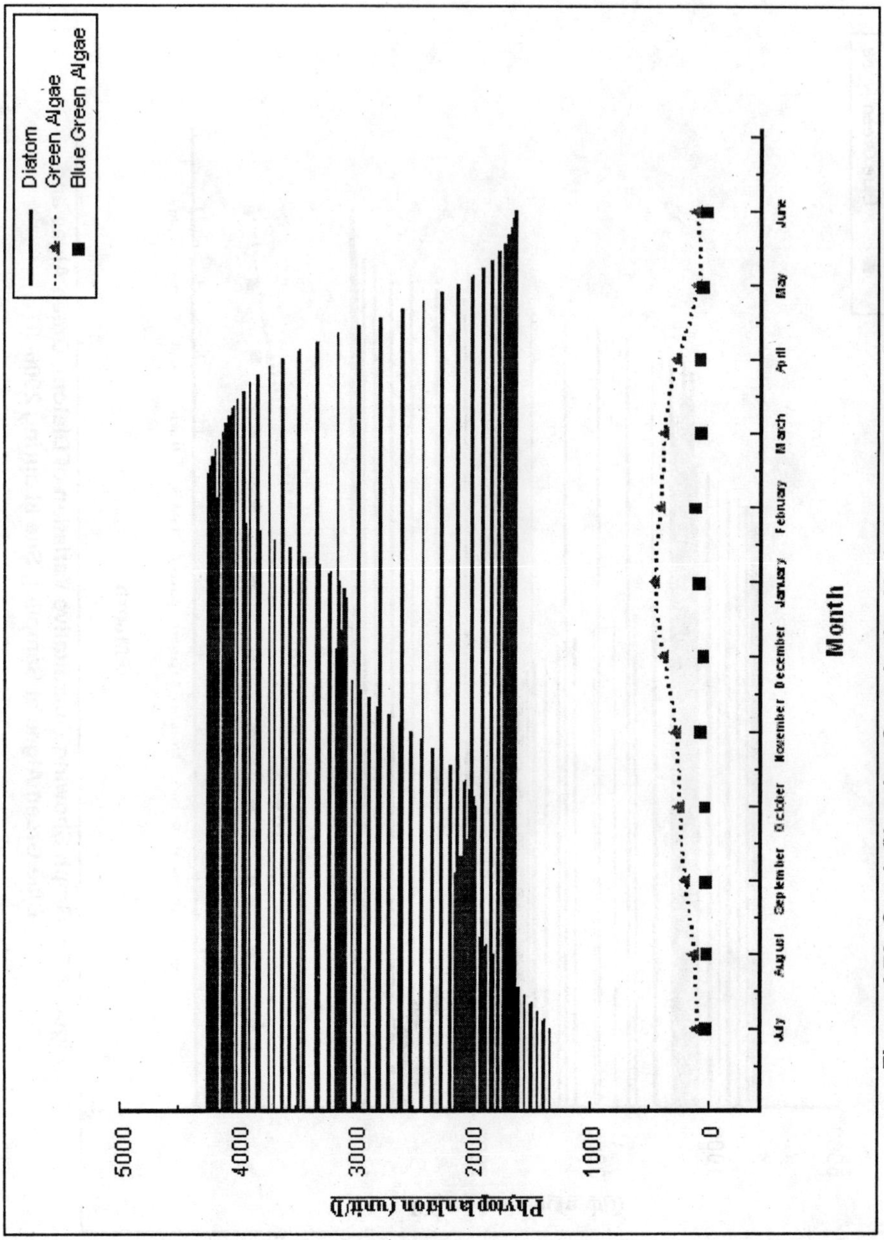

Figure 4.70: Graph Showing Quantitative Variation of Diatom, Green Algae and Blue Green Algae at Sampling Site III during 2007–08

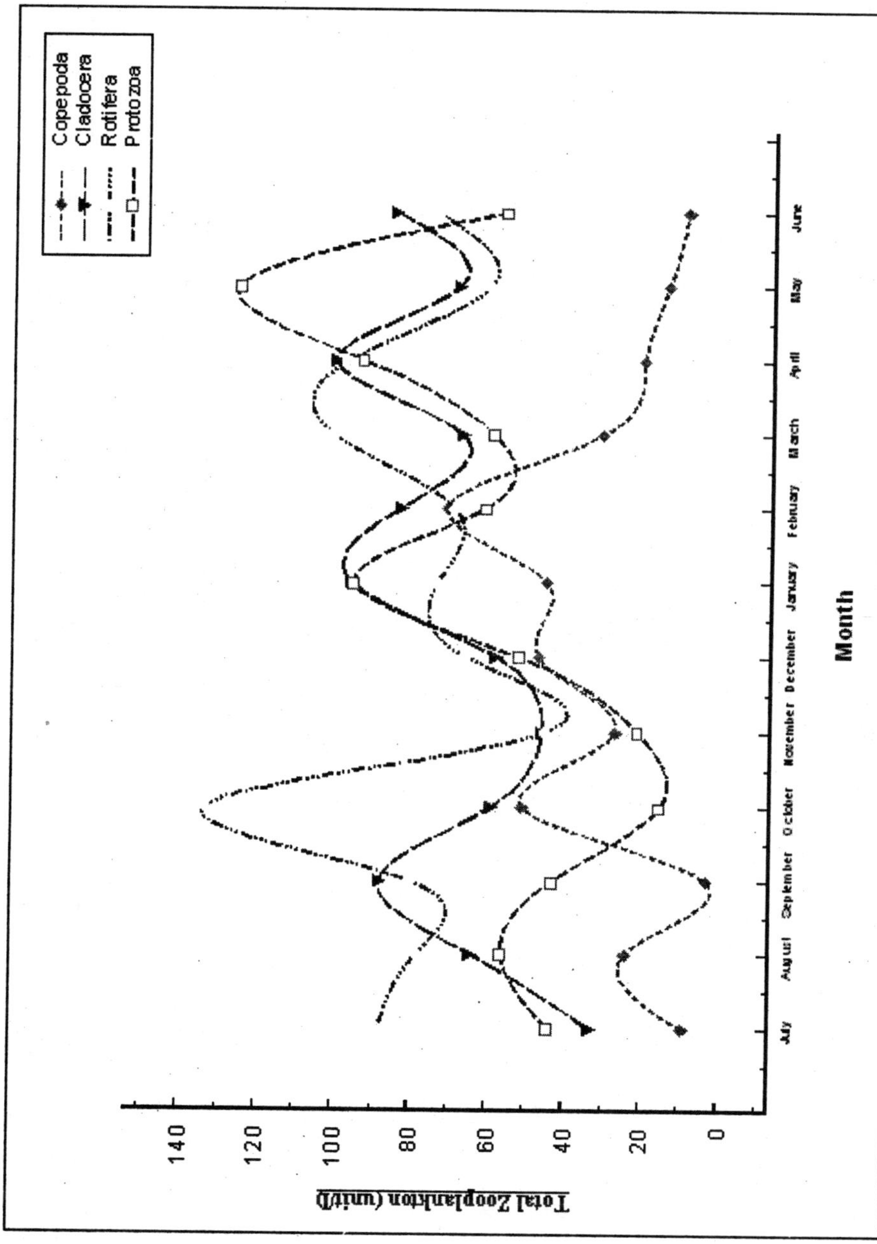

Figure 4:71: Graph Showing Quantitative Variation of Copepoda, Cladocera, Rotifera and Protozoa at Sampling Site III during 2006–07

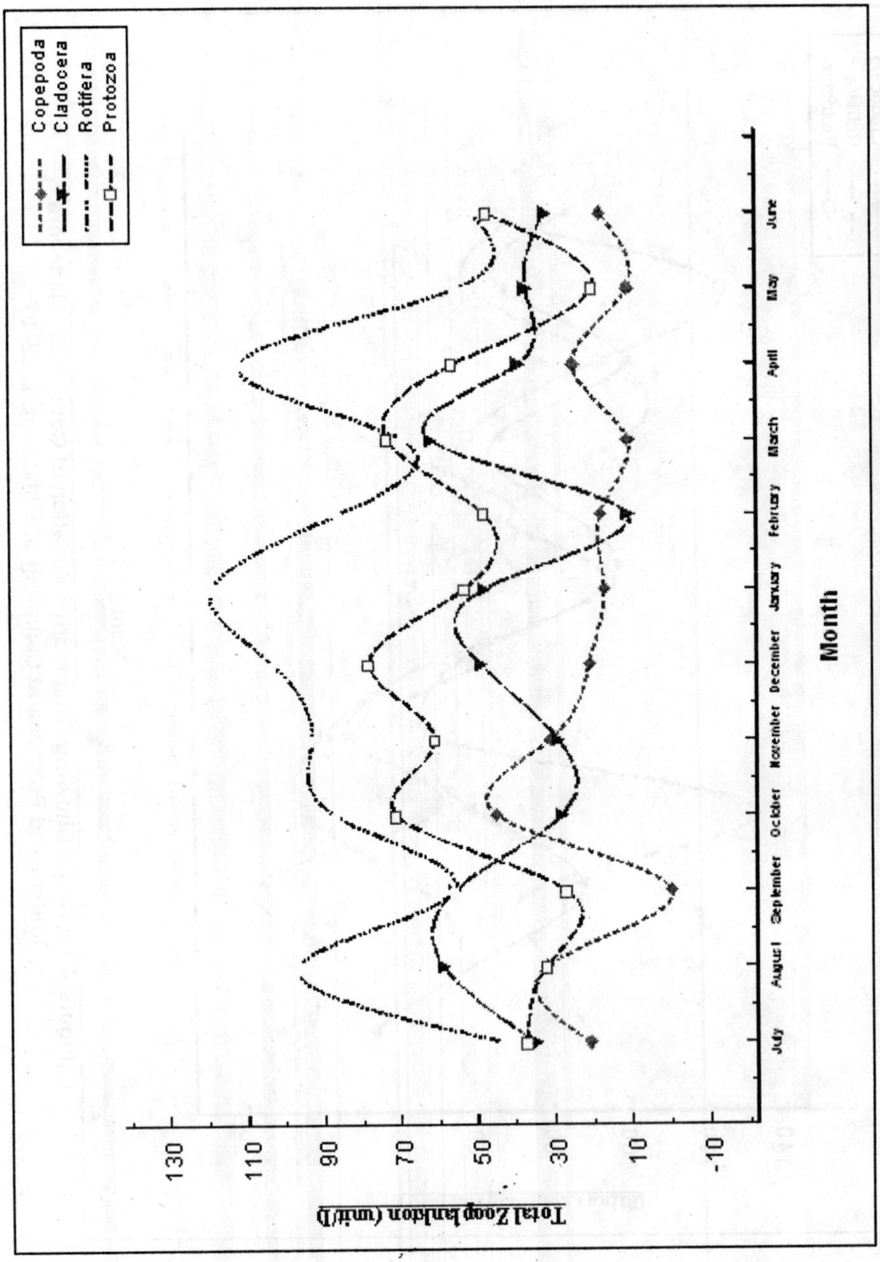

Figure 4.72: Graph Showing Quantitative Variation of Copepoda, Cladocera, Rotifera and Protozoa at Sampling Site III during 2007–08

Sampling Site IV (Koti)

The results of phytoplankton and zoo plankton for this site are shown in Tables 4.33 to 4.36 and 4.44 and Figures 4.73 to 4.76.

Plankton

Quantitative analysis of plankton at this sampling site revealed that the total plankton number per liter of water was maximum 6006.25 unit/l in the month of February in the year 2006-2007 and minimum number was observed 1576.90 unit/l in the month of June in year 2007-2008.

Phytoplankton

Phytoplankton were observed lowest of 1357.90 unit/l in the month of June in year 2007-08 and highest of 5670.25 unit/l in the month of February year 2006-2007. The average number of cells was noted 3508.28 unit/l ±1318.32 in year 2006-2007 and recorded 3447.98 unit/l ±1258.39 in year 2007-2008.

Among phytoplankton Diatoms were dominated and class Blue Green Algae were found least in both the years of study. At this sampling station Diatoms were observed maximum in the month of February of year 2006-2007 (5139.34 unit/l) it is 90.64 per cent of total phytoplankton and noted minimum 1413.39 unit/l (91.12 per cent) in the month of June in the year 2006-07. The average count of cells in 2006-2007 was observed 3192.71unit/l ±1178.80 and in 2007-2008 it was recorded 3142.72 unit/l ±1128.86.

Green algae were recorded to be highest in the month of February year 2007-2008 (472.00 unit/l) it is 8.76 per cent of total phytoplankton and lowest 107.54 unit/l (6.99 per cent) in the month of June in year 2006-2007. The average count of green algae in 2006-2007 was observed 256.88 unit/l ±126.29 and in 2007-2008 it was noted 251.20 unit/l ±124.81.

Blue green algae showed optimum peak during the month of February 2006-2007 (127.53 unit/l) it is 2.25 per cent of total phytoplankton and least peak during the month of August in the year 2007-08 (15.21 unit/l) which is 0.67 per cent of total phytoplankton. In 2006-2007 the average value of Blue green algae was found 58.68 unit/l ±33.13 while in 2007-2008 it was observed 53.65 unit/l ±29.42.

Zooplankton

Among zooplankton protozoa (*Amoeba* and *Arcella*), Rotifera (*Keratella, Notholca* and *Philodima*), Cladocera (*Bosmina* and *Daphnia*) and Copepods (*Cyclops*) were observed. Zooplankton were observed minimum in the month of September 2006-2007 (168.00 unit /l) and maximum 406.00 unit/l in the month of November of the same year. The average value of zooplankton for year 2006-2007 was observed 328.00 unit/l ±75.22 and for year 2007-2008 it was noted 296.44 unit/l ±43.96.

Protozoa

Protozoa were observed maximum 143.00 unit/l in the month of December 2006-2007 and minimum 37.00 unit/l in the month of May 2006-2007. The annual mean recorded for first year 2006-2007 was noted 75.58 unit/l ±33.31 and in year 2007-2008 was observed 77.33 unit/l ±26.24.

Table 4.33: No. and Per cent of Different Group among the Phytoplankton at Sampling Site IV (Koti) during 2006–2007 and 2007–2008

Month	Total Phytoplankton/ Liter of Water		Diatom/Liter of Water		Green Algae/ Liter of Water		Blue Green Algae/ Liter of Water		Percentage of Different Groups					
									Diatoms		Green Algae		Blue Green Algae	
	2006-07	2007-08	2006-07	2007-08	2006-07	2007-08	2006-07	2007-08	2006-07	2007-08	2006-07	2007-08	2006-07	2007-08
July	1691.67	1762.38	1541.39	1624.59	125.28	108.25	25.00	29.54	91.12	92.18	7.40	6.14	1.48	1.68
August	2272.03	2264.84	2139.43	2123.34	111.30	126.29	21.30	15.21	94.16	93.75	4.90	5.57	0.94	0.67
September	3071.99	3164.05	2813.31	2958.34	225.84	182.39	32.84	18.32	91.58	93.50	7.35	5.92	1.07	0.58
October	3364.21	3270.09	3035.40	2973.39	291.00	254.35	37.81	42.35	90.24	90.92	8.64	7.78	1.12	1.30
November	3704.48	3507.23	3299.43	3169.39	322.84	290.39	82.21	47.45	89.07	90.37	8.71	8.28	2.22	1.35
December	4393.76	3995.09	3935.41	3539.34	401.00	387.54	57.35	68.21	89.57	88.59	9.12	9.70	1.31	1.71
January	5063.35	4407.37	4521.45	3931.32	463.50	377.84	78.40	98.21	89.30	89.20	9.15	8.57	1.55	2.23
February	5670.25	5388.95	5139.34	4815.45	403.38	472.00	127.53	101.50	90.64	89.36	9.11	8.76	2.25	1.88
March	4521.87	4934.18	4134.35	4521.34	301.28	351.54	86.24	61.30	91.43	91.63	6.66	7.12	1.91	1.24
April	4188.90	4500.40	3914.34	4215.35	211.21	225.84	63.35	59.28	93.44	93.66	5.05	5.02	1.51	1.32
May	2618.57	2823.37	2425.34	2625.34	118.39	119.82	74.84	78.21	92.62	92.98	4.52	4.24	2.86	2.77
June	1538.28	1357.90	1413.39	1215.45	107.54	118.21	17.35	24.24	91.88	89.51	6.99	8.70	1.13	1.78
Average± SD	3508.28± 1318.32	3447.99± 1258.39	3192.72± 1178.80	3142.72± 1128.86	256.88± 126.29	251.20± 124.81	58.69± 33.13	53.65± 29.42	91.25± 1.61	91.30± 1.90	7.13± 1.72	7.15± 1.74	1.62± 0.59	1.55± 0.61

Table 4.34: Quantitative Analysis for Zooplankton in Different Months of Tehri Reservoir at Sampling Site IV (Koti) during 2006–2007

Month	*Protozoa*		*Rotifera*			*Cladocera*		*Copepoda*	*Total (Unit/l)*
	Amoeba	*Arcella*	*Keratella*	*Notholca*	*Philodima*	*Bosmina*	*Daphnia*	*Cyclops*	
July	+	+	+	+	+	+	+	+	168
August	+	+	+	+	+	+	+	+	292
September	+	+	+	+	+	+	+	+	295
October	+	+	+	+	+	+	+	+	331
November	+	+	+	+	+	+	+	+	406
December	+	+	+	+	+	+	+	+	375
January	+	+	+	+	+	+	+	+	435
February	+	+	+	+	+	+	+	+	336
March	+	+	+	+	+	+	+	+	248
April	+	+	+	+	+	+	+	+	352
May	+	+	+	+	+	+	+	+	232
June	+	+	+	+	+	+	+	+	237

Table 4.35: Quantitative Analysis for Zooplankton in Different Months of Tehri Reservoir at Sampling Site IV (Koti) during 2007–2008

Month	*Protozoa*		*Rotifera*			*Cladocera*		*Copepoda*	*Total (Unit/l)*
	Amoeba	*Arcella*	*Keratella*	*Notholca*	*Philodima*	*Bosmina*	*Daphnia*	*Cyclops*	
July	+	+	+	+	+	+	+	+	216
August	+	+	+	+	+	+	+	+	298
September	+	+	+	+	+	+	+	+	307
October	+	+	+	+	+	+	+	+	366
November	+	+	+	+	+	+	+	+	305
December	+	+	+	+	+	+	+	+	305
January	+	+	+	+	+	+	+	+	337
February	+	+	+	+	+	+	+	+	239
March	+	+	+	+	+	+	+	+	304
April	+	+	+	+	+	+	+	+	274
May	+	+	+	+	+	+	+	+	319
June	+	+	+	+	+	+	+	+	219

Table 4.36: No. of Different Genera among Zooplankton in Sampling Station IV (Koti) during 2006–07 and 2007–08

Months	Protozoa		Rotifera		Cladocera		Copepoda		Total (Unit/liter)	
	2006-07	2007-08	2006-07	2007-08	2006-07	2007-08	2006-07	2007-08	2006-07	2007-08
July	46.00	42.00	56.00	89.00	39.00	70.00	27.00	15.00	168.00	216.00
August	87.00	77.00	80.00	91.00	80.00	96.00	45.00	34.00	292.00	298.00
September	97.00	100.00	89.00	93.00	96.00	65.00	13.00	49.00	295.00	307.00
October	64.00	82.00	142.00	122.00	86.00	117.00	39.00	45.00	331.00	366.00
November	88.00	47.00	137.00	137.00	112.00	74.00	69.00	47.00	406.00	305.00
December	143.00	127.00	103.00	69.00	84.00	70.00	45.00	39.00	375.00	305.00
January	131.00	46.00	140.00	145.00	105.00	83.00	59.00	63.00	435.00	337.00
February	80.00	53.00	140.00	125.00	69.00	40.00	47.00	21.00	336.00	239.00
March	62.00	82.00	117.00	119.00	34.00	66.00	35.00	37.00	248.00	304.00
April	41.00	101.00	145.00	85.00	93.00	41.00	73.00	47.00	352.00	274.00
May	37.00	93.00	98.00	111.00	78.00	86.00	19.00	29.00	232.00	319.00
June	67.00	78.00	83.00	95.00	43.00	19.00	44.00	27.00	237.00	219.00
Average± SD	75.58± 33.33	77.33± 26.24	110.83± 30.15	106.75± 23.16	76.58± 25.72	68.92± 26.45	42.92± 18.26	37.75± 13.46	328.00± 75.22	296.44± 43.96

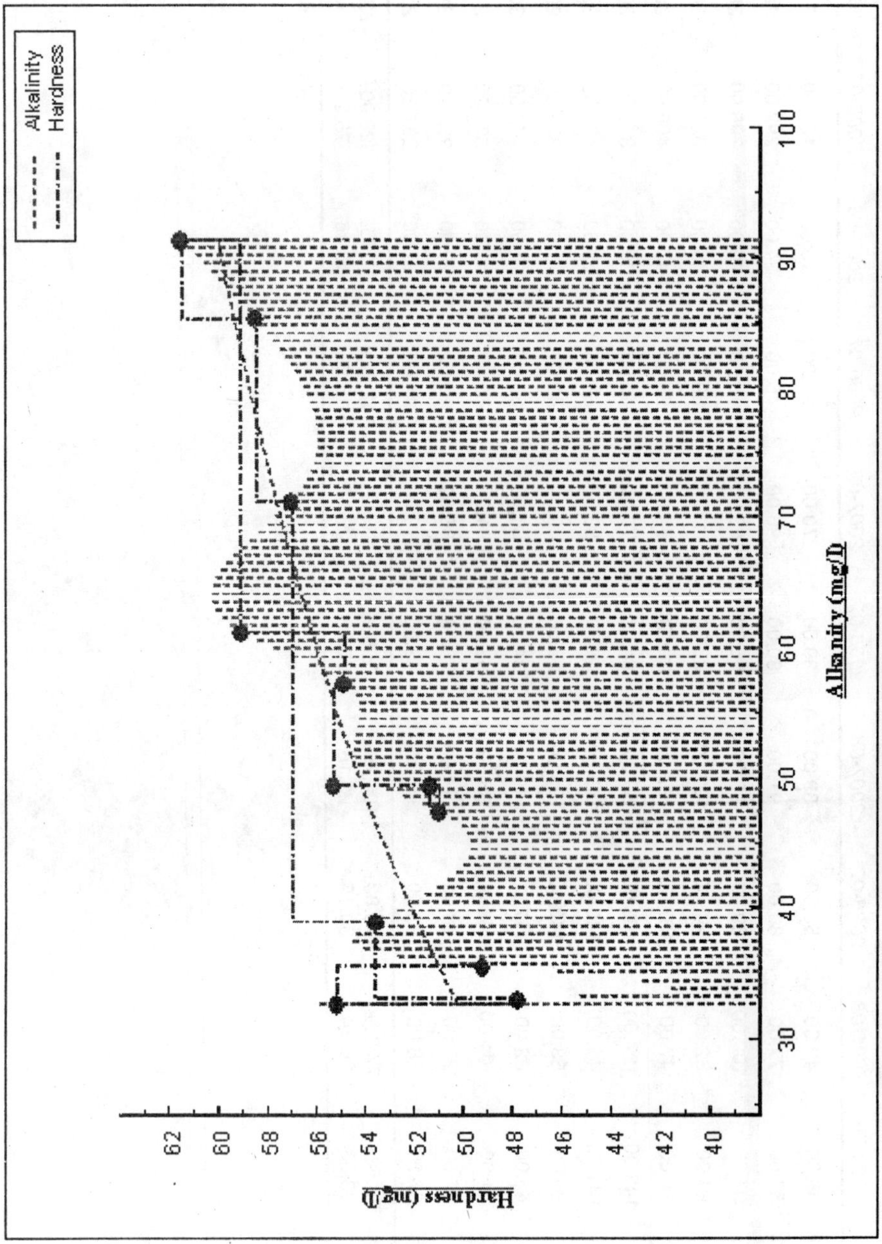

Figure 4.73: Graph Showing Quantitative Variation of Diatom, Green Algae and Blue Green Algae at Sampling Site IV during 2006–07

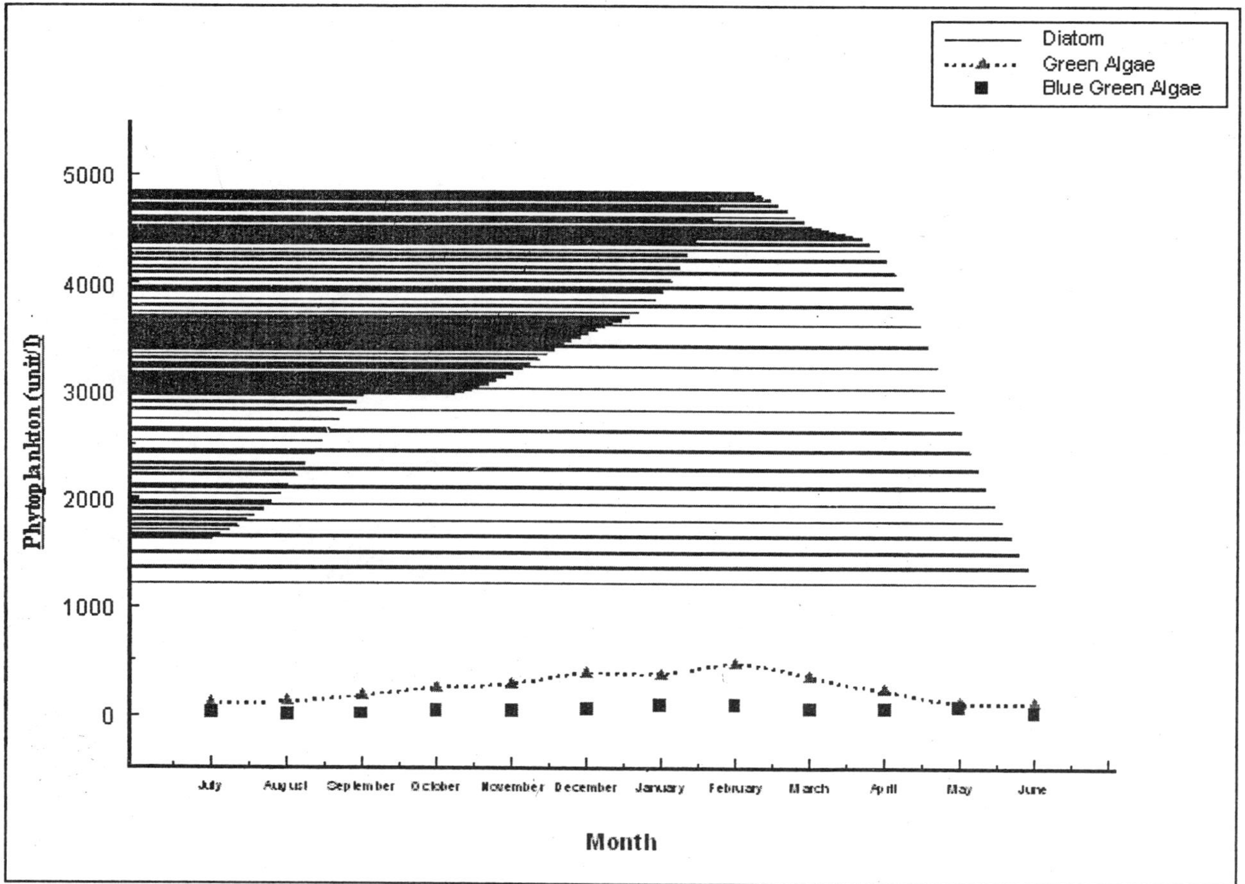

Figure 4.74: Graph Showing Quantitative Variation of Diatom, Green Algae and Blue Green Algae at Sampling Site IV during 2007–08

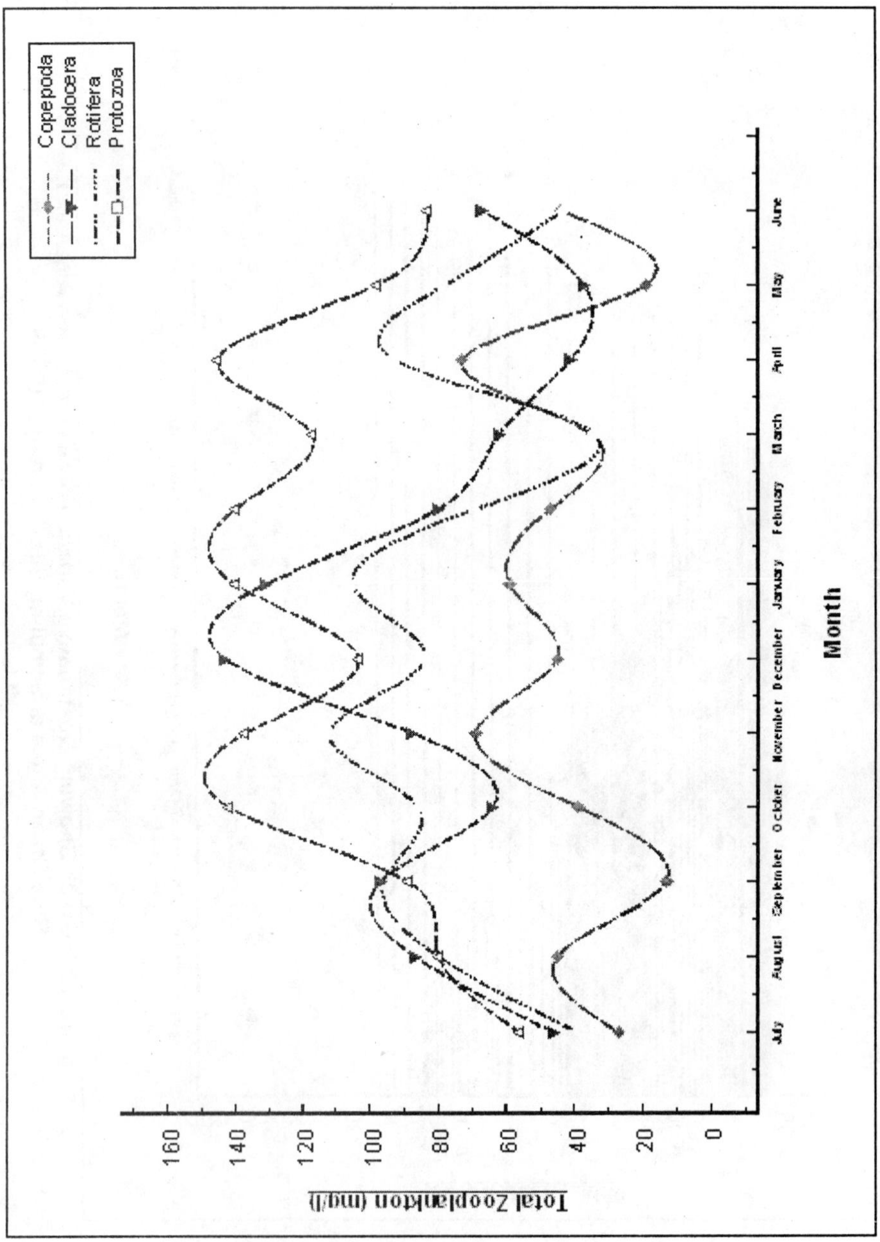

Figure 4.75: Graph Showing Quantitative Variation of Copepoda, Cladocera, Rotifera and Protozoa at Sampling Site IV during 2006–07

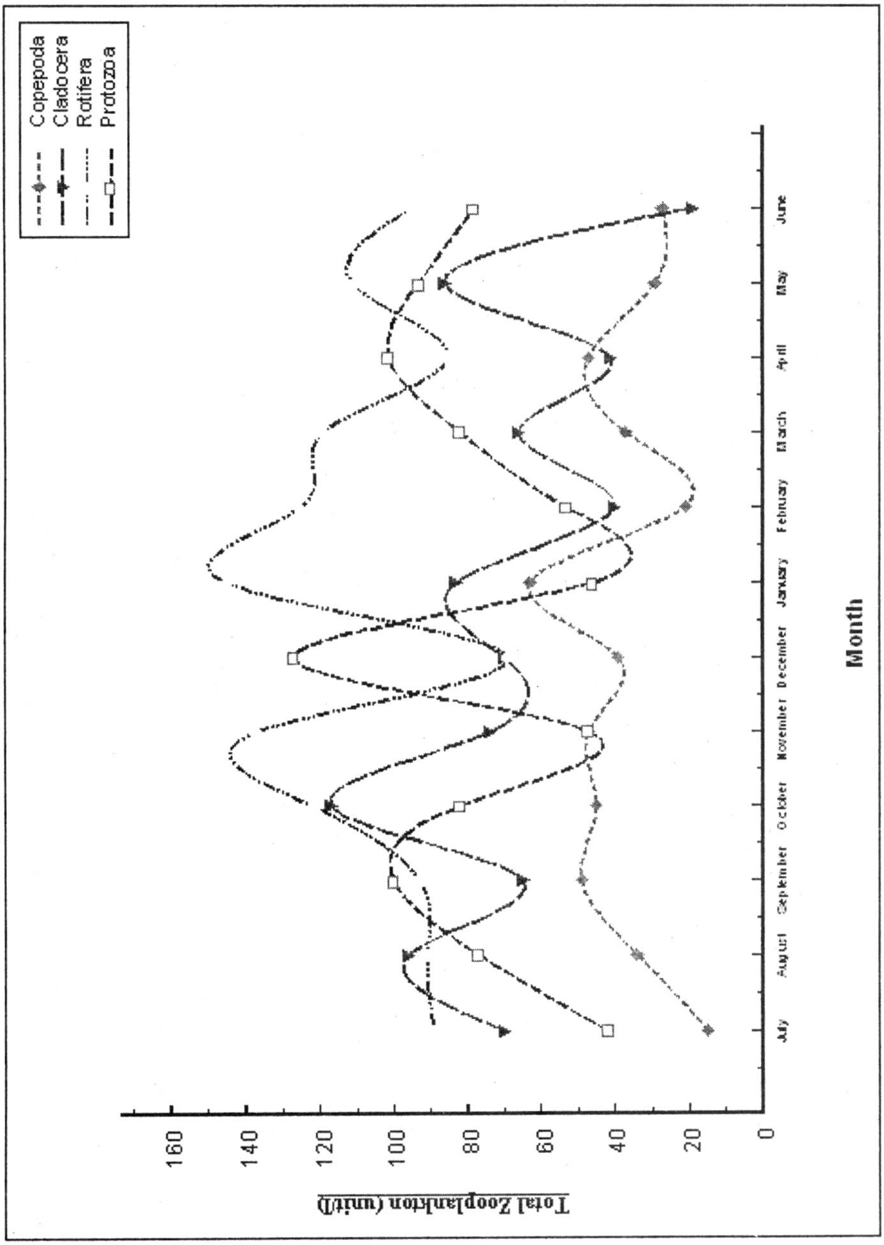

Figure 4.76: Graph Showing Quantitative Variation of Copepoda, Cladocera, Rotifera and Protozoa at Sampling Site IV during 2007–08

Rotifera

The rotifera were found maximum 145.00 unit/l in the month of April 2006-2007 and January 2007-08 and the minimum 56.00 unit/l in the month of July 2006-2007. The annual mean recorded in the first year 2006-2007 was 110.83 unit/l ±30.15 and for second year 2007-2008 it was observed 106.75 unit/l ±23.16.

Cladocera

The Cladocera were observed maximum 117.00 unit/l in the month of October 2007-08 and minimum 19.00 unit/l in the month of June 2007-08. The annual mean recorded for first year 2006-2007 was 76.58 unit/l ±25.72 and for second Year 2007-2008 it was noted 68.92 unit/l ±26.45.

Copepods

Copepods were recorded maximum 73.00 unit/l in the month of April 2006-07 and minimum 13.00 unit/l was found in September month of 2007-08. The annual mean recorded for the first year 2006-2007 was 42.92 unit/l ±18.26 and for second year 2007-2008 it was observed 37.75 unit/l ±13.46.

Sampling Site V (Dam Outlet near Chopra village)

The results of phytoplankton and zooplankton for this site are presented in Tables 4.37 to 4.40 and 4.45 and Figures 4.77 to 4.80.

Plankton

Quantitative analysis of plankton at sampling site showed that the total plankton number per liter of water was maximum 4406.10 unit/l in the month of February in the year 2006-2007 and minimum was 573.93 unit/l in the month of July in year 2006-2007.

Phytoplankton

Phytoplankton were noted lowest of 474.93 unit/l in the month of July in year 2006-07 and highest of 4155.10 unit/l in the month of February year 2006-07. The average number of cells was observed 1922.98 unit/l ±1222.66 in year 2006-2007 and noted 2200.84 unit/l ±1102.66 in year 2007-2008.

In phytoplankton Diatoms were dominated and class Blue Green Algae were found least in both the years of study. At this sampling station Diatoms were observed maximum in the month of February of year 2006-2007 (3710.54 mg/l) it is 89.30 per cent of total phytoplankton and the minimum in the month of July in year 2006-2007 (441.34 unit/l), which is 92.93 per cent of total phytoplankton. The average count of cells in 2006-2007 was observed 1723.88 unit/l ±1114.30 and in 2007-2008 it was found 1972.71 unit/l ±982.57.

Green algae were observed to be highest in the month of February year 2007-08 (384.21 unit/l), it is 12.95 per cent of total phytoplankton and lowest in the month of July in year 2006-2007 (27.24 unit/l), which is 5.73 per cent of total phytoplankton. The average count of green algae in 2006-2007 was 159.00 unit/l ±104.02 and in 2007-2008 it was 179.92 unit/l ±113.05. Blue green algae showed optimum peak during the month of April in the year 2007-08 (101.54 unit/l), it is 3.27 per cent of total phytoplankton and least peak during the month of August in the year 2006-2007

Table 4.37: No. and Per cent of Different Group Among the Phytoplankton at Sampling Site V (Dam Outlet near Chopra Village) during 2006–2007 and 2007–2008

Month	Total Phytoplankton/ Liter of Water		Diatom/Liter of Water		Green Algae/ Liter of Water		Blue Green Algae/ Liter of Water		Percentage of Different Groups					
									Diatoms		Green Algae		Blue Green Algae	
	2006-07	2007-08	2006-07	2007-08	2006-07	2007-08	2006-07	2007-08	2006-07	2007-08	2006-07	2007-08	2006-07	2007-08
July	474.93	574.22	441.34	511.48	27.24	55.34	6.35	8.40	92.93	89.08	5.73	9.46	1.34	1.46
August	599.97	829.48	554.21	734.39	41.54	87.54	4.22	7.55	92.37	88.54	6.92	10.55	0.71	0.91
September	815.16	1365.13	721.54	1259.55	76.24	73.47	17.38	32.11	88.52	92.27	9.35	5.38	2.13	2.35
October	1065.49	1555.76	903.00	1411.21	121.24	114.35	41.25	30.20	84.75	90.71	11.38	7.35	3.87	1.94
November	1332.21	2598.14	1125.41	2411.51	175.45	151.24	31.35	35.39	84.48	92.82	13.17	5.82	2.35	1.36
December	1838.65	3670.50	1580.21	3319.48	201.14	279.52	57.30	71.50	85.94	90.43	10.94	7.62	3.12	1.95
January	2809.64	3531.76	2520.00	3118.94	222.14	325.43	67.54	87.39	89.69	88.32	7.90	9.21	2.41	2.47
February	4155.10	2966.51	3710.54	2529.00	381.21	384.21	63.35	53.30	89.30	85.26	9.18	12.95	1.52	1.79
March	3262.88	2459.9	2925.22	2120.50	291.21	280.10	43.45	59.30	89.65	86.20	8.92	11.39	1.43	2.41
April	3191.78	3102.79	3016.24	2815.84	124.20	185.41	51.34	101.54	94.50	90.75	3.89	5.98	1.60	3.27
May	2491.03	2870.13	2250.29	2625.20	161.39	173.43	79.35	71.50	90.34	91.46	6.48	6.05	3.18	2.49
June	1038.87	885.72	938.59	815.42	85.00	49.00	15.28	21.30	90.35	92.06	8.18	5.53	1.47	2.41
Average± SD	1922.98± 1222.66	2200.84± 1102.66	1723.88± 1114.30	1972.71± 982.57	159.00± 104.02	179.92± 113.05	39.85± 25.05	48.29± 30.63	89.40± 3.13	89.83± 2.39	8.50± 2.58	8.11± 2.56	2.10± 0.93	2.06± 0.63

Table 4.38: Quantitative Analysis for Zooplankton in Different Months of River Bhagirathi at Sampling Site V (Dam Outlet near Chopra village) during 2006–2007

Month	*Protozoa*		*Rotifera*			*Cladocera*		*Copepoda*	*Total (Unit/l)*
	Amoeba	*Arcella*	*Keratella*	*Notholca*	*Philodima*	*Bosmina*	*Daphnia*	*Cyclops*	
July	+	+	+	+	-	+	+	+	99
August	+	-	+	+	+	+	+	+	83
September	+	+	+	+	+	+	+	+	183
October	+	+	+	+	+	+	+	+	187
November	+	+	+	+	+	+	+	+	191
December	+	+	+	+	+	+	+	+	198
January	+	+	+	+	+	+	+	+	383
February	+	+	+	+	+	+	+	+	251
March	+	+	+	+	+	+	+	+	235
April	+	+	+	+	+	+	+	+	186
May	+	+	+	+	+	+	+	+	170
June	+	+	+	+	+	+	+	+	156

Table 4.39: Quantitative Analysis for Zooplankton in Different Months of River Bhagirathi at Sampling Site V (Dam Outlet near Chopra village) during 2007–2008

Month	*Protozoa*		*Rotifera*			*Cladocera*		*Copepoda*	*Total (Unit/l)*
	Amoeba	*Arcella*	*Keratella*	*Notholca*	*Philodima*	*Bosmina*	*Daphnia*	*Cyclops*	
July	+	+	+	+	-	+	+	+	100
August	+	-	+	+	+	+	+	-	217
September	+	+	+	+	+	+	+	+	211
October	+	+	+	+	+	+	+	+	199
November	+	+	+	+	+	+	+	+	253
December	+	+	+	+	+	+	+	+	232
January	+	+	+	+	+	+	+	+	285
February	+	+	+	+	+	+	+	+	203
March	+	+	+	+	+	+	+	+	196
April	+	+	+	+	+	+	+	+	161
May	+	+	-	+	+	+	+	+	130
June	+	+	+	+	+	+	+	+	128

Table 4.40: No. of Different Genera among Zooplankton in Sampling Station V (Dam Outlet near Chopra Village) during 2006–07 and 2007–08

Months	*Protozoa*		*Rotifera*		*Cladocera*		*Copepoda*		*Total (Unit/liter)*	
	2006-07	*2007-08*	*2006-07*	*2007-08*	*2006-07*	*2007-08*	*2006-07*	*2007-08*	*2006-07*	*2007-08*
July	23.00	14.00	46.00	43.00	21.00	26.00	09.00	17.00	99.00	100.00
August	05.00	72.00	42.00	78.00	33.00	52.00	03.00	15.00	83.00	217.00
September	60.00	55.00	69.00	91.00	30.00	53.00	24.00	12.00	183.00	211.00
October	59.00	32.00	62.00	64.00	53.00	70.00	13.00	33.00	187.00	199.00
November	26.00	48.00	80.00	111.00	50.00	57.00	35.00	37.00	191.00	253.00
December	76.00	24.00	62.00	100.00	36.00	61.00	24.00	47.00	198.00	232.00
January	92.00	59.00	159.00	69.00	87.00	109.00	45.00	48.00	383.00	285.00
February	86.00	38.00	82.00	103.00	38.00	31.00	45.00	31.00	251.00	203.00
March	86.00	26.00	58.00	127.00	66.00	43.00	25.00	00.00	235.00	196.00
April	50.00	29.00	77.00	42.00	48.00	62.00	11.00	28.00	186.00	161.00
May	32.00	31.00	78.00	12.00	43.00	62.00	17.00	25.00	170.00	130.00
June	48.00	26.00	73.00	61.00	20.00	32.00	15.00	09.00	156.00	128.00
Average± SD	53.58± 28.15	37.83± 17.09	74.00± 29.74	75.08± 33.18	43.75± 19.04	54.83± 22.12	22.17± 13.65	25.17± 15.01	217.44± 73.19	198.56± 52.04

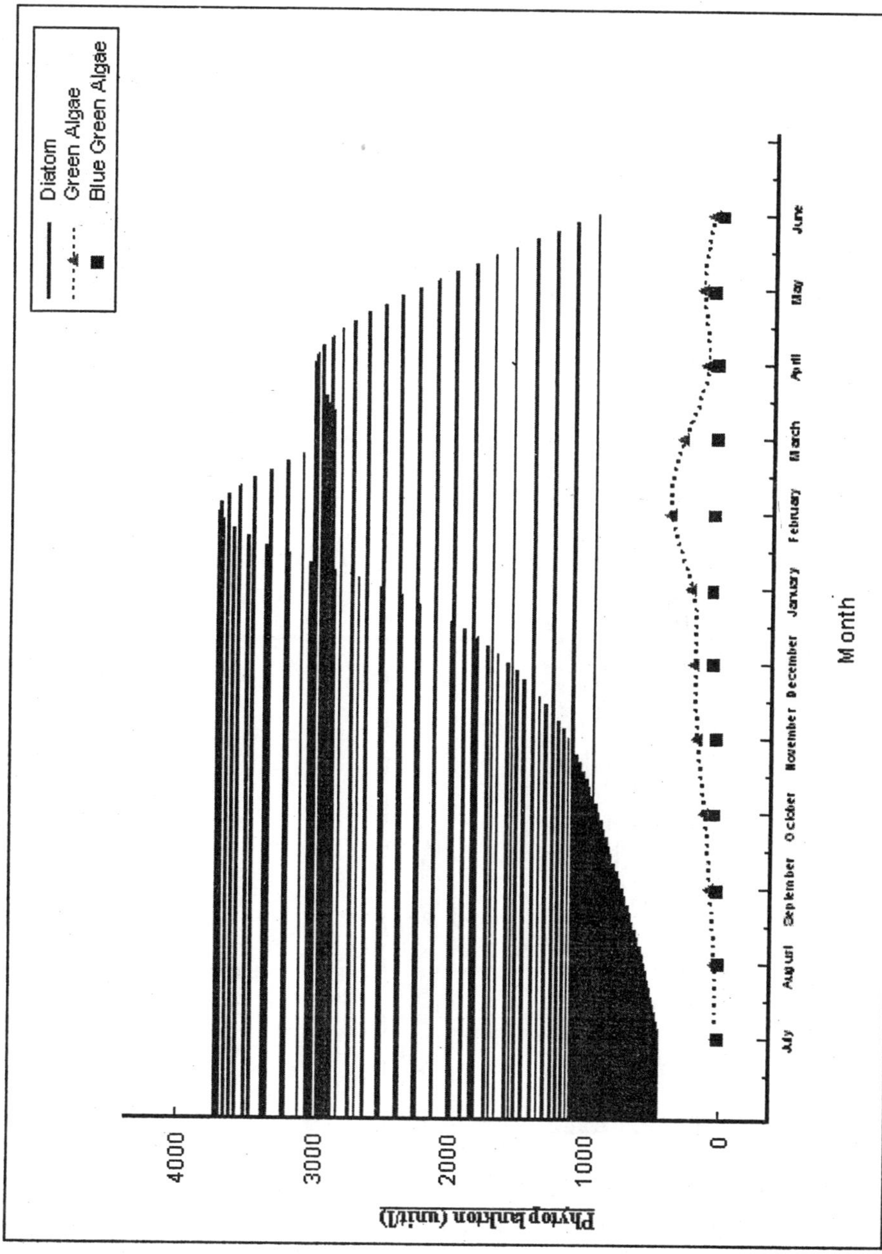

Figure 4.77: Graph Showing Quantitative Variation of Diatom, Green Algae and Blue Green Algae at Sampling Site V during 2006–07

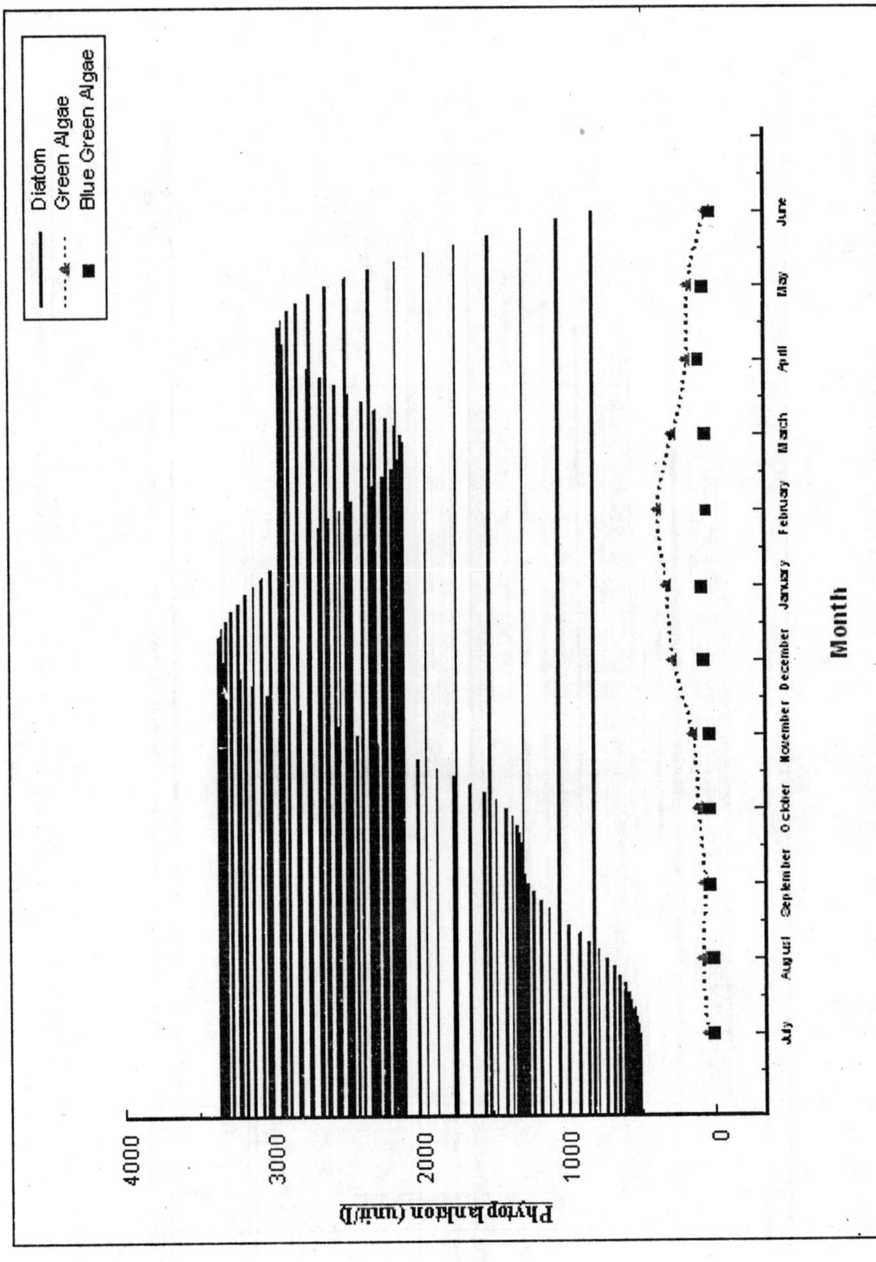

Figure 4.78: Graph Showing Quantitative Variation of Diatom, Green Algae and Blue Green Algae at Sampling Site V during 2007–08

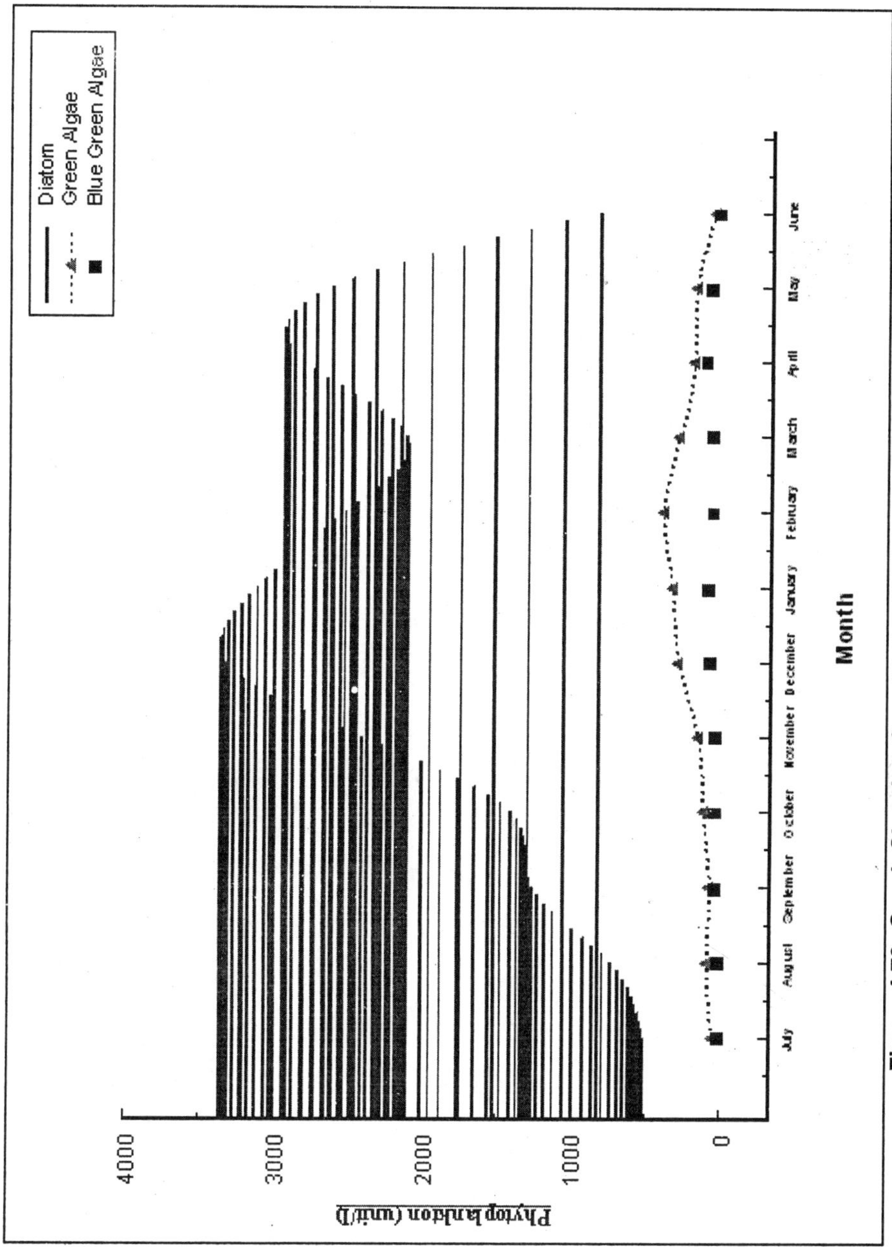

Figure 4.79: Graph Showing Quantitative Variation of Copepoda, Cladocera, Rotifera and Protozoa at Sampling Site V during 2006–07

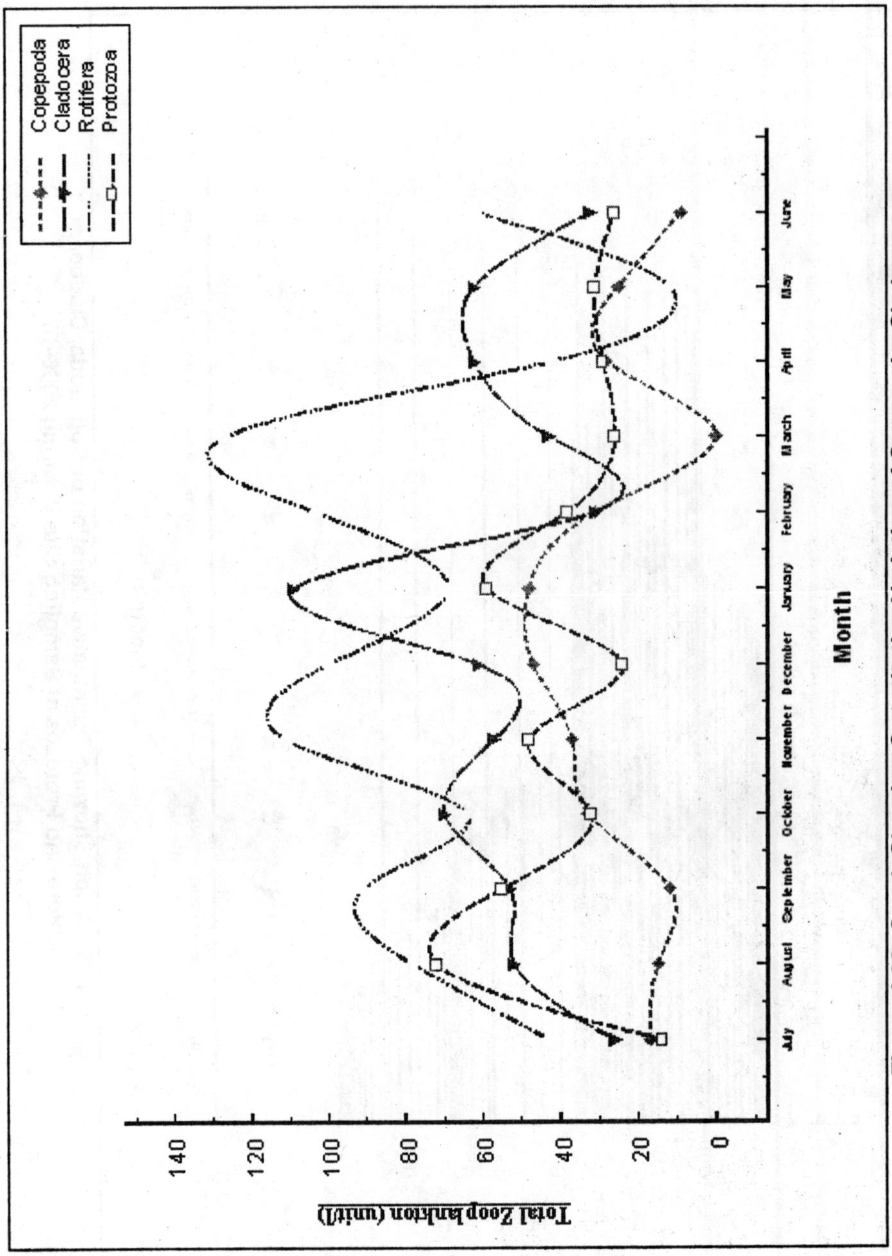

Figure 4.80: Graph Showing Quantitative Variation of Copepoda, Cladocera, Rotifera and Protozoa at Sampling Site V during 2007–08

(4.22 unit/l), which is 0.71 per cent of total phytoplankton. In 2006-2007 the average value of Blue green algae was noted 39.85 unit/l ±25.05 while in 2007-2008 it was recorded 48.29 unit/l ±30.63.

Zooplankton

Among zooplankton protozoa (*Amoeba* and *Arcella*), Rotifera (*Keratella, Notholca* and *Philodima*), Cladocera (*Bosmina* and *Daphnia*) and Copepods (*Cyclops*) were observed. Zooplankton was observed minimum in the month of August year 2006-2007 (83.00 unit/l) and maximum 383.00 unit/l in the month of January of the same year. The average value of zooplankton for year 2006-2007 was observed 217.44 unit/l ±73.19 and for year 2007-2008 it was recorded 198.56 unit/l ±52.04.

Protozoa

Protozoa were found maximum 92.00 unit/l in the month of January in year 2006-07 and minimum 05.00 unit/l in the month of August 2006-2007. The annual mean recorded for first year 2006-07 was 53.58 unit/l ±28.15 and in year 2007-08 it was observed 37.83 unit/l ±17.09.

Rotifera

Rotifera were recorded maximum 159.00 unit/l in the month of January 2006-07 and the minimum 12.00 unit/l in the month of May 2007-08. The annual mean recorded in the first year 2006-07 was observed 74.00 unit/l ±29.74 and for second year 2007-08 it was recorded 75.08 unit/l ±33.18.

Cladocera

Cladocera were found maximum 109.00 unit/l in the month of January 2007-08 and minimum 20.00 unit/l in the month of June 2006-2007. The annual mean recorded for first year 2006-2007 was 43.75 unit/l ±19.04 and for second Year 2007-2008 it was observed 54.83 unit/l ±22.12.

Copepods

Copepods were noted maximum 48.00 unit/l in the month of January in 2007-08 and minimum 00.00 unit/l were found in August month of 2007-08. The annual mean recorded for the first year 2006-2007 was 22.17 unit/l ±13.65 and for second year 2007-2008 it was found 25.17 unit/l ±15.01.

Estimation of Planktonic Biomass

Estimation of phytoplankton assemblage structure can provide a number of useful indications, on the functioning of an aquatic ecosystem. In this section we present the results of different variables of plankton biomass estimation assuming that formation of a huge reservoir in the down stream of a ever flowing hilly river will change the status of plankton in water so biomass estimation is one of the major concerns for the study of impact of reservoir formation at Tehri upon Bhagirathi River.

The observations made during July 2006 to June 2008, are presented in the Tables 4.41 to 4.45 and in Figures 4.81 to 4.95.

Table 4.41: Total Number of Phytoplankton, Zooplankton, Total Plankton and Planktonic Biomass at Sampling Site I (Dharanshu) during 2006–2007 and 2007–2008

Month	*Phytoplankton (Unit/l)*		*Zooplankton (Unit/l)*		*Total Plankton (Unit/l)*		*Planktonic Biomass (mg/l)*	
	2006-07	*2007-08*	*2006-07*	*2007-08*	*2006-07*	*2007-08*	*2006-07*	*2007-08*
July	471.83	616.08	98.00	100.00	569.83	716.08	0.2845	0.3665
August	709.74	672.36	96.00	110.00	805.74	782.36	0.4075	0.3910
September	940.27	1119.11	145.00	163.00	1085.27	1282.11	0.5424	0.6410
October	1266.06	1154.85	202.00	178.00	1468.06	1332.85	0.7340	0.6665
November	1356.58	1408.21	167.00	207.00	1523.58	1615.21	0.7615	0.8090
December	2099.43	2926.00	259.00	302.00	2358.43	3228.00	1.1740	1.6140
January	3299.59	3629.70	224.00	320.00	3523.59	3949.70	1.7565	1.9450
February	5170.00	4220.60	296.00	359.00	5466.00	4579.60	2.7230	2.2450
March	4284.69	4507.75	224.00	270.00	4508.69	4777.75	2.2540	2.3780
April	3738.55	4255.70	155.00	361.00	3893.55	4616.70	1.9465	2.3015
May	2379.58	2696.35	145.00	378.00	2524.58	3074.35	1.2620	1.5485
June	1178.82	1025.35	120.00	142.00	1298.82	1167.35	0.6490	0.5245
Average± SD	2241.26± 1542.89	2352.67± 1516.03	199.11± 60.97	279.67± 98.30	2418.85± 1590.37	2593.51± 1603.33	1.2079± 0.7927	1.2859± 0.7982

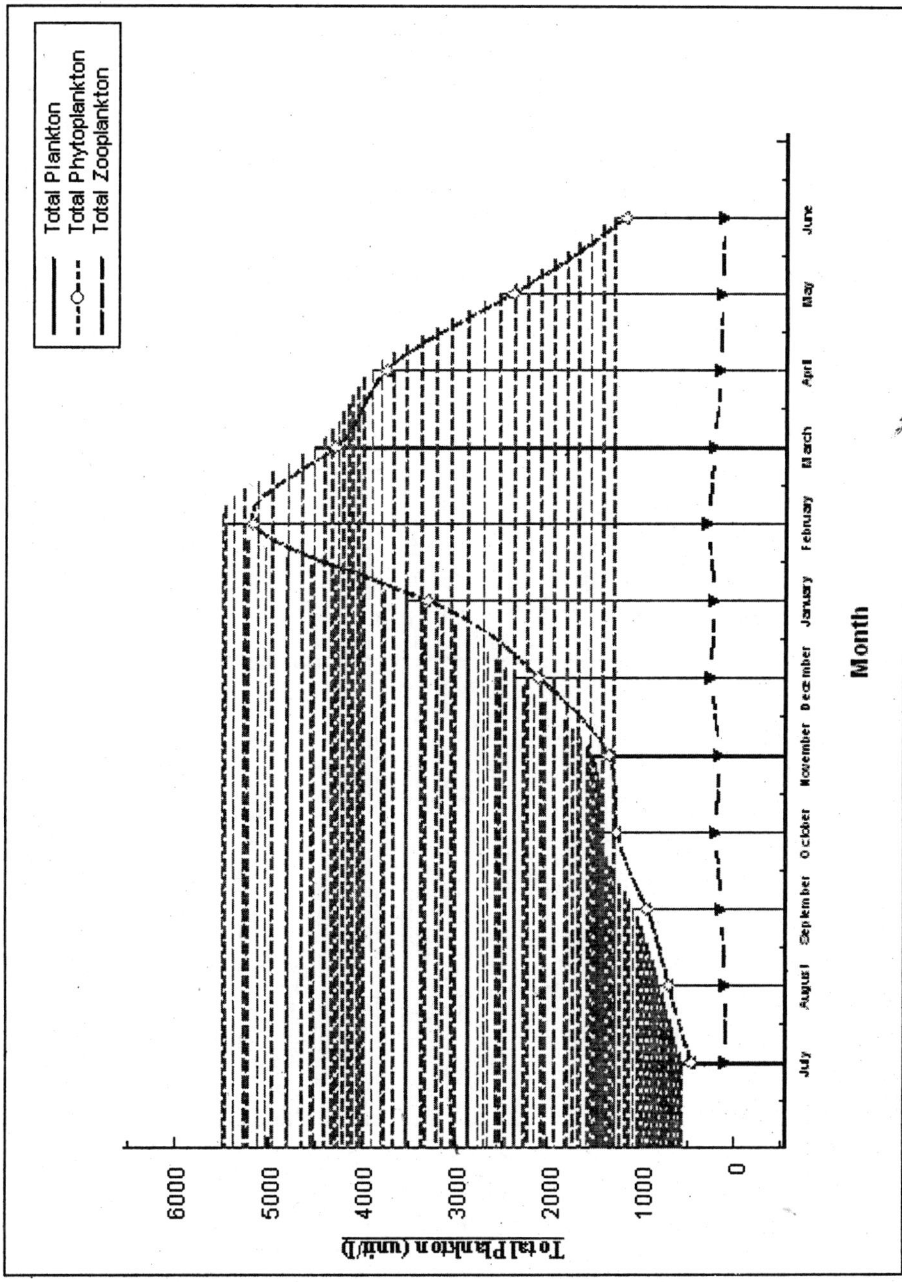

Figure 4.81: Graph Showing Quantitative Variation of Plankton at Sampling Site I during 2006–07

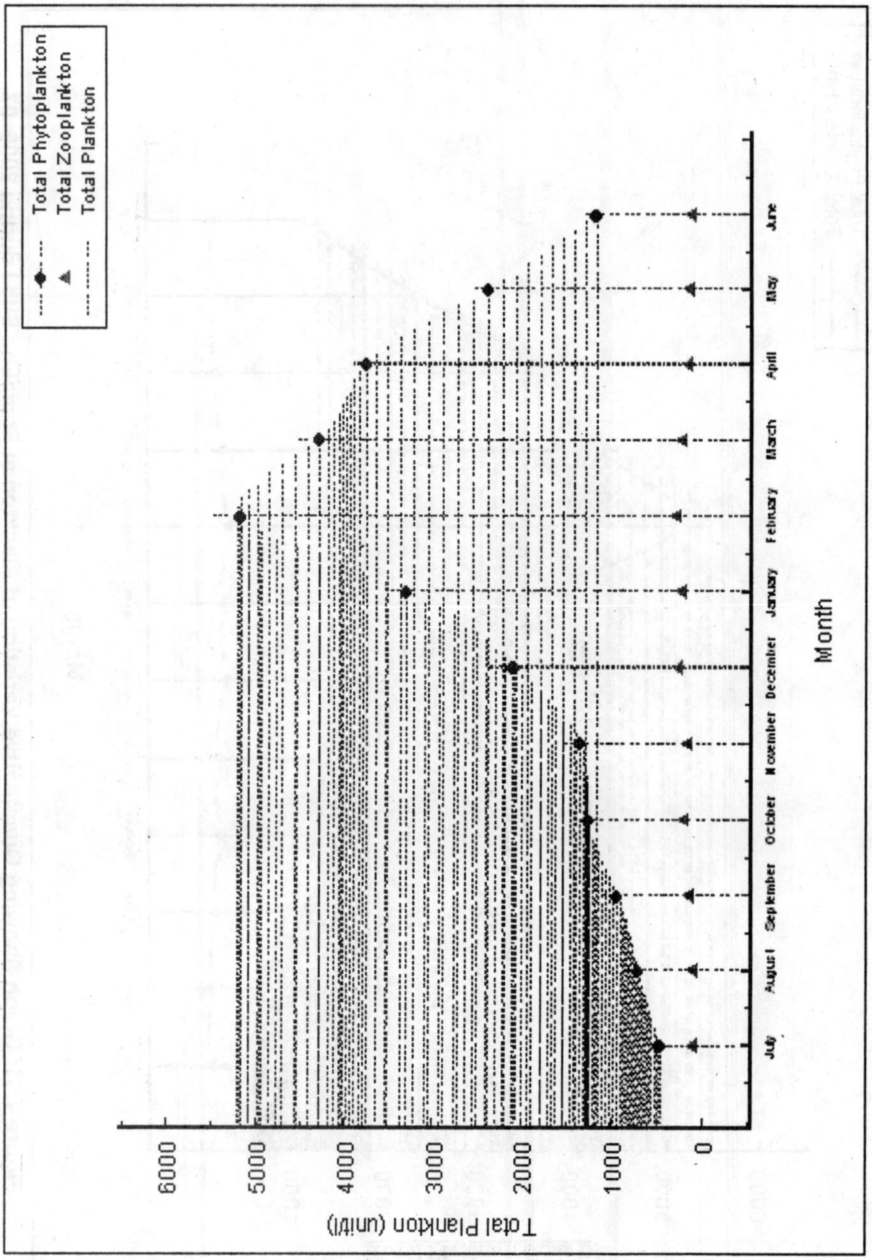

Figure 4.82: Graph Showing Quantitative Variation of Plankton at Sampling Site I during 2007–08

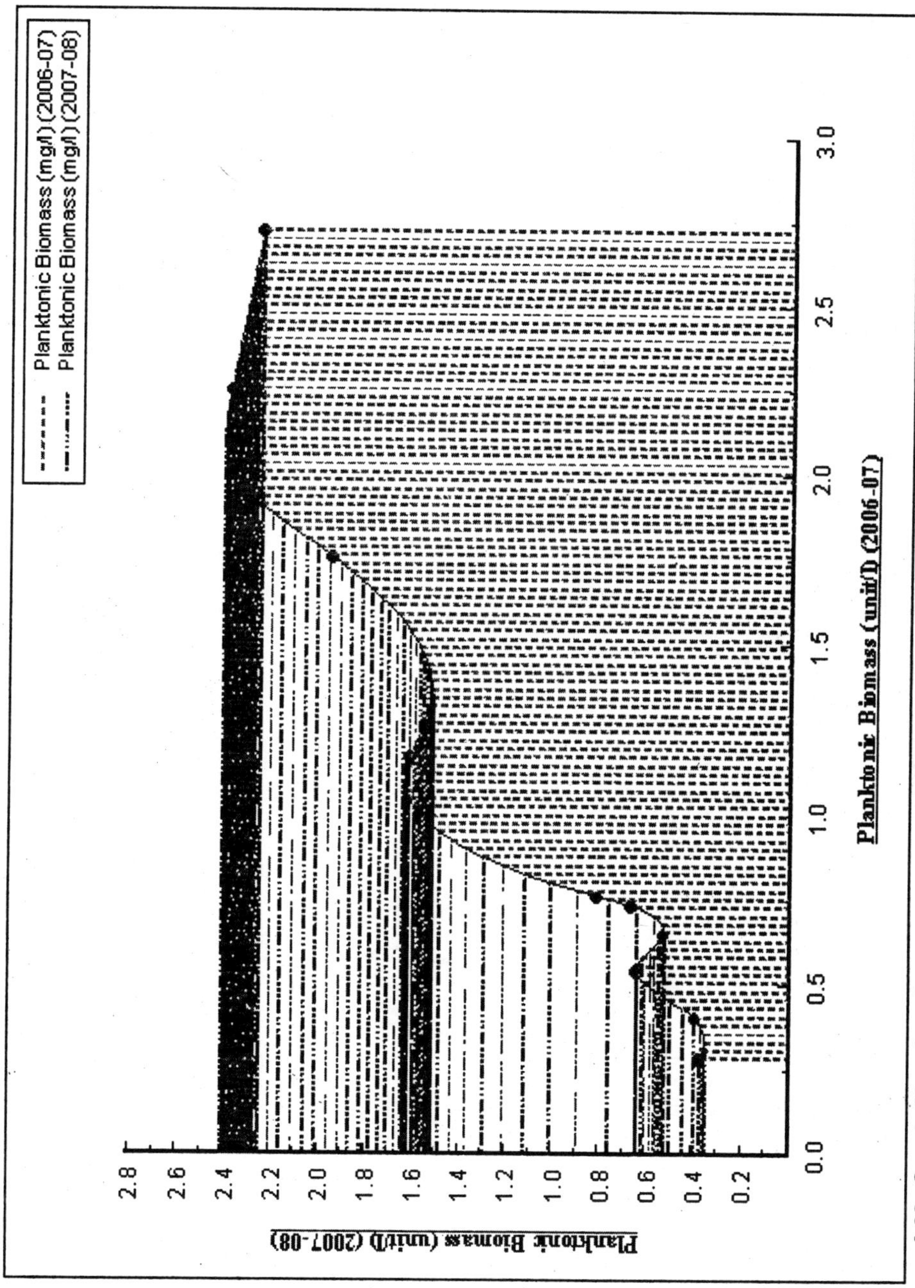

Figure 4.83: Graph Showing Quantitative Variation of Planktonic Biomass at Sampling Site I during 2006–07 and 2007–08

Table 4.42: Total Number of Phytoplankton, Zooplankton, Total Plankton and Planktonic Biomass at Sampling Site II (Chinyalisaur) during 2006–2007 and 2007–2008

Month	*Phytoplankton (Unit/l)*		*Zooplankton (Unit/l)*		*Total Plankton (Unit/l)*		*Planktonic Biomass (mg/l)*	
	2006-07	*2007-08*	*2006-07*	*2007-08*	*2006-07*	*2007-08*	*2006-07*	*2007-08*
July	887.89	799.58	141.00	162.00	1028.89	961.58	0.5570	0.4853
August	1001.74	1188.82	133.00	165.00	1134.74	1353.82	0.5670	0.6535
September	1558.48	1883.84	185.00	170.00	1743.48	2053.84	0.8635	1.085
October	1465.15	1803.14	147.00	140.00	1612.15	1943.14	0.8145	0.965
November	1969.47	2117.05	193.00	143.00	2162.47	2260.05	1.1455	1.135
December	2578.25	2643.42	191.00	155.00	2769.25	2798.42	1.4970	1.4295
January	3722.49	4301.90	264.00	209.00	3986.49	4510.90	1.9455	2.3780
February	5439.08	4705.96	224.00	208.00	5663.08	4913.96	2.8295	2.5295
March	4668.73	4670.98	233.00	287.00	4901.73	4957.98	2.5450	2.4200
April	4267.99	4487.04	200.00	188.00	4467.99	4675.04	2.3545	2.3895
May	3232.80	3025.08	217.00	212.00	3449.80	3237.08	1.7245	1.6290
June	1227.64	1045.17	160.00	182.00	1387.64	1227.17	0.6545	0.6455
Average± SD	2668.31± 1562.82	2722.67± 1483.70	203.22± 38.45	191.56± 38.84	2858.98± 1596.14	2907.75± 1512.01	1.4582± 0.0814	1.4787± 0.07718

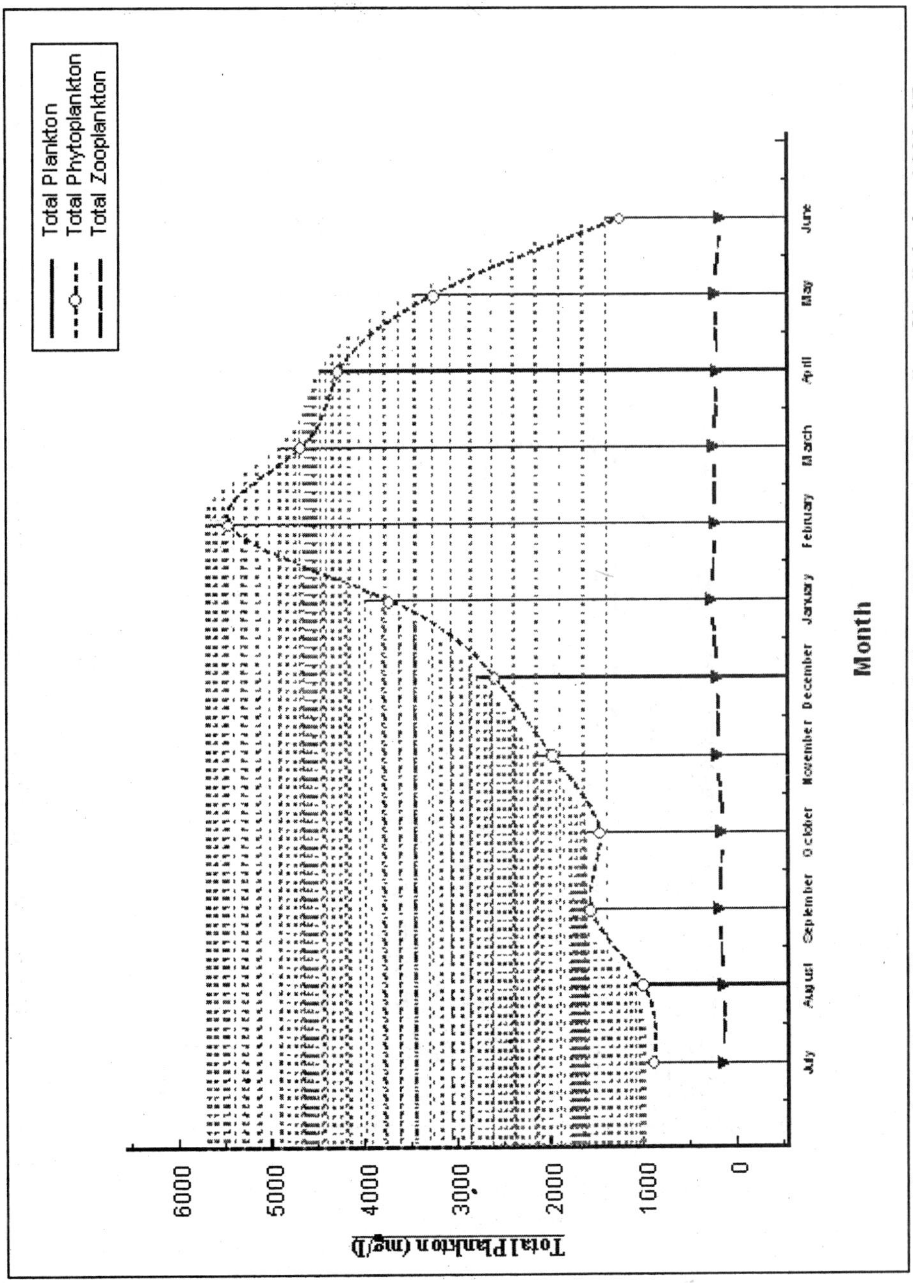

Figure 4.84: Graph Showing Quantitative Variation of Plankton at Sampling Site II during 2006–07

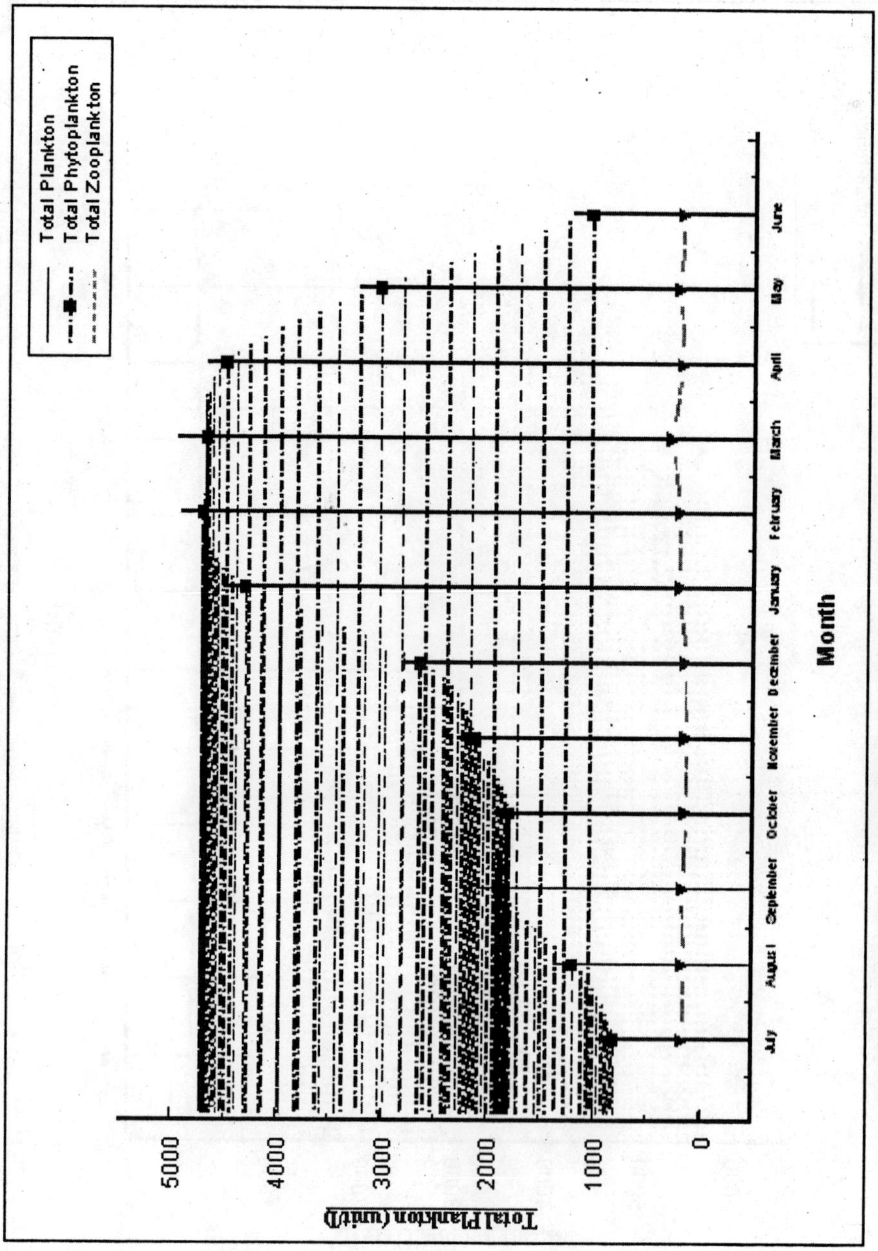

Figure 4.85: Graph Showing Quantitative Variation of Plankton at Sampling Site II during 2007–08

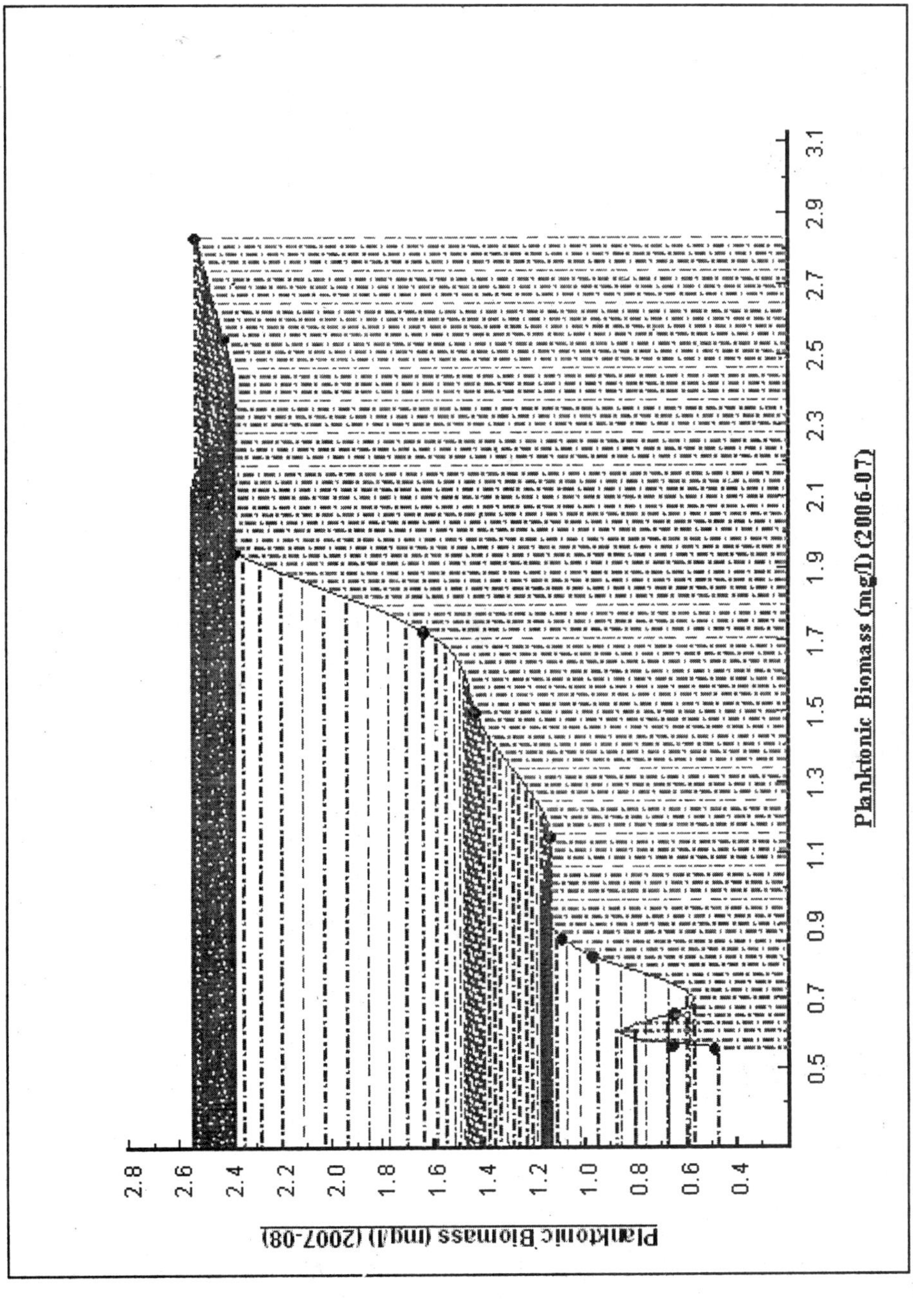

Figure 4.86: Graph Showing Quantitative Variation of Planktonic Biomass at Sampling Site II during 2006–07 and 2007–08

Table 4.43: Total Number of Phytoplankton, Zooplankton, Total Plankton and Planktonic Biomass at Sampling Site III (Bhaldiyana) during 2006–2007 and 2007–2008

Month	*Phytoplankton (Unit/l)*		*Zooplankton (Unit/l)*		*Total Plankton (Unit/l)*		*Planktonic Biomass (mg/l)*	
	2006-07	*2007-08*	*2006-07*	*2007-08*	*2006-07*	*2007-08*	*2006-07*	*2007-08*
July	1477.05	1457.39	174.00	136.00	1651.05	1593.39	0.8455	0.7554
August	1860.32	1967.08	222.00	220.00	2082.32	2187.08	0.9350	1.0290
September	2570.59	2370.99	214.00	139.00	2784.59	2509.99	1.5290	1.2455
October	2461.57	2256.50	258.00	231.00	2719.57	2487.50	1.4585	1.3947
November	2749.21	2868.96	139.00	214.00	2888.21	3082.96	1.4975	1.5586
December	3010.19	3581.83	223.00	251.00	3233.19	3832.83	1.6975	1.9673
January	4510.05	3692.05	309.00	237.00	4819.05	3929.05	2.4255	2.0569
February	4740.31	4665.02	284.00	163.00	5024.31	4828.02	2.4850	2.4451
March	4032.84	4581.33	258.00	215.00	4290.84	4796.33	2.1856	2.3694
April	3925.10	3973.99	310.00	232.00	4235.10	4205.99	2.0569	2.0472
May	1588.98	2284.80	267.00	123.00	1855.98	2407.80	0.8976	1.1894
June	1051.81	1770.97	224.00	148.00	1275.81	1918.97	0.653	0.9857
Average± SD	2831.50± 1237.03	2955.91± 1105.77	252.44± 49.22	201.56± 44.63	3071.67± 1267.50	3148.33± 1127.51	1.5555± 0.6353	1.5870± 0.0572

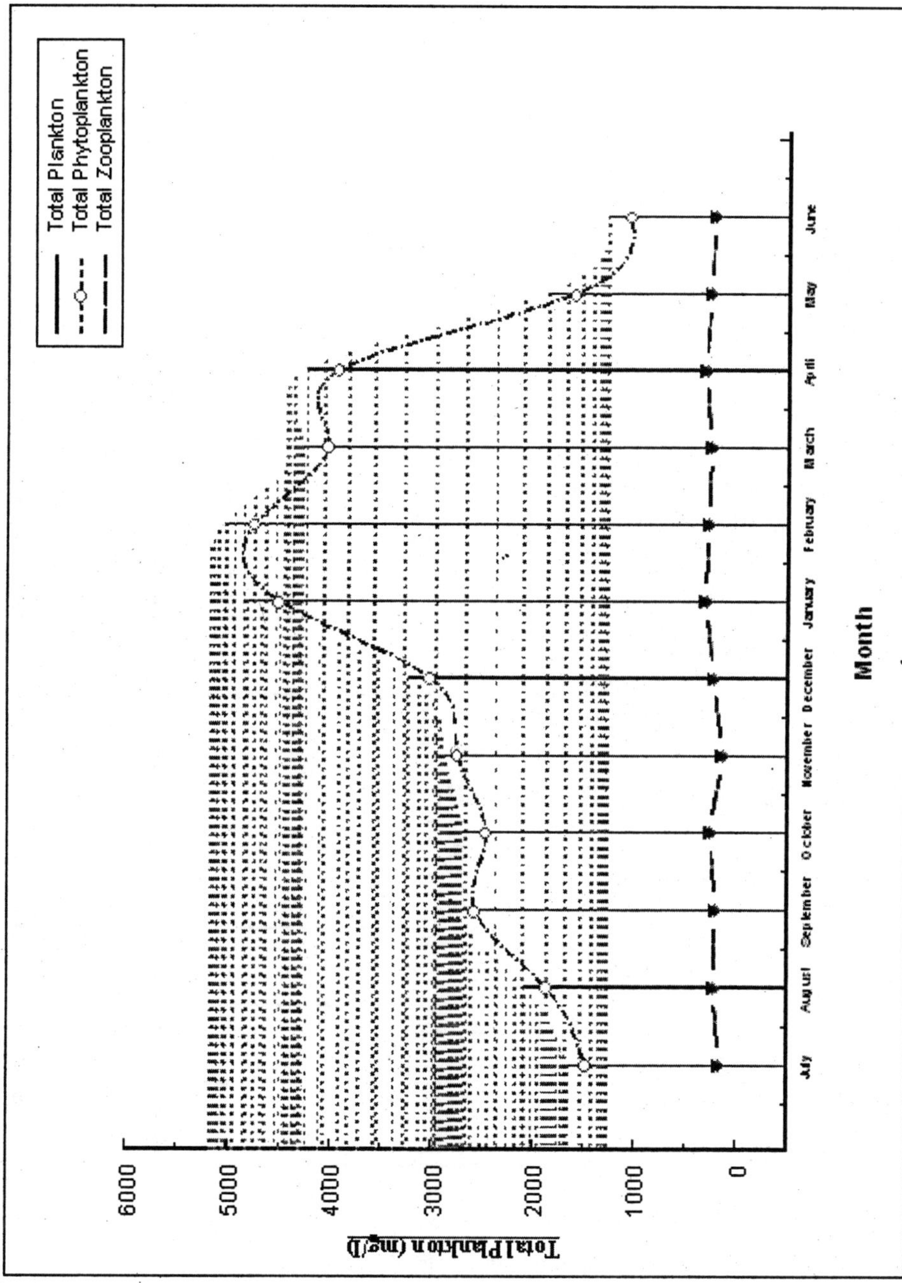

Figure 4.87: Graph Showing Quantitative Variation of Plankton at Sampling Site III during 2006–07

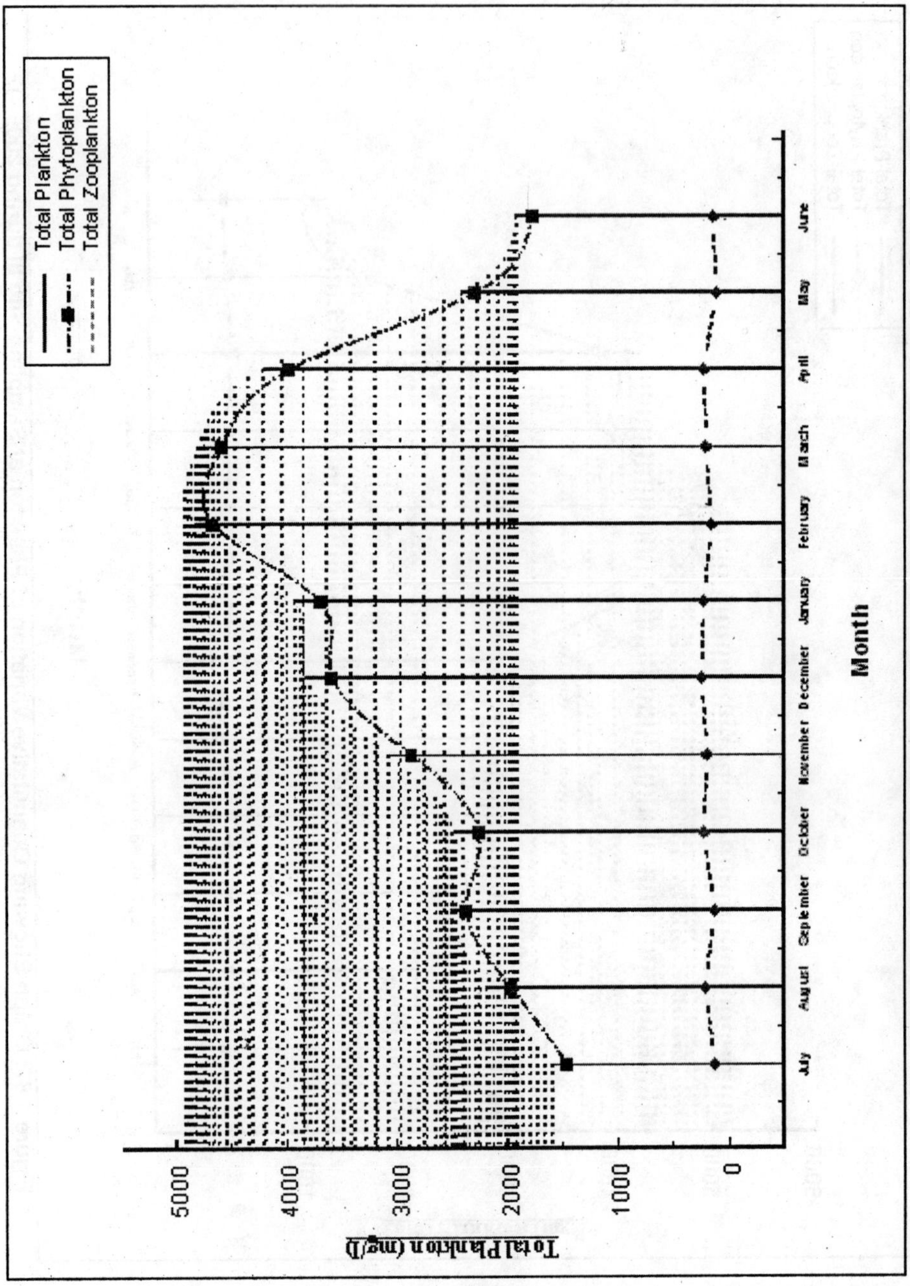

Figure 4.88: Graph Showing Quantitative Variation of Plankton at Sampling Site III during 2007–08

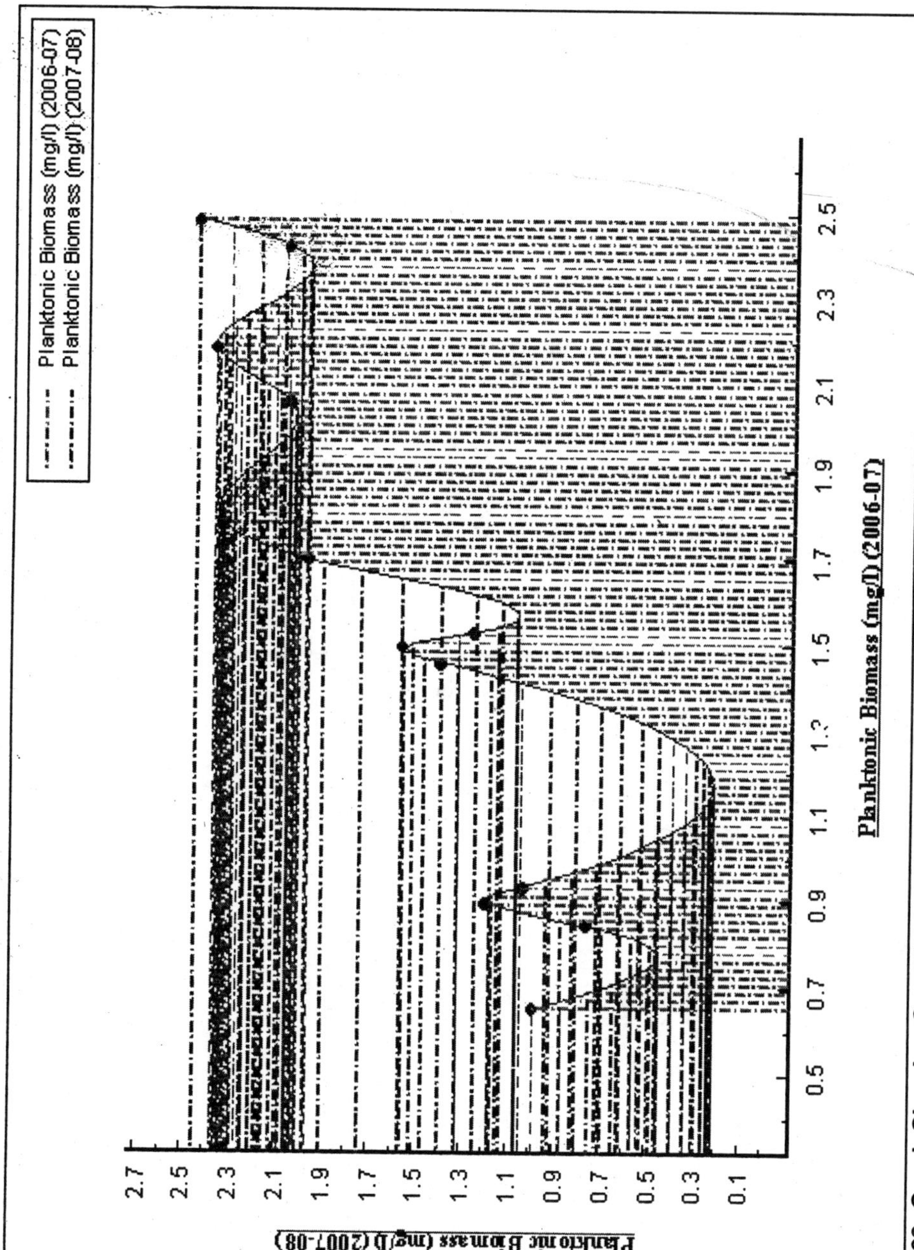

Figure 4.89: Graph Showing Quantitative Variation of Planktonic Biomass at Sampling Site III during 2006–07 and 2007–08

Table 4.44: Total Number of Phytoplankton, Zooplankton, Total Plankton and Planktonic Biomass at Sampling Site IV (Koti) during 2006–2007 and 2007–2008

Month	Phytoplankton (Unit/l)		Zooplankton (Unit/l)		Total Plankton (Unit/l)		Planktonic Biomass (mg/l)	
	2006-07	2007-08	2006-07	2007-08	2006-07	2007-08	2006-07	2007-08
July	1691.67	1762.38	168.00	216.00	1859.67	1978.38	0.8547	0.9567
August	2272.03	2264.84	292.00	298.00	2564.03	2562.84	1.2378	1.2578
September	3071.99	3164.05	295.00	307.00	3366.99	3471.05	1.6549	1.6525
October	3364.21	3270.09	331.00	366.00	3695.21	3636.09	1.7678	1.7159
November	3704.48	3507.23	406.00	305.00	4110.48	3812.23	2.1390	1.8842
December	4393.76	3995.09	375.00	305.00	4768.76	4300.09	2.3455	2.0931
January	5063.35	4407.37	435.00	337.00	5498.35	4744.37	2.6549	2.3458
February	5670.25	5388.95	336.00	239.00	6006.25	5627.95	2.9780	2.7582
March	4521.87	4934.18	248.00	304.00	4769.87	5238.18	2.3085	2.6973
April	4188.90	4500.40	352.00	274.00	4540.90	4774.40	2.1150	2.2450
May	2618.57	2823.37	232.00	319.00	2850.57	3142.37	1.3866	1.6469
June	1538.28	1357.90	237.00	219.00	1775.28	1576.90	0.8743	0.8245
Average± SD	3508.28± 1318.32	3447.99± 1258.39	328.00± 75.72	296.44± 43.96	3817.20± 1373.94	3738.74± 1272.61	1.8598± 0.6813	1.8398± 0.6254

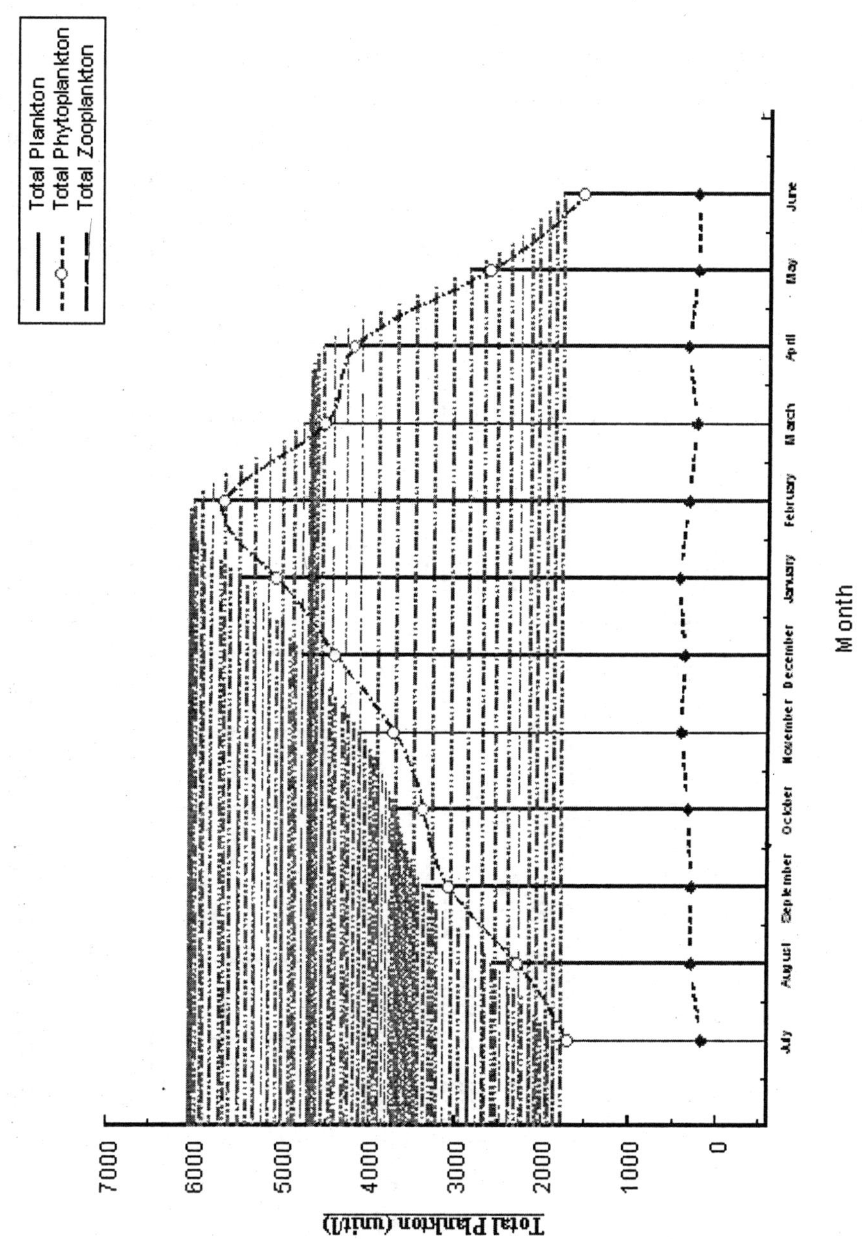

Figure 4.90: Graph Showing Quantitative Variation of Plankton at Sampling Site IV during 2006–07

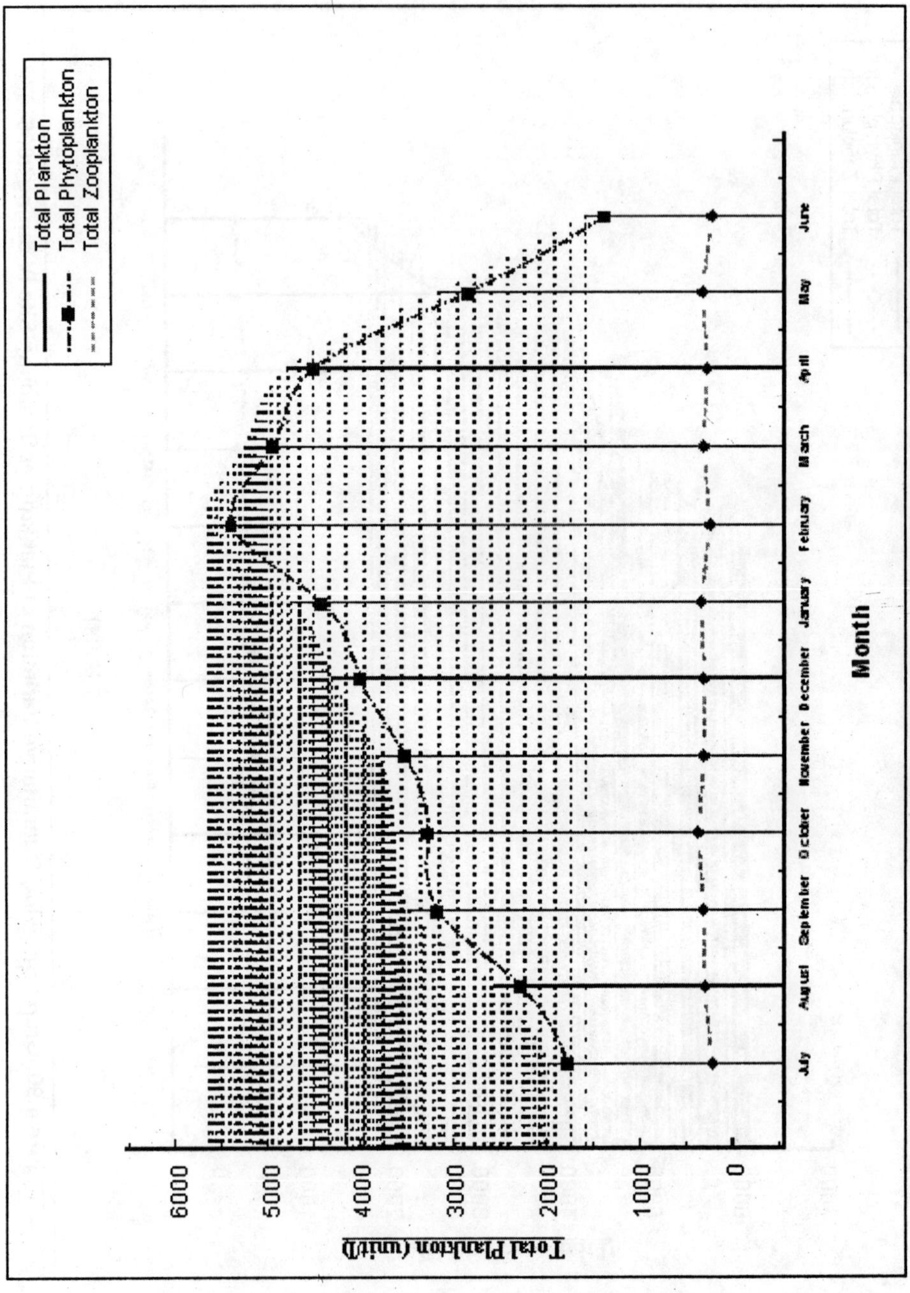

Figure 4.91: Graph Showing Quantitative Variation of Plankton at Sampling Site IV during 2007–08

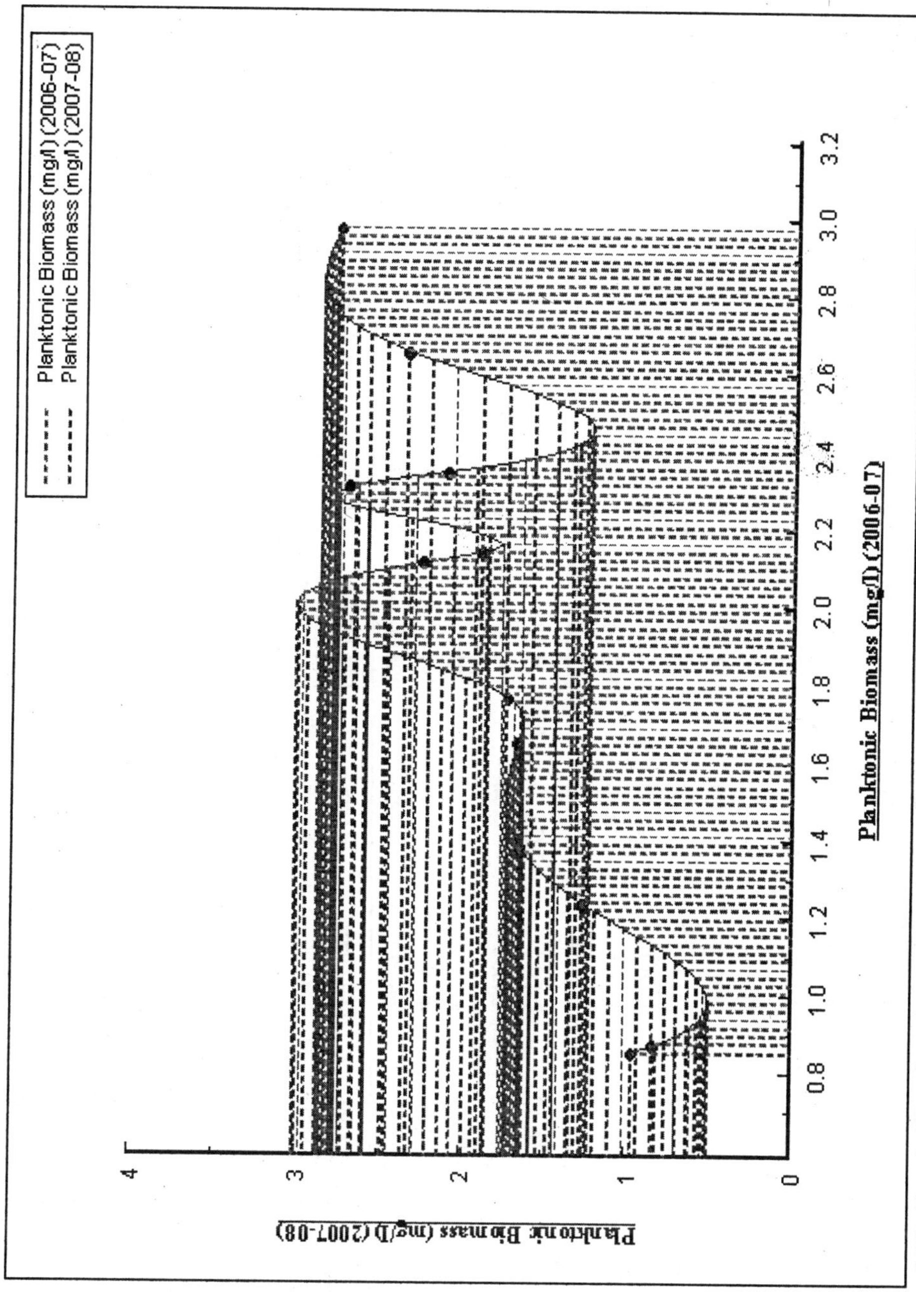

Figure 4.92: Graph Showing Quantitative Variation of Planktonic Biomass at Sampling Site IV during 2006–07 and 2007–08

Table 4.45: Total Number of Phytoplankton, Zooplankton, Total Plankton and Planktonic Biomass at Sampling Site V (Dam outlet near Chopra village) during 2006–2007 and 2007–2008

Month	*Phytoplankton (Unit/l)*		*Zooplankton (Unit/l)*		*Total Plankton (Unit/l)*		*Planktonic Biomass (mg/l)*	
	2006-07	*2007-08*	*2006-07*	*2007-08*	*2006-07*	*2007-08*	*2006-07*	*2007-08*
July	474.93	574.22	99.00	100.00	573.93	674.22	0.2843	0.3247
August	599.97	829.48	83.00	217.00	682.97	1046.48	0.3371	0.4920
September	815.16	1365.13	183.00	211.00	998.16	1576.13	0.4947	0.7495
October	1065.49	1555.76	187.00	199.00	1252.49	1754.76	0.59124	0.7842
November	1332.21	2598.14	191.00	253.00	1523.21	2851.14	0.7549	1.3729
December	1838.65	3670.50	198.00	232.00	2036.65	3902.50	0.9935	1.8514
January	2809.64	3531.76	383.00	285.00	3192.64	3816.76	1.4996	1.7829
February	4155.10	2966.51	251.00	203.00	4406.10	3169.51	2.1863	1.5595
March	3262.88	2459.9	235.00	196.00	3497.88	2655.90	1.8874	1.2445
April	3191.78	3102.79	186.00	161.00	3377.78	3263.79	1.7493	1.7683
May	2491.03	2870.13	170.00	130.00	2661.03	3000.13	1.4838	1.6154
June	1038.87	885.72	156.00	128.00	1194.87	1013.72	0.6746	0.5471
Average± SD	1922.98± 1222.66	2200.84± 1102.66	217.44± 73.19	198.56± 52.04	2116.48± 1273.34	2393.75± 1130.52	1.0781± 0.6541	1.1744± 0.5627

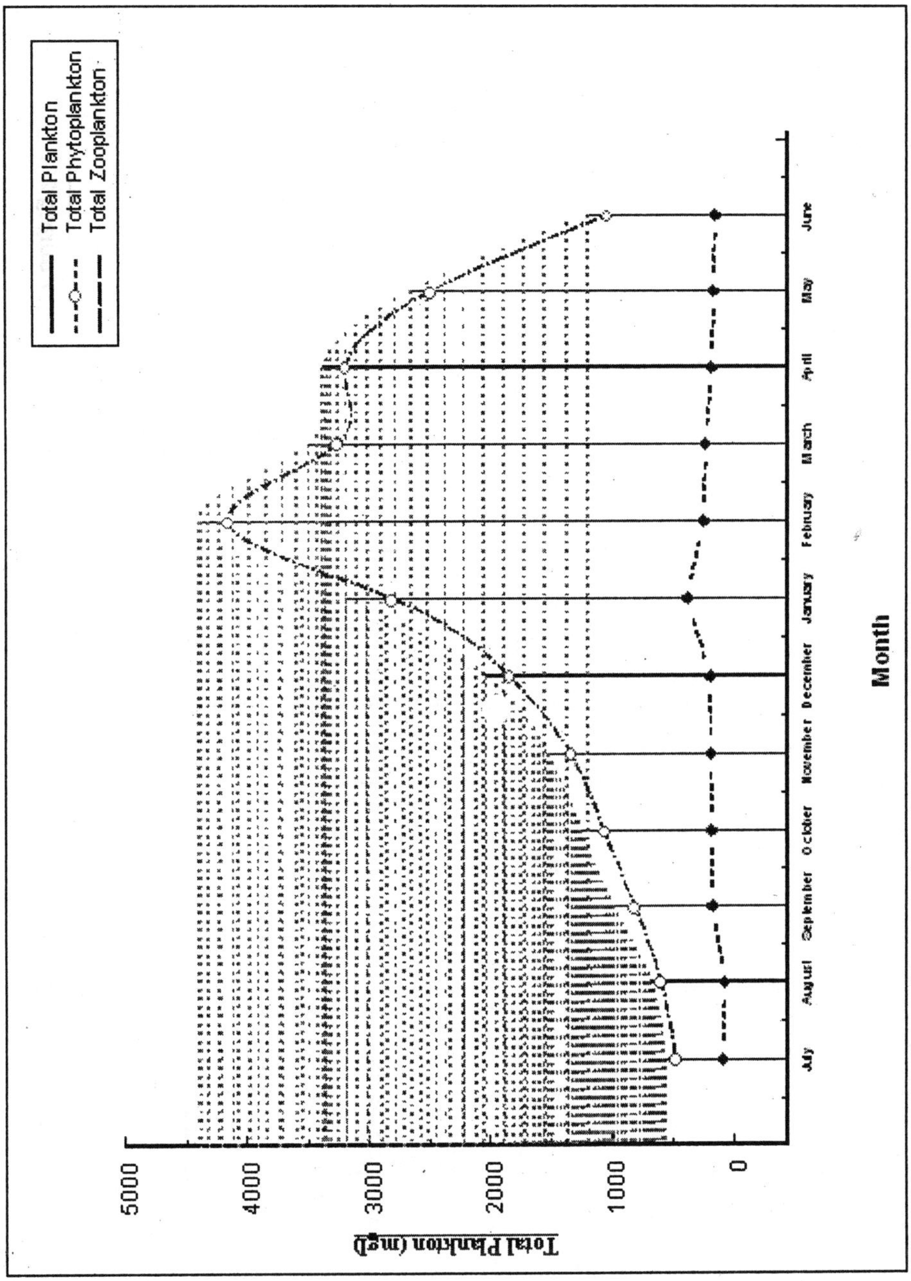

Figure 4.93: Graph Showing Quantitative Variation of Plankton at Sampling Site V during 2006–07

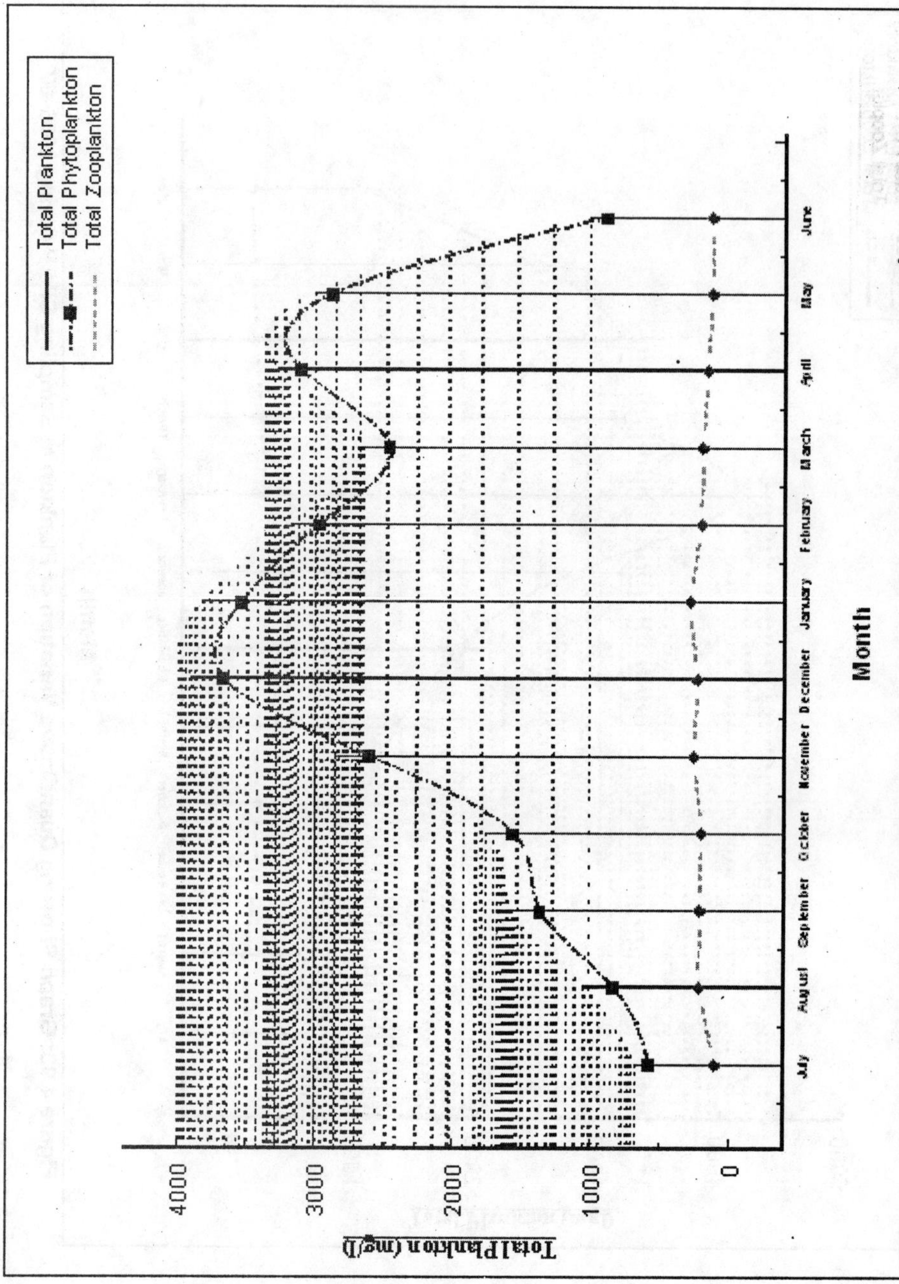

Figure 4.94: Graph Showing Quantitative Variation of Plankton at Sampling Site V during 2007–08

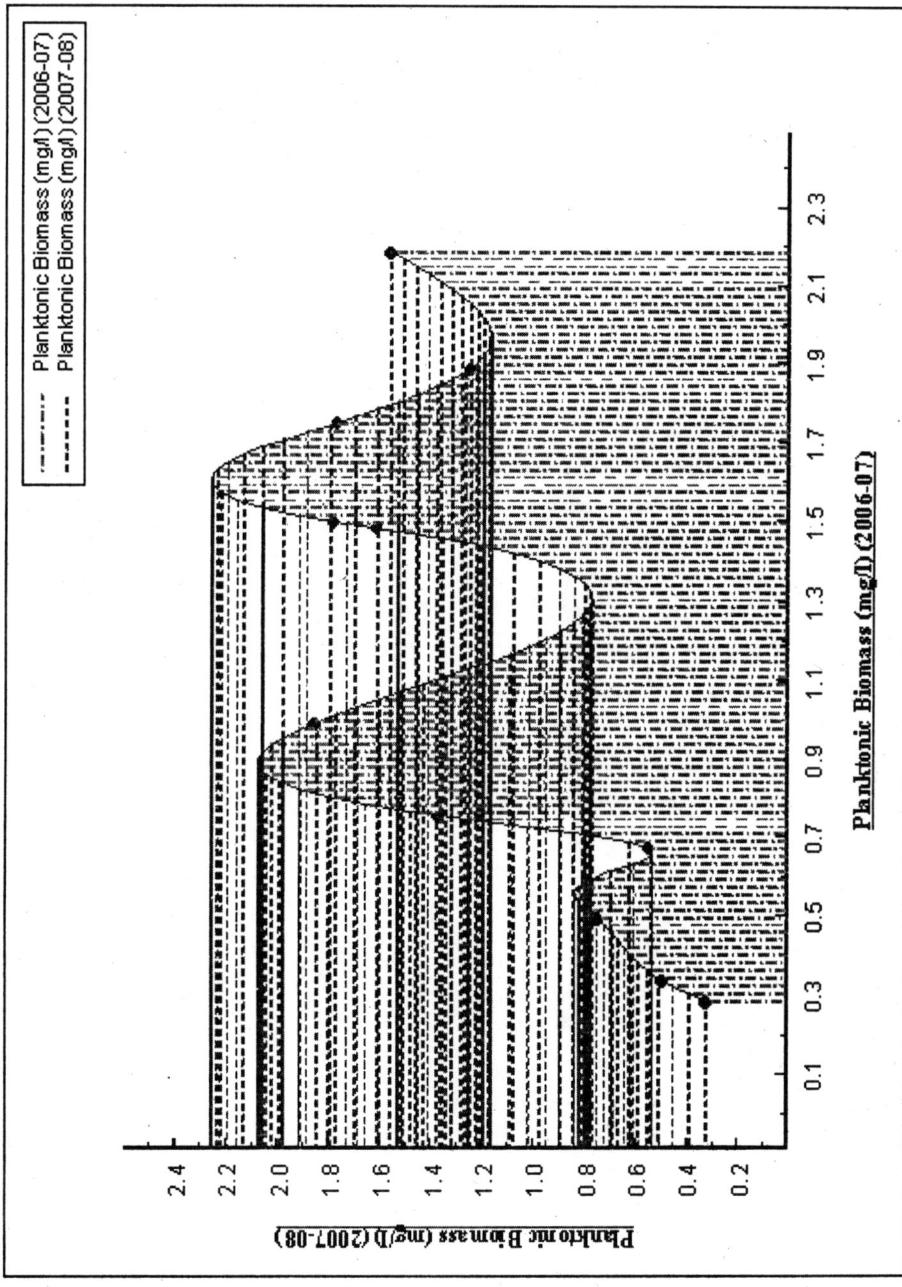

Figure 4.95: Graph Showing Wuantitative Variation of Planktonic Biomass at Sampling Site V during 2006–07 and 2007–08

Table 4.46: Systematic List of Fishes

Phylum	Vertebrata
Sub phylum	Craniata
Super class	Gnathostomata
Series	Pisces
Class	Teleostomi
Sub class	Aetinopterygii
Order	Cyprinoidei
Sub order	Cyprinoidei

Sl.No.	*Name of Fish*	*Upstream*	*Reservoir*	*Downstream*
	Family: Cyprinidae			
1.	*Puntius hexasticus* (McClelland)	+	–	+
2.	*Puntius chilinoides* (McClelland)	+	–	+
3.	*Tor putitora* (Ham) endangered	+	+	+
4.	*Tor tor* (Ham) endangered	+	+	+
5.	*Cyprinus caprio* var *communis* (Linn)	+	+	+
6.	*Cyprinus caprio* var *specularis* (Linn)	+	+	+
7.	*Schizothorax richardsoni* (Gray)	+	–	+
8.	*Schizothorax sinatus* (Heckel)	+	–	+
9.	*Schizothorax plagiostomus* (Heckel)	+	–	+
10.	*Schizothorax curvifrons* (Heckel)	+	+	+
11.	*Schizothoraichthys progastus* (McClelland)	+	+	+
	Family: Cobitdae			
12.	*Noemacheilus montanus* (McClelland)		+	– –
13.	*Noemacheilus rupicola* (McClelland)	+	–	–
14.	*Noemacheilus butansis* (Gunther)	+	–	–
15.	*Noemacheilus beavani* (Gunther)	+	–	–
16.	*Noemacheilus denisoni* (Day)	+	–	–
17.	*Noemacheilus multifasciatus* (Day)	+	–	–
18.	*Noemacheilus zonatus* (McClelland)	+	–	–
	Division: SiluriFamily: Sisoridae			
19.	*Glyptothorax pectinopterus* (McClelland)	+	–	–
20.	*Pseudecheneis sulcatus* (McClelland)	+	–	–
	Order: ClupeiformesFamily: Salmonidae			
21.	*Salmo trutta fario* (Linn)	+	-	-

Sampling Site I (Dharanshu)

Planktonic biomass at sampling site I was found maximum 2.7236 mg/l in the month of February in year 2006-2007 and minimum of 0.2845 mg/l in the month of

July in year 2006-2007. Annual average of planktonic biomass in year 2006-07 was noted 1.2079 mg/l ±0.7927 and in year 2007-2008 observed 1.2859 mg/l ±0.7982.

Sampling Site II (Chinyalisaur)

Planktonic biomass at sampling site II was observed maximum 2.8295 mg/l in the month of February in year 2006-2007 and minimum 0.4853 mg/l in the month of July in year 2007-2008. Annual average of planktonic biomass in year 2006-2007 was recorded 1.4582 mg/l ± 0.08142 and in year 2007-2008 found 1.4787 mg/l ±0.07718.

Sampling Site III (Bhaldiyana)

Planktonic biomass at sampling site III was recorded maximum 2.4850 mg/l in the month of February in year 2006-2007 and minimum 0.653 mg/l in the month of June in year 2006-2007. Annual average of planktonic biomass in year 2006-2007 was observed 1.5555 mg/l ±0.6353 and in year 2007-08 noted1.5870 mg/l ±0.0572.

Sampling Site IV (Koti)

Planktonic biomass at sampling site IV was observed maximum 2.9780 mg/l in the month of February in year 2006-2007 and minimum 0.8245 mg/l in the month of June in year 2007-2008. Annual average of planktonic biomass in year 2006-2007 was noted 1.8598 mg/l ±0.6813 and in next year 2007-2008 observed 1.8398 mg/l ±0.6254.

Sampling Site V (Dam Outlet Near Chopra village)

Planktonic biomass at sampling site V was observed maximum 2.1863 mg/l in the month of February in year 2006-2007 and minimum 0.2843 mg/l in the month of July in year 2006-2007. Annual average of planktonic biomass in year 2006-2007 was observed 1.0781 mg/l ±0.6541 and in next year 2007-2008 found 1.1744 mg/l ±0.5627.

Observation for Fish Diversity

Fishes were collected from various spots of River Bhagirathi and reservoir all through the sampling range. The River contained rich fish diversity and population but reservoir is very poor in fish diversity, 21 species of fishes belonging to four families were collected from River Bhagirathi (sampling site I and II) while there were only 6 species observed and collected from reservoir (sampling site III and IV). 11 species of fishes belonging to one family were found in dam outlet during the study period. A systematic list of fish found during the course of study is given in Table 4.46.

Chapter 5
Discussion

Water is one of the most important of all natural resources. It is vital for all living organisms and major ecosystem as well as for human health, food production and economic development. It is having a remarkable tendency to dissolve other substances. Therefore water bodies offer a convenient option for the disposal of domestic, agricultural and industrial wastes. Due to increased urbanization and industrialization the consumption of electricity has been increased very much. To meet this increased requirement of electricity various hydro-electric projects have been constructed upon hilly rivers. All of which can significantly affect the natural physical, chemical and biological characteristics of river waters.

Reservoirs usually operate as simple units, but each reservoir can behave differently from physical, chemical and biological point of view (Rodgher *et al.*, 2005). Therefore it was purposeful to study the effect of Tehri dam on aquatic ecology of River Bhagirathi at Tehri Garhwal. In this chapter we will discuss the result obtained during our study period which has already been mentioned in the previous chapter.

Temperature is an important abiotic factor in the aquatic ecosystem (Francis *et al.*, 2007). The maximum monthly average value of temperature for both the years in upstream (sampling site I and II) was recorded 19.24 °C±1.34 in the month of September and minimum 6.95 °C±0.55 in the month of February while the maximum monthly average value of temperature (surface water) for both the years in reservoir (sampling site III and IV) was recorded 29.08 °C±0.79 in the month of July and minimum 12.30 °C±1.14 in the month of February. The maximum monthly average value of temperature for both the years in downstream (sampling site V) was recorded 18.70 °C±0.60 in the month of June and minimum 9.70 °C±1.20 in the month of January. The annual average value of temperature for both the years in upstream, reservoir and downstream ranged between 13.10 °C±4.24 to 22.04°C±5.60.

Maximum average value of temperature was recorded from the surface water of reservoir and minimum was recorded from upstream. Increased average temperature of surface water in reservoir is due to thermal stratification in reservoir water. Upper layer of water in reservoir remains in the direct contact of sunlight. No significant upwelling recorded in the surface water of reservoir so water present at upper surface remains stagnant and is directly affected by atmospheric temperature. During late summers when temperature of water in river is decreased due to melting of glaciers temperature of reservoir remains unchanged by this cooling effect because cold water entering in reservoir from upstream goes down in the deeper parts of reservoir because of having more density (total solids) and low temperature. Water coming out from reservoir also has comparatively lower temperature because it comes from deeper parts of reservoir up to which the effect of thermal stratification is not reached. Temperature of downstream water is more than upstream but lower than that of reservoir. Study of water temperature of surface water in reservoir indicates that it is determined only by atmospheric temperature and sun light while temperature of river is determined by melting of glaciers, raining and seasonal changes in air temperature. Similar results with reference to reservoir and river were shown by Hannan *et al.* (1979), Harding (1992). Same variation in Tongue River reservoir was stated by Whalen *et al.* (1982) and for Aiba Reservoir by Atobatele and Ugwumba (2008). In this study mean average temperature of surface water in Tehri Reservoir was recorded 22.04 °C, Idowu *et al.* (2004) observed the mean average temperature, 25.25 °C for Alau lake. Similar temperature trend was observed by Jekatierynczuk-Rudczyk *et al.* (2002) in Siemianowka Reservoir.

The seasonal variation observed in this study for river system was also observed by Chakrabarty *et al.* (1959) in Yamuna River, Badola and Singh (1981) in river Alaknanda, Khanna *et al.* (1999) in Ganga River at Rishikesh and in River Ganga at Haridwar by Khanna and Bhutiani (2003).

Any matter get dissolved in the solvent only when the size of the molecule of the solute is large enough to get accommodate in the space present between the molecules of solvent and water has a specific capacity to dissolve any matter in it but this dissolving limit rises with rising temperature.

In the present study the maximum monthly average value of total dissolved solids in upstream (sampling site I and II) observed was 372.45 mg/l±113.30 in the month of August and minimum value was recorded 70.90 mg/l±1.83 in the month of April. In the same way maximum monthly average value of total dissolved solids in reservoir (sampling site III and IV) observed was 140.97 mg/l±48.35 in the month of July and minimum value was recorded 52.72 mg/l±5.17 in the month of March. In the present study the maximum monthly average value of total dissolved solids in downstream (sampling site V) observed was 223.40 mg/l ±7.10 in the month of July and minimum value was recorded 72.30 mg/l±1.10 in the month of April. The annual average in upstream, reservoir and downstream for both the years varies from 79.35 mg/l ±25.98 to 151.10 mg/l ±100.06. Amount of dissolved solids is affected by the temperature they are positively correlated and the value of correlation shows the following trend site wise

Site I > Site II > Site III > Site IV > Site V, the regression equation for this study is; Temp. = 11.4 + 0.0499 TDS (P= 0.016), from this equation it is also clear that they are positively correlated.

TDS trend observed by Jekatierynczuk-Rudczyk *et al.* (2002) in Siemianowka Reservoir was similar to the findings of this study. The average of TDS in Tehri Reservoir in this study recorded was 79.35 mg/l while TDS within the Kinangop Reservoirs observed by Mwaura (2006) was between 40 and 60 mg/l. Gaudet and Melack (1981) reported a mean of 91.00 mg/l in the Lower Malewa Reservoir.

Amount of suspended solids in river system is affected by various factors like geographical location of the river, velocity of stream, volume of water and season. In the present study the maximum monthly average value of total suspended solids in upstream (sampling site I and II) observed was 2293.09 mg/l±631.33 in the month of August and minimum value was recorded 75.82 mg/l±26.23 in the month of April.

The maximum monthly average value of total suspended solids for both the years, in reservoir (sampling site III and IV) was recorded 909.77 mg/l ±365.23 in the month of July while minimum 34.13 mg/l ±15.51 in the month of March. The maximum monthly average value of total suspended solids for both the years, in downstream (sampling site V) was recorded 1362.90 mg/l ±48.10 in the month of July while minimum 20.99 mg/l ±4.62 in the month of April. The annual average value of total suspended solids in upstream, reservoir and downstream for both the years (2006-07 and 2007-08) fluctuated from 323.96 mg/l ±335.72 to 780.98 mg/l ±682.11.

Due to the construction of dam the suspended solids present in water are getting deposited in reservoir. It ultimately affects ecology of not only reservoir but also that of downstream. Early work on the influence of inorganic sediment on aquatic life has been reviewed by Cordone and Kelley (1961). Correlation observed in this study between suspended solid and velocity of the River is positive and variation in its value is as follow

Site I > Site II > Site V, site III and IV didn't showed any relation as water velocity at these two sites was zero.

Total solids are the sum of total suspended solids and total dissolved solids and from the above discussion it is clear that amount of total solid will be higher in both upper and downstream than the reservoir. The maximum monthly average of total solids in upstream (sampling site I and II) was recorded 2665.54mg/l ±744.33 in the month of August due to heavy rain in upper catchments area resulting soil erosion increasing solids in water. The minimum total solids were recorded 146.72 mg/l ±28.06 in the month of April in upstream.

The maximum monthly average of total solids in reservoir (sampling site III and IV) was recorded 1050.74 mg/l ±408.10 in the month of July while minimum value of total solids obtained 86.85mg/l ±20.65 in the month of March. The maximum monthly average of total solids in downstream (sampling site V) was recorded 1586.30 mg/l ±55.20 in the month of July while minimum value of total solids obtained 93.29 mg/l ±5.72 in the month of April. The annual average in upstream, reservoir and downstream varies from 403.31 mg/l ±360.49 to 930.21 mg/l ±778.91.

The reduction in amount of total solids in downstream than that of upstream shows that the process of sedimentation is taking place in reservoir. Sedimentation is causing siltation resulting in to submerging the habitat of benthic fauna. Same trend were obtained by Gaudet and Melack (1981) in the Lower Malewa Reservoir.

Turbidity is the measure of visibility in the water column. The maximum monthly average value of turbidity for both the years recorded in upstream (sampling site I and II) during the study period was 291.23 JTU±97.79 in the month of August and the minimum value recorded was 0.00 JTU±0.00 in the month of April. The maximum monthly average value of turbidity for both the years recorded in reservoir (sampling site III and IV) during the study period was 46.53 JTU±16.56 in the month of July the minimum value recorded was 1.60 JTU±0.41 in the month of April. The maximum monthly average value of turbidity for both the years recorded in downstream (sampling site V) during the study period was 152.75 JTU±17.75 in the month of July and the minimum value recorded was 0.40 JTU±0.40 in the month of April. The average value of turbidity in upstream, reservoir and downstream fluctuated between 16.01JTU±15.80 to 59.15±91.79.

Lesser turbidity in reservoir is due to sedimentation of suspended solids. The water of river becomes highly turbid during monsoon while there is no significant fluctuation in the turbidity of reservoir. The turbidity and total solids are closely related to each other. The regression equation for turbidity and total solids found for this study was, Turbidity =–22.8 + 0.0918 TS, (P= 0.00). Turbidity is an important characteristic to know the physical status of the river. The suspended particles, soil particles, discharged effluent, decomposed organic matters, total dissolved solids as well as the microscopic organisms increase the turbidity of water, which interferes with the light penetration. The increased turbidity during monsoon season was due to heavy soil erosion in the nearby catchments. Surface runoffs and domestic waste mainly contribute to the increased turbidity in the river. In River Ganga minimum value for turbidity was recorded in winter and summer season while maximum in monsoon period (Bilgrami and Duttamunshi, 1985). The low value of turbidity in lake water was observed by Randolph *et al.* (1984). Trend similar to this study for reservoir was observed by Whalen *et al.* (1982) in Tongue River reservoir.

The velocity is considered as an important spatial character for any river system and lay very important role in regulation of the ecology of the particular system. It regulate the transport includes advection and dispersion of water quality constituents. Advective flow carries water quality constituents downstream with the water while dispersion causes further mixing and dilution between the regions of high concentrations and the regions of the low concentrations. Velocity for river system depends upon the slope of its path and volume of water flowing in it.

During our study between July 2006 to June 2008, the maximum monthly average value of velocity recorded in upstream (sampling site I and II) was 1.91 m/s±0.07 in the month of July and minimum value was 0.51 m/s±0.02 in the month of March, while the maximum value of velocity observed for surface water of the sampling site reservoir (sampling site III and IV) was 0.00 m/s±0.00 through out both the years during the study period. The velocity of water at reservoir is zero because here river flow influence is nil and water is purely standing. The maximum monthly average value of velocity recorded in downstream (sampling site V) was 1.68 m/s±0.03 in the month of July and minimum value was 0.52 m/s±0.02 in the month of March and April. The average value of velocity in upstream, reservoir and downstream for both the years varies from 0.00 m/s±0.00 to 1.07 m/s±0.47.

Velocity of river starts to increase in the month of May due to the melting of glaciers and starts to decrease soon after monsoon. Same trend in the velocity in different hilly rivers was obtained by Khanna (1993), Khanna and Bhutiani (2003) in River Ganga, Badola and Singh (1981) in river Alaknanda.

pH is the scale of intensity of acidity or alkalinity of water and this scale gives the concentration of hydrogen ions present in water. Most of the biochemical and biological processes in water are pH dependent (Minns, 1989).

In this study the maximum monthly average value of pH recorded in upstream (sampling site I and II) was 7.72±0.28 in the month of November and minimum value of pH recorded was 7.15±0.02 in the month of June. In this study the maximum monthly average value of pH recorded in reservoir (sampling site III and IV) was 7.63±0.20 in the month of December and minimum value of pH recorded was 7.12±0.05 in the month of July while the maximum monthly average value of pH recorded in downstream (sampling site V) was 7.76±0.19 in the month of December and minimum value of pH recorded was 7.11±0.02 in the month of July. The average value in upstream, reservoir and downstream in both the years of study fluctuated between 7.32±0.17 to 7.40±0.22.

The water quality assessment showed the existence of slightly alkaline water pH in the reservoirs which is quite similar to other reservoirs in different parts of the world including Bangladesh (Haldar *et al.*, 1992), Sri Lanka (De Silva, 1992). According to Moehl and Davies (1993) in North America the typical pH range in reservoirs is usually 5-10, Randolph *et al.* (1984), Harding (1992), Mwaura (2006), Francis *et al.* (2007) found the value of pH variation from 6.38 to 8.09 in Andoni brackishwater River system. Prapurna and Shashikanth (2002) found the pH in alkaline trend throughout in Hussain Sagar Lake of Hyderabad.

Oxygen in water goes from the atmosphere due to upwelling and due to photosynthetic activity of phytoplankton present in water column. Oxygen present in water column support the aquatic life, it acts as a health indicator of aquatic system. As the pollution level increases its value decreases.

The maximum monthly average value of dissolved oxygen in upstream (sampling site I and II) for both the years was recorded 11.10±0.57 in the month of January and minimum monthly average value of dissolved oxygen recorded was 7.98 mg/l±0.48 in the month of July. The maximum monthly average value of dissolved oxygen in reservoir (sampling site III and IV) for both the years was recorded 11.53 mg/l±0.93 in the month of January and minimum monthly average value of dissolved oxygen recorded was 8.85 mg/l±0.93 in the month of July. While the maximum monthly average value of dissolved oxygen in downstream (sampling site V) for both the years was recorded 10.75 mg/l±0.05 in the month of December and minimum

Table 5.1: Average of Physical Parameters in Upstream (Sampling sites I and II) during July 2006 to June 2008

Month	*Temp. (°C)*	*TS (mg/l)*	*TDS (mg/l)*	*TSS (mg/l)*	*Velocity (m/s)*	*Turbidity (JTU)*
July	15.28±0.26	1937.65±11.56	318.45±5.50	1619.21±7.04	1.91±0.07	178.28±5.50
August	18.11±1.40	2665.54±744.33	372.45±113.30	2293.09±631.33	1.77±0.78	291.23±97.79
September	19.24±1.34	1531.23±289.50	216.21±12.68	1315.02±277.15	1.55±0.91	120.65±20.88
October	16.68±0.56	1072.81±118.07	151.34±47.34	943.98±74.42	1.21±0.72	48.08±15.33
November	11.53±0.47	780.08±70.00	101.15±6.68	678.93±63.36	1.10±0.24	16.00±4.60
December	10.22±0.40	571.42±130.38	95.52±4.85	475.90±125.53	0.83±0.05	6.30±3.75
January	7.35±0.62	307.04±89.04	86.29±3.74	220.75±86.20	0.86±0.01	4.45±1.84
February	6.95±0.55	225.62±76.24	78.12±6.26	147.49±70.22	0.66±0.03	2.43±0.63
March	8.53±0.45	159.32±24.92	75.04±0.95	84.28±25.04	0.51±0.02	1.43±1.71
April	11.98±0.30	146.72±28.06	70.90±1.83	75.82±26.23	0.56±0.01	0.00±0.00
May	16.13±0.05	790.09±68.09	117.43±12.17	672.66±56.24	0.78±0.04	15.88±4.35
June	15.18±0.05	975.02±115.33	130.33±16.70	844.69±98.67	1.13±0.05	25.15±4.22
Average±SD	13.10±4.24	930.21±778.91	151.10±100.06	780.98±682.11	1.07±0.47	59.15±91.79

Table 5.2: Average of Chemical Parameters in Upstream (Sampling sites I and II) during July 2006 to June 2008

Month	*DO (mg/l)*	*COD (mg/l)*	*BOD (mg/l)*	*Alkalinity (mg/l)*	*pH*	*Hardness (mg/l)*	*Chloride (mg/l)*
July	7.98±0.48	8.08±0.55	2.81±0.21	34.56±3.27	7.20±0.07	43.23±1.59	3.41±0.28
August	8.59±1.06	6.67±1.55	2.50±0.59	32.64±4.34	7.19±0.06	38.58±4.55	3.86±0.30
September	9.15±0.32	5.47±0.44	2.20±0.12	36.38±8.97	7.18±0.13	39.81±8.03	2.01±0.36
October	9.33±0.45	5.92±0.90	2.13±0.34	45.79±6.54	7.21±0.13	49.07±1.19	1.58±0.18
November	10.08±0.41	4.40±0.64	1.68±0.27	56.12±10.04	7.72±0.28	55.54±4.30	1.44±0.16
December	10.15±0.85	4.32±1.22	1.71±0.49	65.94±14.28	7.66±0.17	61.93±2.82	1.73±0.39
January	11.10±0.57	3.26±0.85	1.28±0.40	87.19±6.15	7.65±0.26	64.60±1.82	2.11±0.26
February	10.95±0.27	3.21±0.46	1.28±0.19	71.22±15.64	7.63±0.21	61.43±0.13	1.94±0.22
March	10.58±0.19	4.01±0.34	1.44±0.12	55.57±6.80	7.53±0.24	57.54±1.18	2.37±0.25
April	9.15±0.25	5.87±0.22	2.16±0.09	46.56±9.03	7.47±0.13	54.19±1.33	2.52±0.26
May	10.63±0.74	3.43±0.94	1.43±0.39	43.74±5.93	7.22±0.08	51.12±1.11	2.32±0.21
June	10.18±0.39	5.49±0.49	1.81±0.11	40.61±4.46	7.15±0.02	47.99±1.90	2.96±0.16
Average±SD	9.82±0.94	5.01±1.44	1.87±0.47	51.38±15.85	7.40±0.22	52.08±8.29	2.36±0.71

Table 5.3: Average of Physical Parameters in Reservoir (Sampling sites III and IV) during July 2006 to June 2008

Month	*Temp. (°C)*	*TS (mg/l)*	*TDS (mg/l)*	*TSS (mg/l)*	*Velocity (m/s)*	*Turbidity (JTU)*
July	29.08±0.79	1050.74±408.10	140.97±48.35	909.77±365.23	0.00±0.00	46.53±16.56
August	27.35±1.13	989.31±342.85	118.97±31.79	870.35±316.51	0.00±0.00	45.45±15.32
September	26.18±2.14	776.34±365.22	94.55±17.24	681.79±349.88	0.00±0.00	27.95±7.81
October	25.83±1.77	730.45±276.11	88.50±10.92	641.95±266.85	0.00±0.00	25.33±5.24
November	21.78±0.72	417.31±219.98	78.00±14.18	339.31±208.83	0.00±0.00	15.70±7.05
December	17.38±0.79	194.90±95.39	68.93±14.24	125.98±85.80	0.00±0.00	8.45±6.01
January	12.48±2.48	134.16±49.61	65.57±14.31	68.59±40.34	0.00±0.00	5.88±5.55
February	12.30±1.14	121.85±35.22	61.87±15.09	59.99±23.54	0.00±0.00	4.50±3.51
March	19.95±0.93	86.85±20.65	52.72±5.17	34.13±15.51	0.00±0.00	1.90±1.41
April	21.43±3.35	92.01±14.04	54.82±3.41	37.20±10.70	0.00±0.00	1.60±0.41
May	24.58±1.79	99.85±17.13	61.92±7.86	37.93±11.97	0.00±0.00	3.95±2.96
June	26.15±1.37	145.95±56.98	65.39±4.51	80.57±53.81	0.00±0.00	4.95±4.75
Average±SD	22.04±5.60	403.31±360.49	79.35±25.98	323.96±335.72	0.00±0.00	16.01±15.80

Table 5.4: Average of Chemical Parameters in Reservoir (Sampling sites III and IV) during July 2006 to June 2008

Month	*DO (mg/l)*	*COD (mg/l)*	*BOD (mg/l)*	*Alkalinity (mg/l)*	*pH*	*Hardness (mg/l)*	*Chloride (mg/l)*
July	8.85±0.93	6.74±1.51	2.35±0.53	36.50±3.40	7.12±0.05	45.56±4.13	2.39±0.39
August	9.36±0.53	5.42±0.66	2.07±0.21	30.83±3.80	7.30±0.19	42.53±6.17	2.12±0.52
September	9.00±0.94	5.86±1.12	2.31±0.49	43.83±11.69	7.19±0.07	44.14±3.74	1.78±0.52
October	9.43±0.37	5.89±0.45	2.10±0.16	55.78±9.48	7.34±0.16	53.65±3.56	1.68±0.11
November	10.63±0.13	3.78±0.35	1.42±0.10	70.05±5.82	7.47±0.17	61.42±2.56	1.18±0.14
December	10.58±0.66	4.23±0.45	1.74±0.16	80.15±4.07	7.63±0.20	64.10±2.16	1.19±0.05
January	11.53±0.93	2.79±1.17	1.02±0.45	92.28±1.70	7.61±0.24	65.61±2.59	1.54±0.37
February	11.08±0.20	3.08±0.45	1.21±0.12	80.53±5.21	7.55±0.17	62.35±2.33	1.47±0.17
March	10.68±0.69	4.13±0.91	1.43±0.35	55.29±6.61	7.54±0.13	58.04±1.17	1.47±0.24
April	9.88±0.62	4.94±0.75	1.83±0.31	54.80±5.71	7.43±0.22	57.23±1.94	1.88±0.15
May	10.03±0.26	4.09±0.41	1.72±0.14	45.70±4.03	7.22±0.12	53.47±2.57	1.93±0.33
June	9.40±0.16	6.32±0.34	2.10±0.08	45.05±3.96	7.27±0.14	52.93±4.06	2.30±0.27
Average±SD	10.04±0.82	4.77±1.23	1.77±0.41	57.56±18.39	7.39±0.17	55.08±7.50	1.74±0.38

Table 5.5: Average of Physical Parameters in Downstream (Sampling site V) during July 2006 to June 2008

Month	*Temp. (°C)*	*TS (mg/l)*	*TDS (mg/l)*	*TSS (mg/l)*	*Velocity (m/s)*	*Turbidity (JTU)*
July	17.55±0.05	1586.30±55.20	223.40±7.10	1362.90±48.10	1.68±0.03	152.75±17.75
August	18.00±0.00	1191.92±987.42	205.92±115.42	986.00±872.00	1.08±0.83	102.60±77.40
September	17.60±0.60	1081.85±150.54	119.95±11.55	961.90±138.99	1.58±0.08	75.80±20.70
October	17.15±0.35	915.21±164.92	109.96±2.24	805.25±162.68	1.03±0.23	62.25±4.05
November	16.30±1.00	676.12±24.88	102.61±1.31	573.52±26.19	0.75±0.15	20.55±3.55
December	12.80±0.60	454.80±57.30	98.50±1.00	356.30±58.30	1.05±0.45	13.65±1.45
January	9.70±1.20	176.10±75.10	85.30±4.00	90.80±71.10	1.20±0.70	5.40±0.10
February	10.70±1.80	127.32±32.25	78.40±0.00	48.92±32.25	0.53±0.08	2.35±1.15
March	14.50±2.60	157.27±47.07	80.35±5.05	76.92±42.02	0.52±0.01	2.40±0.90
April	14.55±1.35	93.29±5.72	72.30±1.10	20.99±4.62	0.52±0.02	0.40±0.40
May	18.15±1.96	461.42±64.03	90.72±8.58	370.70±55.45	0.61±0.05	14.40±1.90
June	18.70±0.60	798.00±7.00	109.00±2.50	689.00±9.50	0.73±0.08	23.00±4.50
Average±SD	15.48±2.91	643.30±463.04	114.70±46.87	528.60±421.84	0.94±0.38	39.63±46.57

Table 5.6: Average of Chemical Parameters in Downstream (Sampling site V) during July 2006 to June 2008

Month	*DO (mg/l)*	*COD (mg/l)*	*BOD (mg/l)*	*Alkalinity (mg/l)*	*pH*	*Hardness (mg/l)*	*Chloride (mg/l)*
July	8.13±0.57	7.62±1.06	4.84±1.73	32.50±3.00	7.11±0.02	49.84±0.64	3.20±0.00
August	9.95±0.15	4.45±0.21	3.16±1.51	31.95±0.75	7.19±0.02	55.07±0.09	2.99±0.57
September	9.90±0.20	4.45±0.35	3.26±1.54	36.25±3.15	7.30±0.10	50.96±3.17	2.32±0.19
October	9.35±0.15	5.99±0.27	3.94±1.78	39.70±0.80	7.18±0.08	54.77±1.17	1.62±0.52
November	9.95±0.15	4.74±0.52	3.47±1.79	72.35±1.15	7.38±0.23	58.16±1.16	1.02±0.03
December	10.75±0.05	3.18±0.06	2.26±0.98	78.90±6.40	7.76±0.19	61.85±3.36	1.60±0.41
January	9.60±1.20	4.67±1.56	2.80±0.32	89.90±1.40	7.43±0.03	63.81±2.31	2.27±0.29
February	9.70±0.40	4.47±0.19	3.14±1.14	73.75±12.55	7.35±0.05	59.35±0.25	2.14±0.37
March	9.60±0.50	5.30±0.93	3.91±2.31	59.80±2.50	7.45±0.25	56.20±1.33	2.15±0.05
April	8.90±0.40	6.18±0.20	4.26±2.12	55.35±5.85	7.33±0.18	55.21±0.09	2.60±0.10
May	9.60±0.60	4.31±0.80	2.89±0.62	50.85±3.35	7.20±0.01	53.71±2.71	2.75±0.15
June	9.15±0.15	6.67±0.02	4.45±2.21	45.90±3.60	7.15±0.05	53.83±2.43	3.03±0.08
Average±SD	9.55±0.62	5.17±1.18	3.52±0.73	55.60±18.67	7.32±0.17	56.06±3.95	2.31±0.63

Table 5.7a: Average of Biological Parameters in Upstream (Sampling sites I and II) during July 2006 to June 2008

Month	*Protozoa (Unit/l)*	*Rotifera (Unit/l)*	*Cladocera (Unit/l)*	*Copepoda (Unit/l)*	*Total Zooplankton (Unit/l)*
July	30.00±17.94	46.25±6.99	35.75±9.50	13.25±4.79	125.25±31.51
August	17.50±9.57	55.50±9.15	38.25±19.31	14.75±4.03	126.00±30.14
September	25.75±15.11	86.00±16.51	39.25±10.87	14.75±17.33	165.75±16.60
October	33.50±11.36	69.75±10.81	46.25±19.60	17.25±20.69	166.75±28.72
November	56.25±5.06	70.50±26.49	30.50±15.89	20.25±14.08	177.50±28.35
December	67.50±10.15	80.75±25.84	53.75±25.06	24.75±9.74	226.75±66.15
January	66.75±24.47	87.25±12.12	71.25±43.18	29.00±13.29	254.25±49.60
February	49.75±7.14	131.50±41.73	64.00±27.66	26.50±9.98	271.75±69.63
March	67.50±15.97	107.00±25.07	53.75±15.52	25.25±18.14	253.50±29.92
April	46.50±19.12	115.00±52.24	41.00±14.72	23.50±11.12	226.00±91.99
May	72.50±26.06	104.50±71.23	38.75±17.50	22.25±10.50	238.00±98.94
June	45.50±8.85	55.00±18.24	37.50±12.48	13.00±4.97	151.00±26.36
Average ± SD	48.25±18.47	84.08±26.38	45.83±12.37	20.38±5.61	198.54±52.23

Table 5.7b: Average of Biological Parameters in Upstream (Sampling sites I and II) during July 2006 to June 2008

Month	*Diatom (Unit/l)*	*Green Algae (Unit/l)*	*Blue Green Algae (Unit/l)*	*Total Phytoplankton (Unit/l)*	*Planktonic Biomass (mg/l)*
July	631.68±182.43	53.65±26.84	8.52±2.69	693.85±186.34	0.4233±0.1214
August	810.28±231.51	74.79±16.55	8.10±3.68	893.17±246.04	0.5048±0.1270
September	1226.68±416.87	118.84±12.01	29.90±7.63	1375.43±427.03	0.7830±0.2420
October	1187.80±252.28	186.36±24.82	48.15±16.97	1422.30±284.50	0.7950±0.1285
November	1447.03±372.67	218.76±11.96	47.04±9.33	1712.83±386.85	0.9628±0.2059
December	2107.15±336.11	373.07±35.39	81.58±12.87	2561.78±343.20	1.4286±0.1861
January	3203.80±383.98	444.02±39.09	90.60±11.90	3738.42±417.20	2.0063±0.2633
February	4360.15±549.98	451.57±32.65	72.19±16.68	4883.91±535.94	2.5818±0.2566
March	4116.05±143.31	345.73±39.79	71.33±11.53	4533.04±182.35	2.3993±0.1200
April	3859.44±246.14	243.07±58.53	84.80±32.08	4187.32±317.50	2.2480±0.2042
May	2554.66±367.23	190.97±8.25	87.82±8.07	2833.45±374.61	1.5410±0.1994
June	764.05±439.42	86.96±15.42	18.23±6.64	1119.25±99.33	0.6184±0.0627
Average ± SD	2189.07±1389.04	232.315±141.56	54.02±31.54	2496.23±1510.82	1.3577±0.7854

Table 5.8a: Average of Biological Parameters in Reservoir (Sampling sites III and IV) during July 2006 to June 2008

Month	*Protozoa (Unit/l)*	*Rotifera (Unit/l)*	*Cladocera (Unit/l)*	*Copepoda (Unit/l)*	*Total Zooplankton (Unit/l)*
July	42.25±3.86	69.00±23.14	44.25±17.34	18.00±7.75	173.50±32.88
August	63.00±24.37	86.50±9.04	74.75±16.76	33.75±8.66	258.00±42.80
September	66.75±37.26	79.50±16.58	76.25±18.84	16.25±22.53	238.75±78.28
October	58.00±29.61	121.00±24.10	72.50±37.97	45.00±4.90	296.50±62.70
November	54.25±27.94	103.00±43.88	65.25±36.27	43.50±19.07	266.00±115.40
December	100.00±42.29	85.25±20.53	65.25±15.17	38.00±11.83	288.50±66.97
January	81.25±39.60	119.25±32.83	83.00±25.02	46.00±20.82	329.50±81.98
February	60.50±14.06	105.00±33.08	50.75±31.98	39.25±24.85	255.50±73.31
March	69.00±10.55	101.50±23.12	57.25±15.65	28.50±11.93	256.25±36.75
April	72.75±28.85	109.75±25.97	68.25±32.76	41.25±24.20	292.00±51.15
May	68.75±48.77	81.00±27.72	67.25±21.47	18.25±7.89	235.25±82.93
June	62.00±13.44	75.75±18.61	44.75±28.57	24.50±14.93	207.00±40.06
Average ±SD	66.54±14.36	94.71±17.42	64.13±12.48	32.69±11.16	258.06±41.98

Table 5.8b: Average of Biological Parameters in Reservoir (Sampling sites III and IV) during July 2006 to June 2008

Month	*Diatom (Unit/l)*	*Green Algae (Unit/l)*	*Blue Green Algae (Unit/l)*	*Total Phytoplankton (Unit/l)*	*Planktonic Biomass (mg/l)*
July	1209.97±465.89	113.32±10.55	23.84±4.44	1597.12±152.96	0.8531±0.0823
August	1951.37±212.26	121.90±7.66	17.80±3.09	2091.07±209.41	1.1149±0.1584
September	2563.32±384.58	205.39±18.31	24.45±6.27	2794.41±384.30	1.5205±0.1925
October	2529.16±553.26	274.26±25.05	34.68±6.80	2838.09±560.79	1.5842±0.1851
November	2820.14±490.21	320.03±46.77	67.31±14.62	3207.47±469.56	1.7698±0.2990
December	3291.84±591.94	400.57±17.75	52.81±13.14	3745.22±591.61	2.0259±0.2695
January	3884.83±566.82	448.32±48.45	85.06±9.98	4418.21±563.34	2.3708±0.2469
February	4579.88±479.95	423.56±33.45	112.70±13.25	5116.13±492.01	2.6667±0.2499
March	4102.35±367.37	341.38±33.77	73.80±10.38	4517.56±370.86	2.3902±0.2185
April	3840.33±286.90	234.48±29.94	72.31±13.20	4147.10±261.93	2.1160±0.0910
May	2147.31±530.73	112.71±8.47	68.91±9.23	2328.93±540.94	1.2801±0.3164
June	1296.51±299.86	109.13±8.88	24.11±4.82	1429.74±303.43	0.8343±0.1384
Average ±SD	2851.42±1103.64	258.75±128.30	54.81±29.98	3185.92±1208.70	1.7105±0.6160

Table 5.9a: Average of Biological Parameters in Downstream (Sampling site V) during July 2006 to June 2008

Month	*Protozoa (Unit/l)*	*Rotifera (Unit/l)*	*Cladocera (Unit/l)*	*Copepoda (Unit/l)*	*Total Zooplankton (Unit/l)*
July	18.50±6.36	44.50±2.12	23.50±3.54	13.00±5.66	99.50±0.71
August	38.50±47.38	60.00±25.46	42.50±13.44	9.00±8.49	150.00±94.75
September	57.50±3.54	80.00±15.56	41.50±16.26	18.00±8.48	197.00±19.80
October	45.50±19.09	63.00±1.41	61.50±12.02	23.00±14.14	193.00±8.49
November	37.00±15.56	95.50±21.92	53.50±4.95	36.00±1.41	222.00±43.84
December	50.00±36.77	81.00±26.87	48.50±17.68	35.50±16.26	215.00±24.04
January	75.50±23.33	114.00±63.64	98.00±15.56	46.50±2.12	334.00±69.30
February	62.00±33.94	92.50±14.85	34.50±4.95	38.00±9.90	227.00±33.94
March	56.00±42.43	92.50±48.79	54.50±16.26	12.50±17.68	215.50±27.58
April	39.50±14.84	59.50±24.75	55.00±9.90	19.50±12.02	173.50±17.68
May	31.50±0.71	45.00±46.67	52.50±13.44	21.00±5.66	150.00±28.28
June	37.00±15.56	67.00±8.49	26.00±8.49	12.00±4.24	142.00±19.80
Average ± SD	45.71±15.38	74.54±21.56	49.29±19.41	23.67±12.31	193.21±59.01

Table 5.9b: Average of Biological Parameters in Downstream (Sampling site V) during July 2006 to June 2008

Month	*Diatom (Unit/l)*	*Green Algae (Unit/l)*	*Blue Green Algae (Unit/l)*	*Total Phytoplankton (Unit/l)*	*Planktonic Biomass (mg/l)*
July	476.41±49.60	41.29±19.87	7.38±1.45	524.58±70.21	0.3045±0.0286
August	644.30±127.41	64.54±32.53	5.89±2.35	714.73±162.29	0.4146±0.1095
September	990.55±380.43	74.86±1.96	24.75±10.42	1090.15±388.89	0.6221±0.1801
October	1157.11±359.36	117.80±4.87	35.73±7.81	1310.63±346.67	0.6877±0.1364
November	1768.46±909.41	163.35±17.12	33.37±2.86	1965.18±895.15	1.0639±0.4370
December	2449.85±1229.85	240.33±55.42	64.40±10.04	2754.58±1295.31	1.4225±0.6066
January	2819.47±423.51	273.79±73.04	77.47±14.04	3170.70±510.62	1.6413±0.2003
February	3119.77±835.47	382.71±2.12	58.33±7.11	3560.81±840.46	1.8729±0.4432
March	2522.86±569.02	285.66±7.86	51.38±11.21	2861.39±567.79	1.5660±0.4546
April	2916.04±141.70	154.81±43.28	76.44±35.50	3147.29±62.93	1.7588±0.1334
May	2437.75±265.10	167.41±8.51	75.43±5.55	2680.58±268.06	1.5496±0.0930
June	877.01±87.09	67.00±25.46	18.29±4.26	962.30±108.29	0.6109±0.0901
Average ± SD	1848.30±970.44	169.46±106.47	44.70±26.70	2061.91±1079.68	1.1262±0.5706

monthly average value of dissolved oxygen recorded was 8.13 mg/l±0.57 in the month of July.

Table 5.10: Average of Phytoplankton, Zooplankton and Planktonic Biomass in Upstream (Sampling sites I and II) during July 2006 to June 2008

Month	*Phytoplankton (Unit/l)*	*Zooplankton (Unit/l)*	*Planktonic Biomass (mg/l)*
July	693.85±186.34	125.25±31.51	0.4233±0.1214
August	893.17±246.04	126.00±30.14	0.5048±0.1270
September	1375.43±427.03	165.75±16.60	0.7830±0.2420
October	1422.30±284.50	166.75±28.72	0.7950±0.1285
November	1712.83±386.85	177.50±28.34	0.9628±0.2059
December	2561.78±343.20	226.75±66.15	1.4286±0.1861
January	3738.42±417.20	254.25±46.60	2.0063±0.2633
February	4883.91±535.94	271.75±69.63	2.5818±0.2566
March	4533.04±182.35	253.50±29.92	2.3993±0.1200
April	4187.32±317.50	226.00±91.99	2.2480±0.2042
May	2833.45±374.61	238.00±98.94	1.5410±0.1994
June	1119.25±99.33	151.00±26.36	0.6184±0.0627
Average±SD	2496.23±1510.82	198.54±52.23	1.3576±0.7854

Table 5.11: Average of Phytoplankton, Zooplankton and Planktonic Biomass in Reservoir (Sampling sites III and IV) during July 2006 to June 2008

Month	*Phytoplankton (Unit/l)*	*Zooplankton (Unit/l)*	*Planktonic Biomass (mg/l)*
July	1597.12±152.96	173.50±32.88	0.8531±0.0823
August	2091.07±209.41	258.00±42.80	1.1149±0.1584
September	2794.41±384.30	238.75±78.28	1.5205±0.1925
October	2838.09±560.79	296.50±62.70	1.5842±0.1851
November	3207.47±469.56	266.00±115.40	1.7698±0.2990
December	3745.22±591.61	288.50±66.96	2.0259±0.2695
January	4418.25±563.34	329.50±81.98	2.3708±0.2469
February	5116.11±452.01	255.50±73.30	2.6667±0.2499
March	4517.56±370.86	256.25±36.75	2.3902±0.2185
April	4147.10±261.93	292.00±51.15	2.1160±0.0910
May	2328.93±540.94	235.25±82.93	1.2801±0.3164
June	1429.74±303.43	207.00±40.06	0.8344±0.1383
Average±SD	3185.92±1208.70	258.06±41.98	1.7105±0.6160

Table 5.12: Average of Phytoplankton, Zooplankton and Planktonic Biomass in Downstream (Sampling site V) during July 2006 to June 2008

Month	*Phytoplankton (Unit/l)*	*Zooplankton (Unit/l)*	*Planktonic Biomass (mg/l)*
July	524.58±70.21	99.50±0.71	0.3045±0.0286
August	714.73±162.29	150.00±94.75	0.4146±0.1095
September	1090.14±388.89	197.00±19.80	0.6221±0.1801
October	1310.63±346.67	193.00±8.49	0.6877±0.1364
November	1965.18±895.15	222.00±43.84	1.0639±0.4370
December	2754.58±1295. 31	215.00±24.04	1.4225±0.6066
January	3170.70±510.62	334.00±69.30	1.6413±0.2003
February	3560.81±840.46	227.00±332.94	1.8729±0.4432
March	2861.39±567.79	215.50±27.58	1.5660±0.4546
April	3147.29±62.93	173.50±27.58	1.7588±0.0134
May	2680.58±268.06	150.00±17.68	1.5496±0.0931
June	962.30±108.29	142.00±19.80	0.6109±0.0902
Average±SD	2061.91±1089.68	193.21±59.01	1.1262±0.5706

The value of dissolved oxygen reduced gradually from May onward due to turbidity, which retarded the photosynthetic activity of aquatic flora for river system. Chopra *et al.* (1994) and Sharma (1986) also got the same result and reasoned that low temperature in November to February increases the oxygen retaining capacity of water.

This trend was also observed by Badola and Singh (1981) in the river Alaknanda. Khanna (1993) and Khanna and Chugh (2004) also reported the same trend in the River Ganga at Haridwar. Same observations were also given by Pandey *et al.* (1993) in Kosi River. The value of dissolved oxygen was observed higher in reservoir than the river because reservoir water is standing, because of which clarity is more which resulted in more photosynthetic activity, ultimately more value of dissolved oxygen. In the reservoir the yearly trend was same as in case of river but fluctuation was relatively low. Seasonal trend observed in this study was similar to the findings of Nicholas and Gordan (1972) for Cherokee reservoir, Nix (1974) for De Gray Reservoir, Gnilka (1975) for Center Hill Reservoir.

The average value of DO in upstream, reservoir and downstream for both the years varied from 9.55 mg/l±0.62 to 10.04 mg/l±0.82. Maximum value of DO obtained in reservoir while minimum value recorded from downstream. It is a temperature, turbidity and velocity dependent parameter. Increase in temperature decreases the dissolved oxygen but increased velocity increases the upwelling in water column resulting more mixing of oxygen in water and ultimately more value of dissolved oxygen.

Turbidity and dissolved oxygen are negatively related, more the turbid water means more dissolved matter in water so lesser is the space available for oxygen to

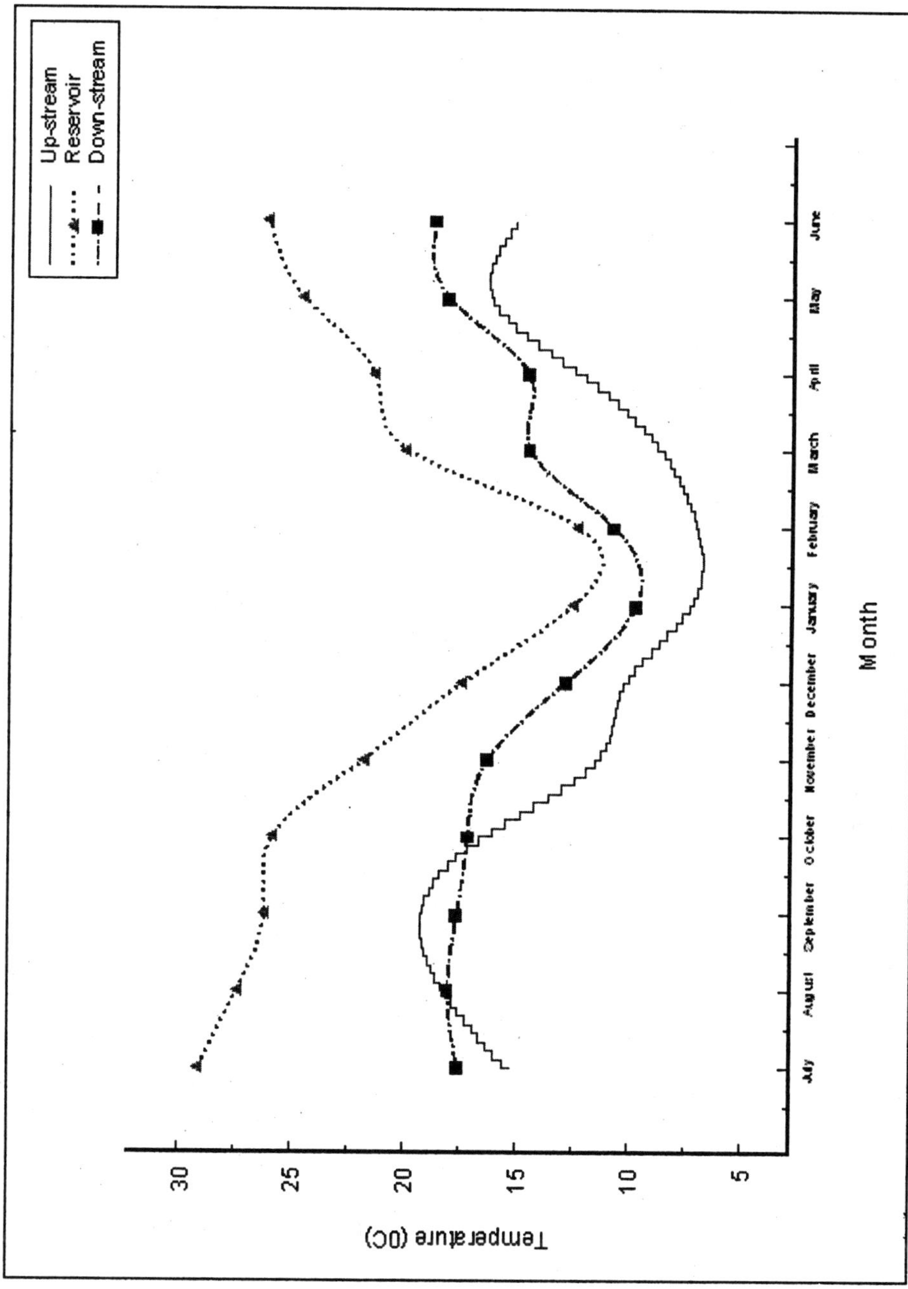

Figure 5.1: Comparative Variation of Temperature between Upstream, Reservoir and Downstream during 2006–08

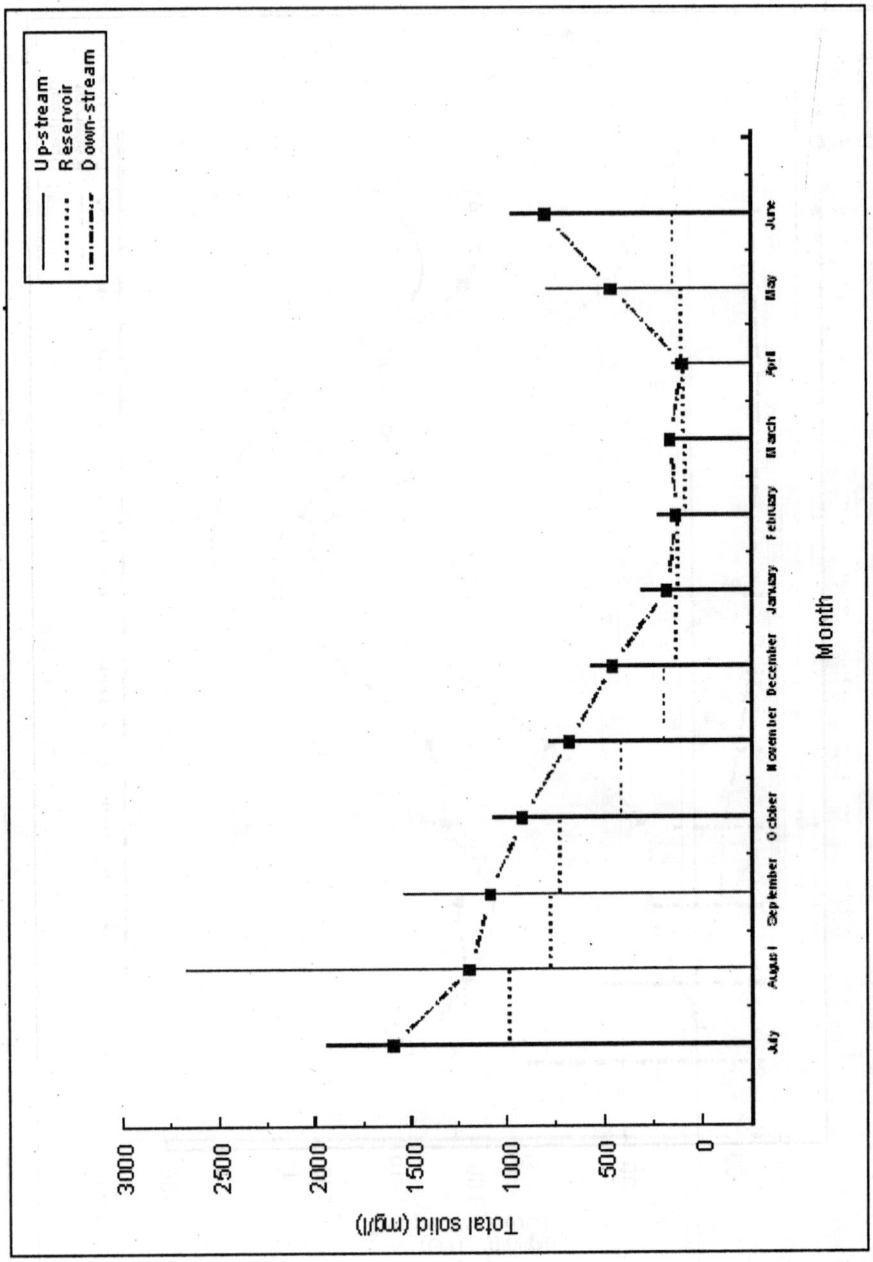

Figure 5.2: Comparative Variation of Total Solid between Upstream, Reservoir and Downstream during 2006–08

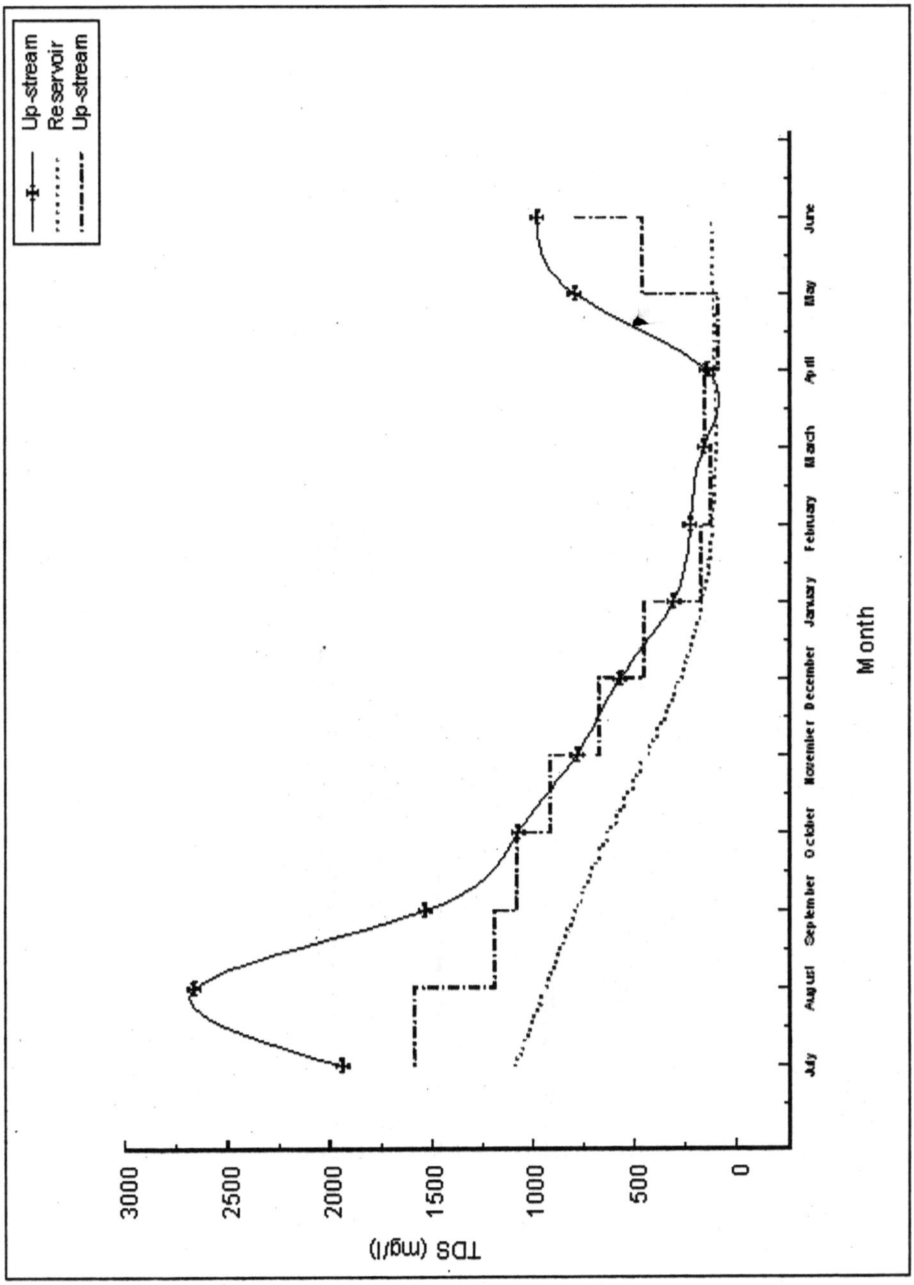

Figure 5.3: Comparative Variation of TDS between Upstream, Reservoir and Downstream during 2006–08

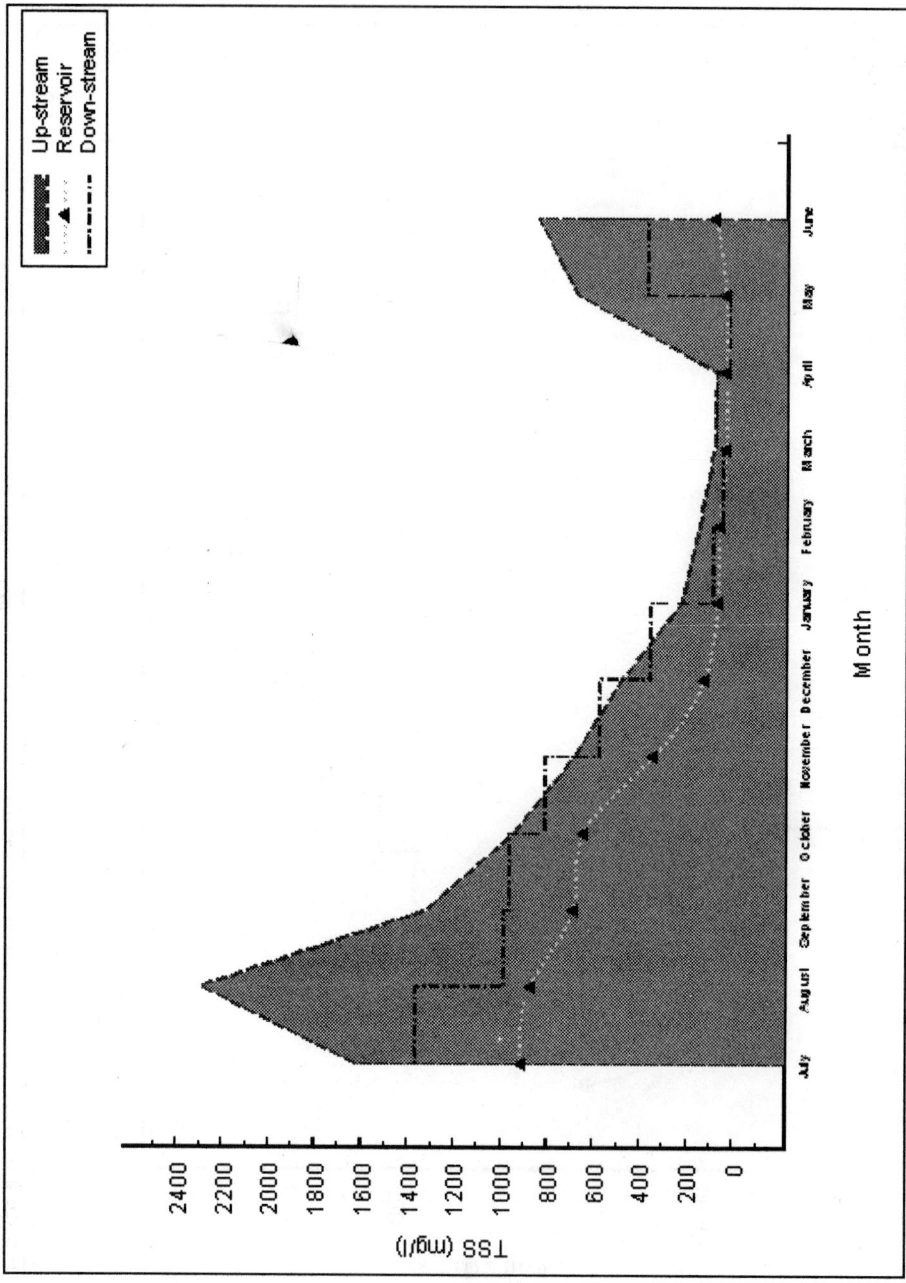

Figure 5.4: Comparative Variation of TSS between Upstream, Reservoir and Downstream during 2006–08

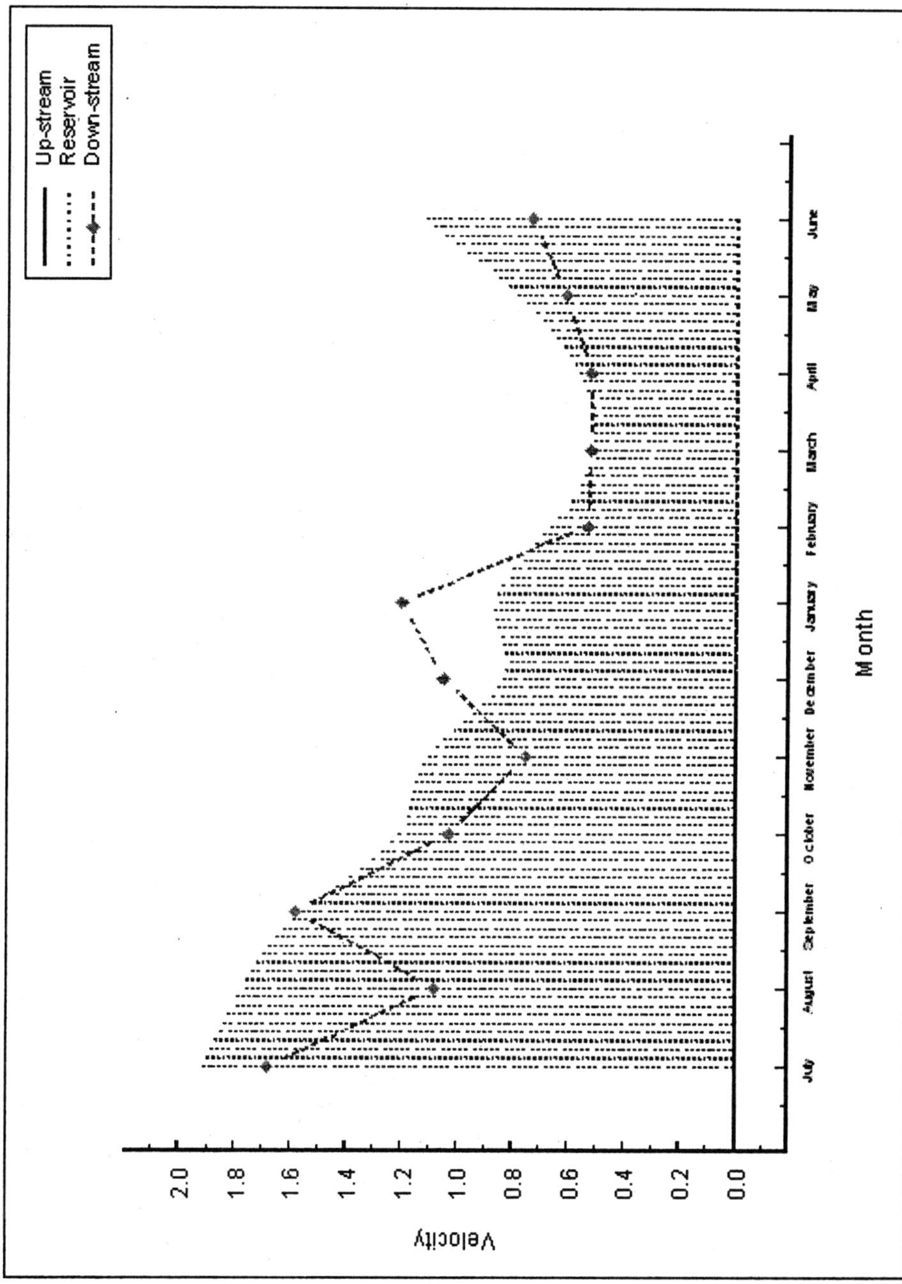

Figure 5.5: Comparative Variation of Velocity between Upstream, Reservoir and Downstream during 2006–08

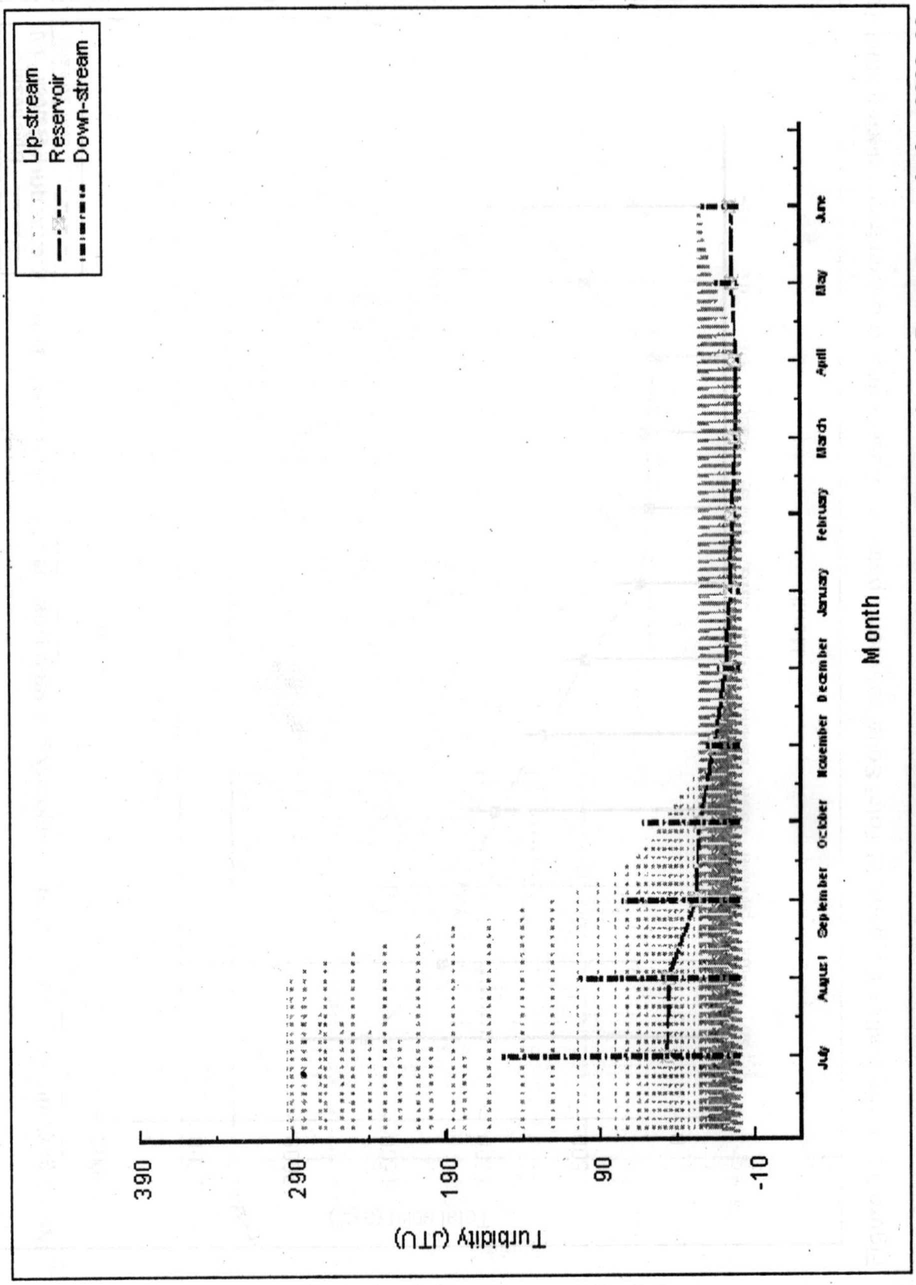

Figure 5.6: Comparative Variation of Turbidity between Upstream, Reservoir and Downstream during 2006–08

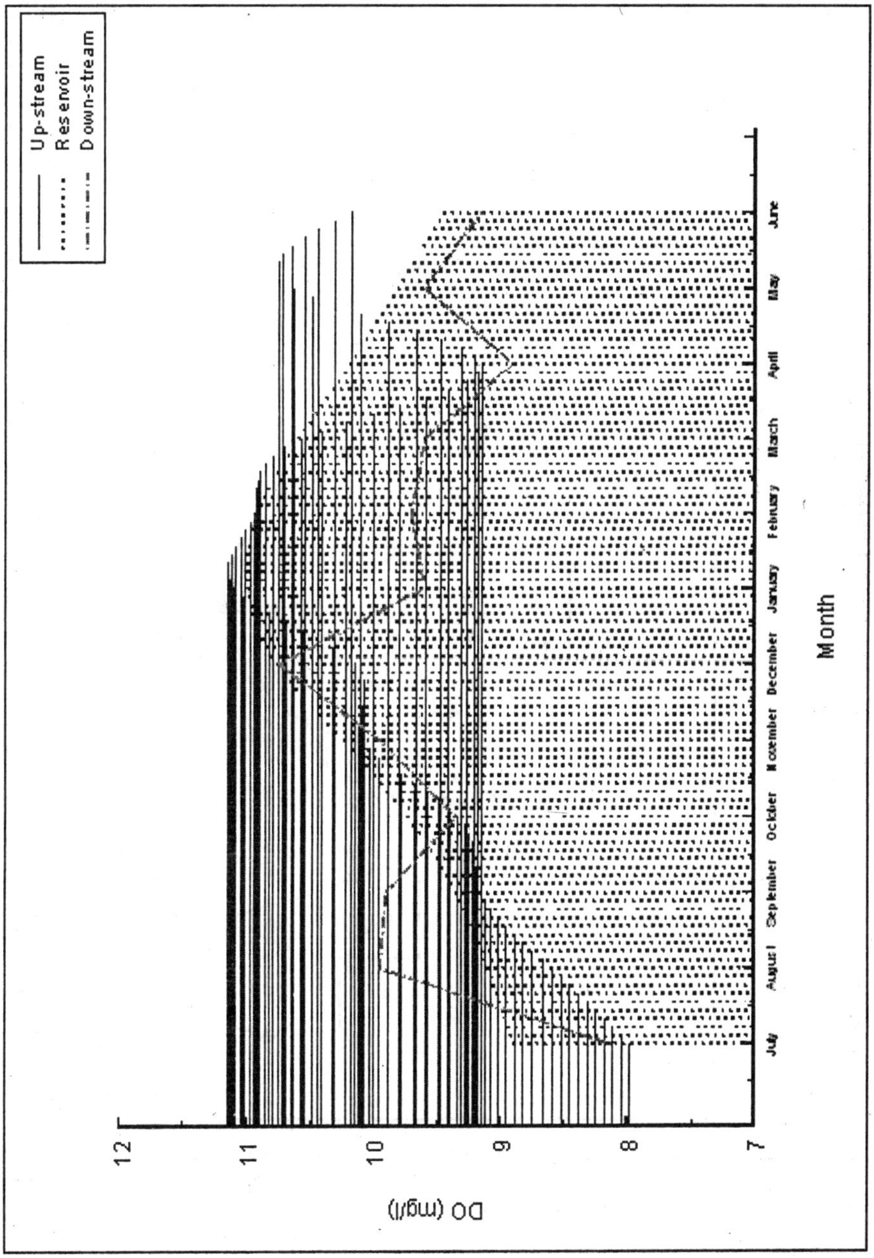

Figure 5.7: Comparative Variation of DO between Upstream, Reservoir and Downstream during 2006–08

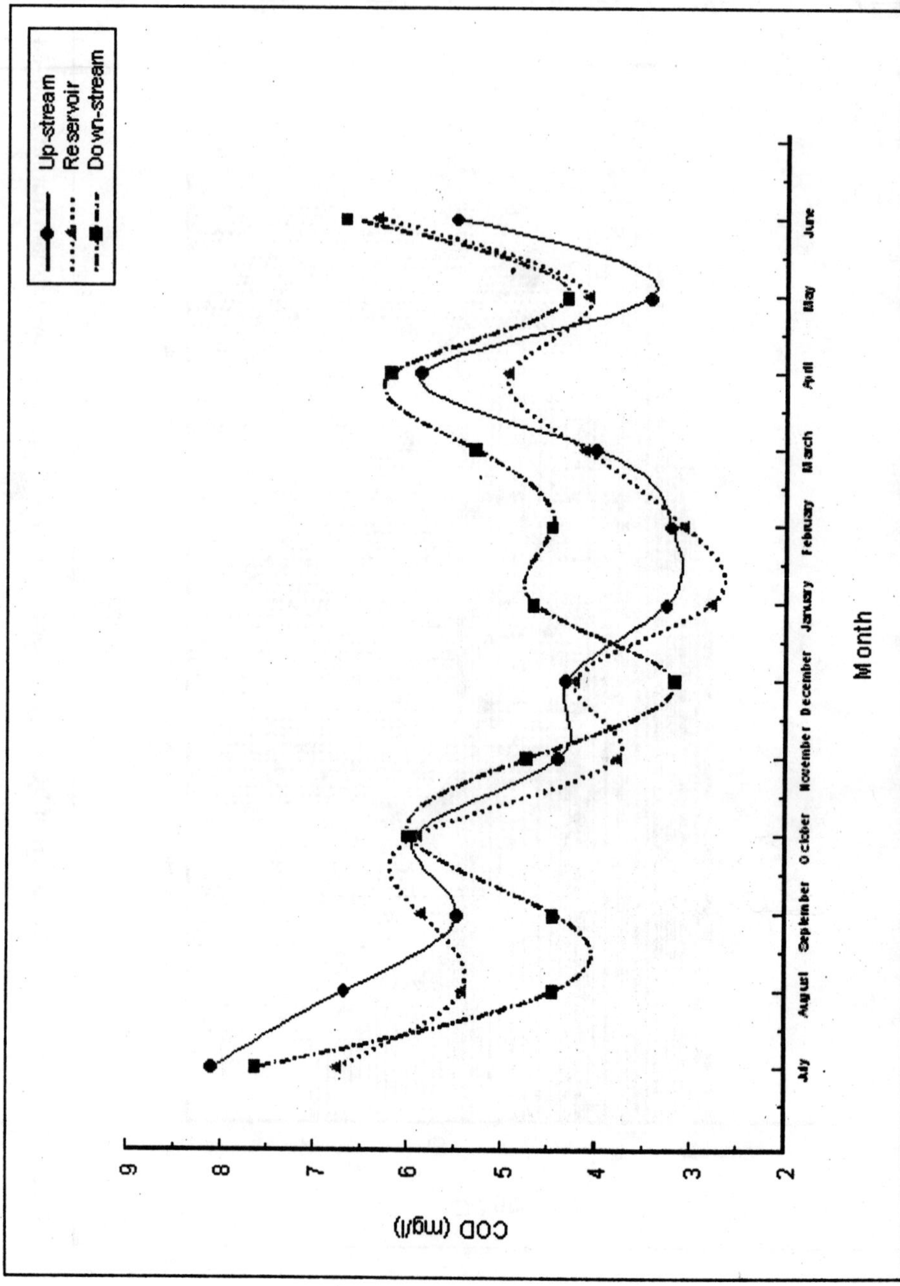

Figure 5.8: Comparative Variation of COD between Upstream, Reservoir and Downstream during 2006–08

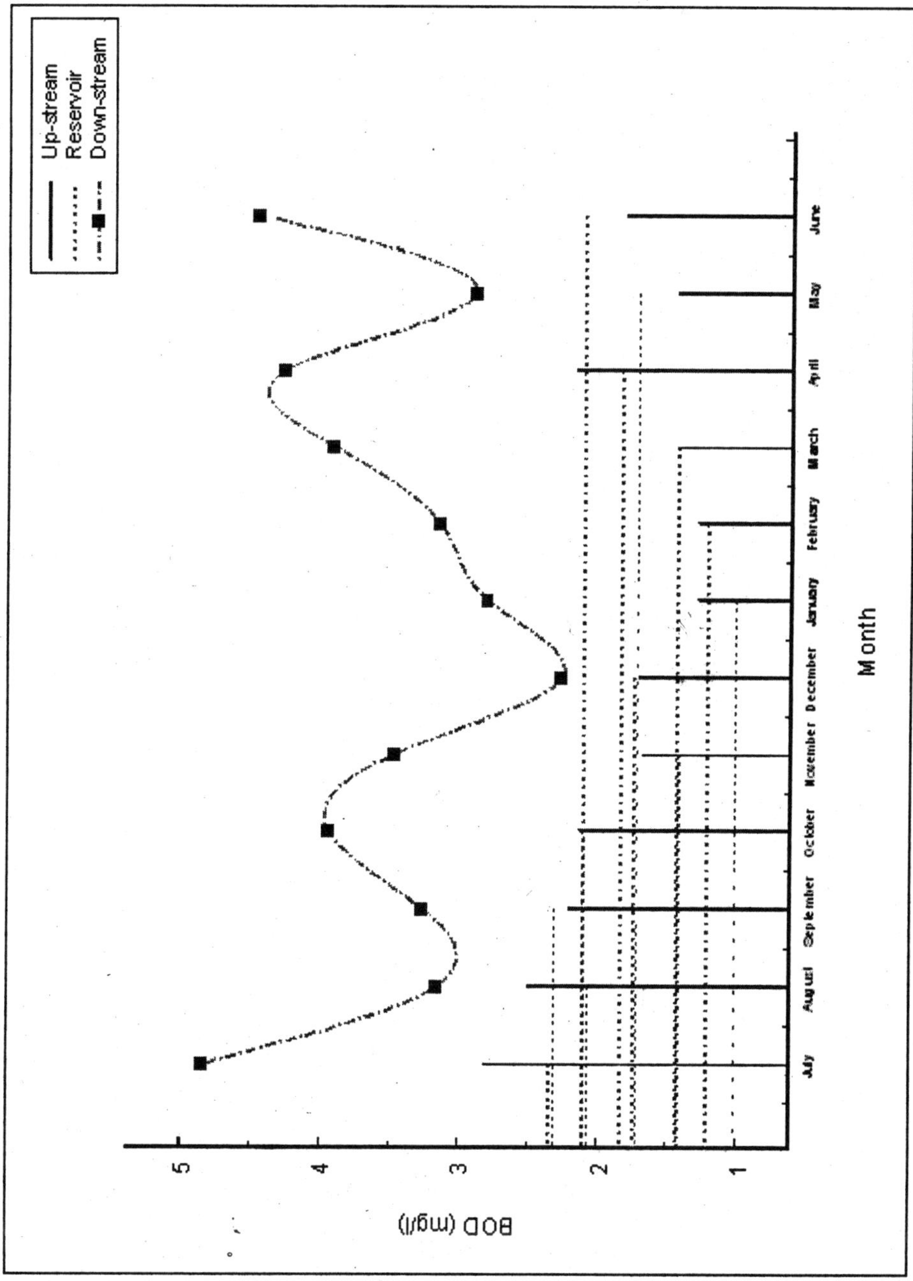

Figure 5.9: Comparative Variation of BOD between Upstream, Reservoir and Downstream during 2006–08

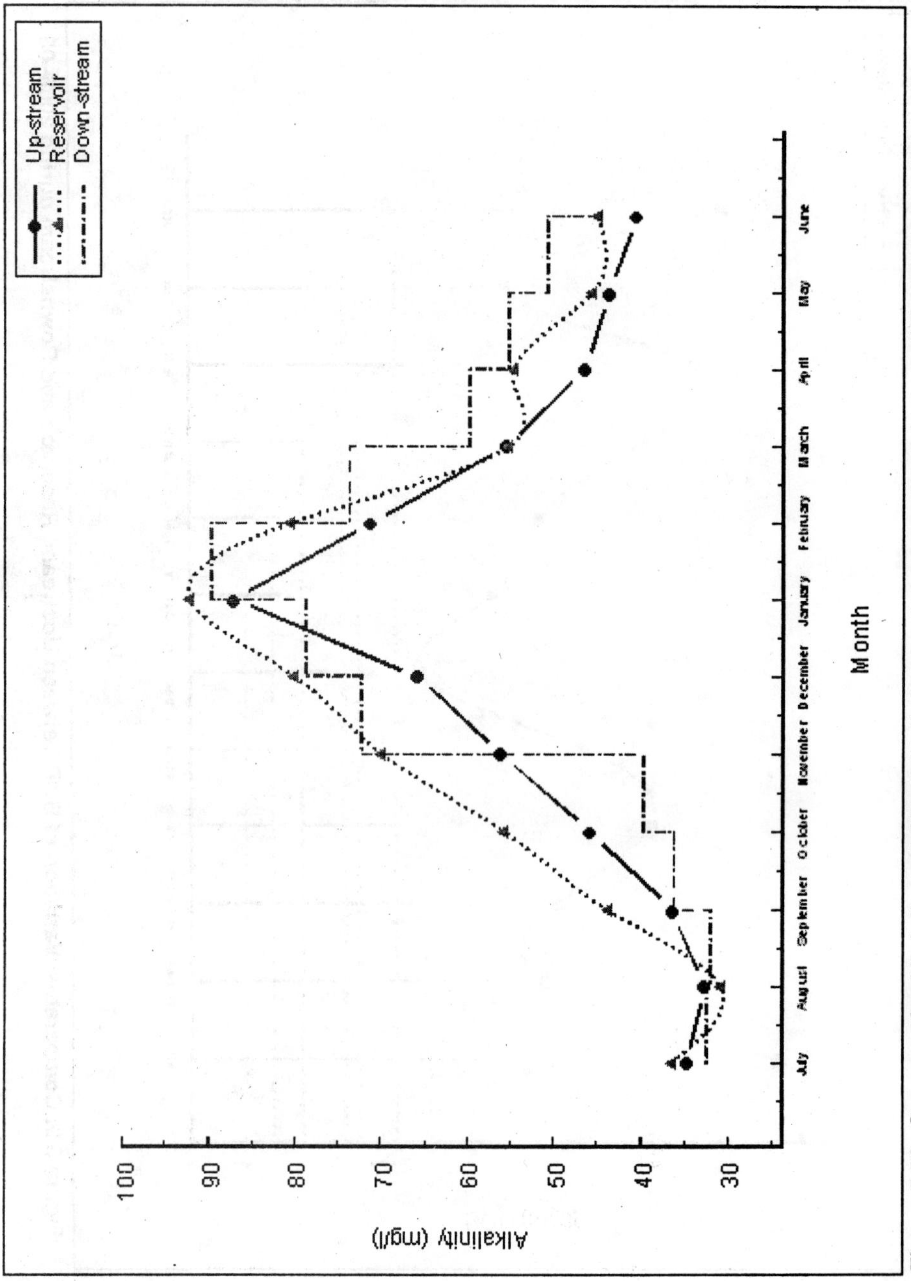

Figure 5.10: Comparative Variation of Alkalinity between Upstream, Reservoir and Downstream during 2006–08

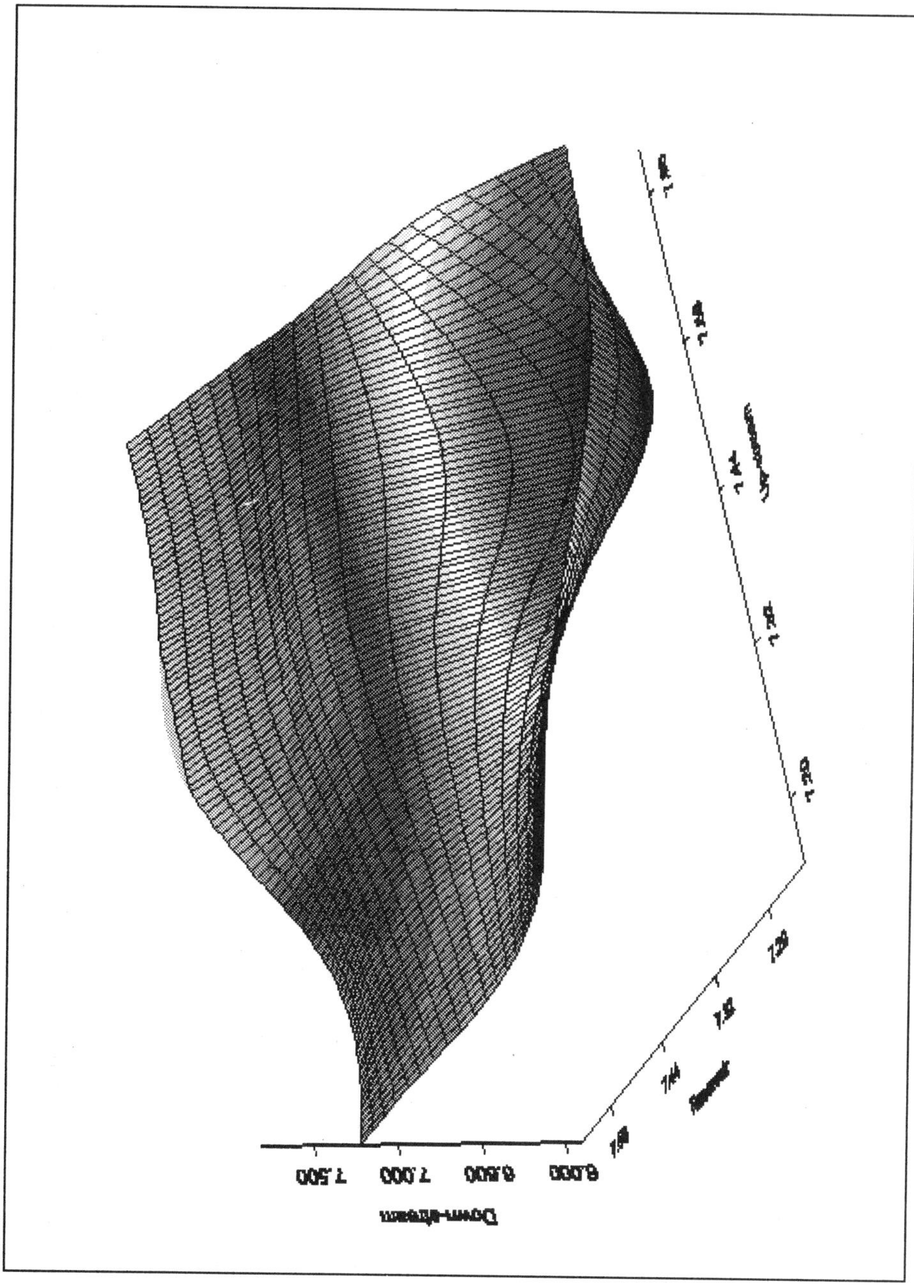

Figure 5.11: Comparative Variation of pH between Upstream, Reservoir and Downstream during 2006–08

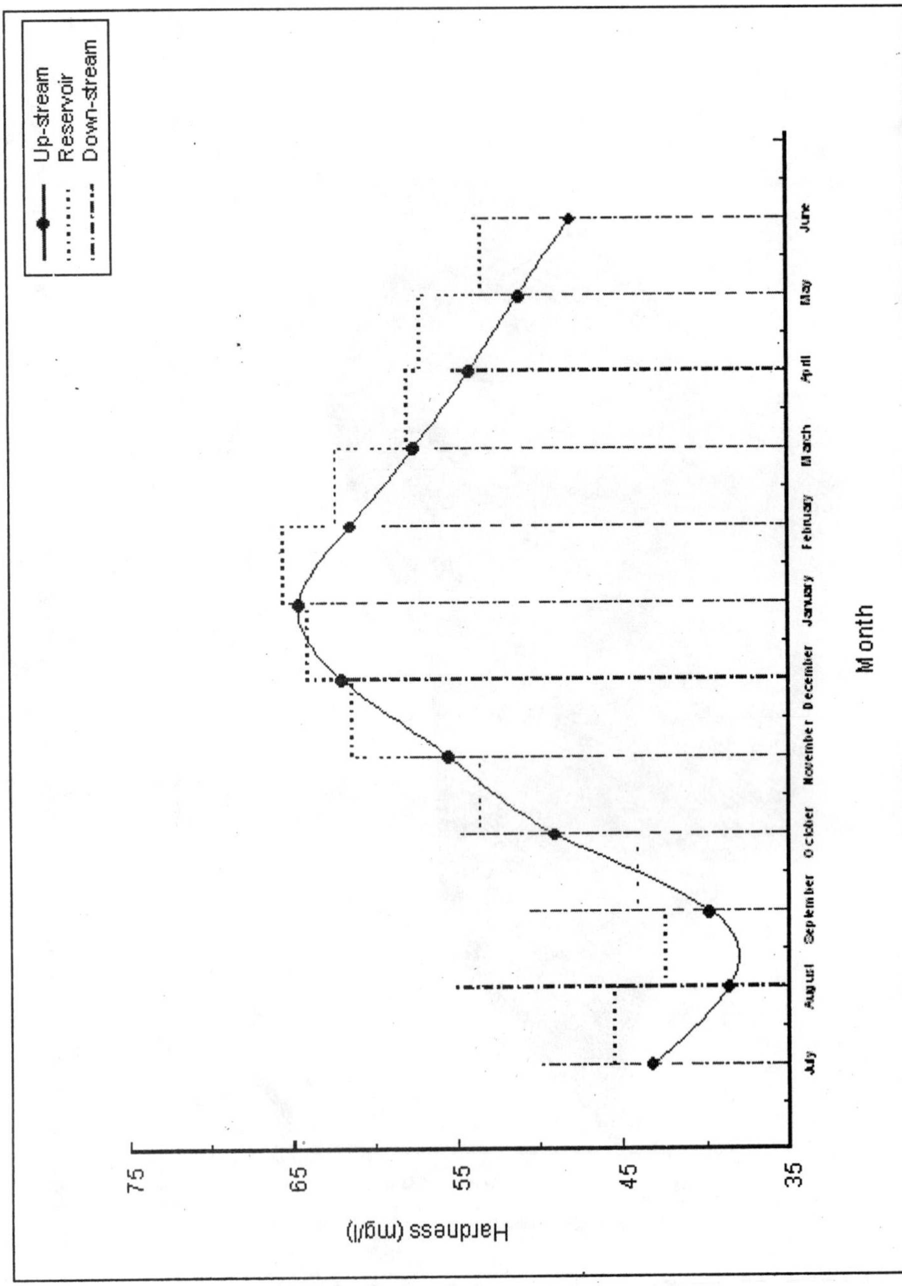

Figure 5.12: Comparative Variation of Hardness between Upstream, Reservoir and Downstream during 2006–08

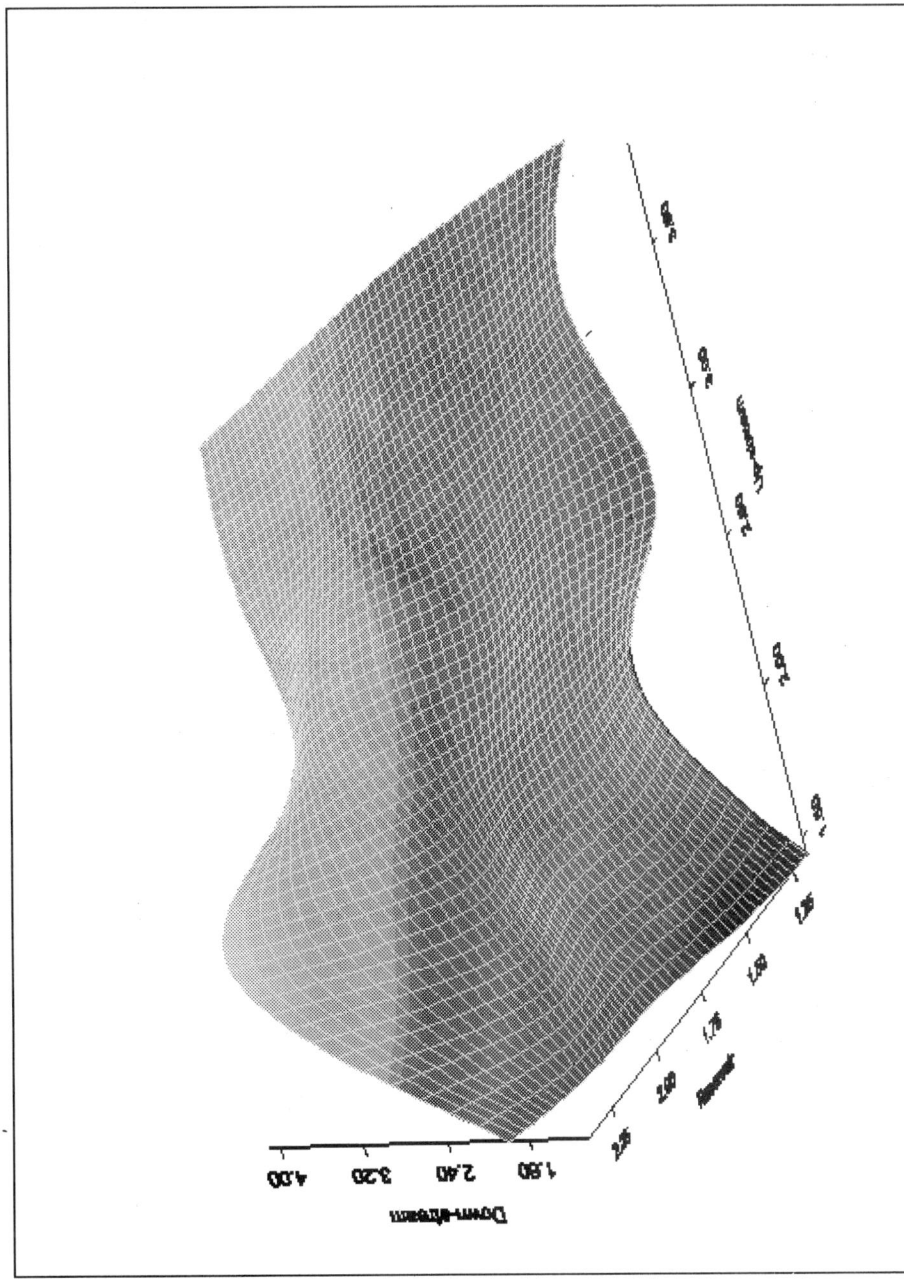

Figure 5.13: Comparative Variation of Chloride between Upstream, Reservoir and Downstream during 2006–08

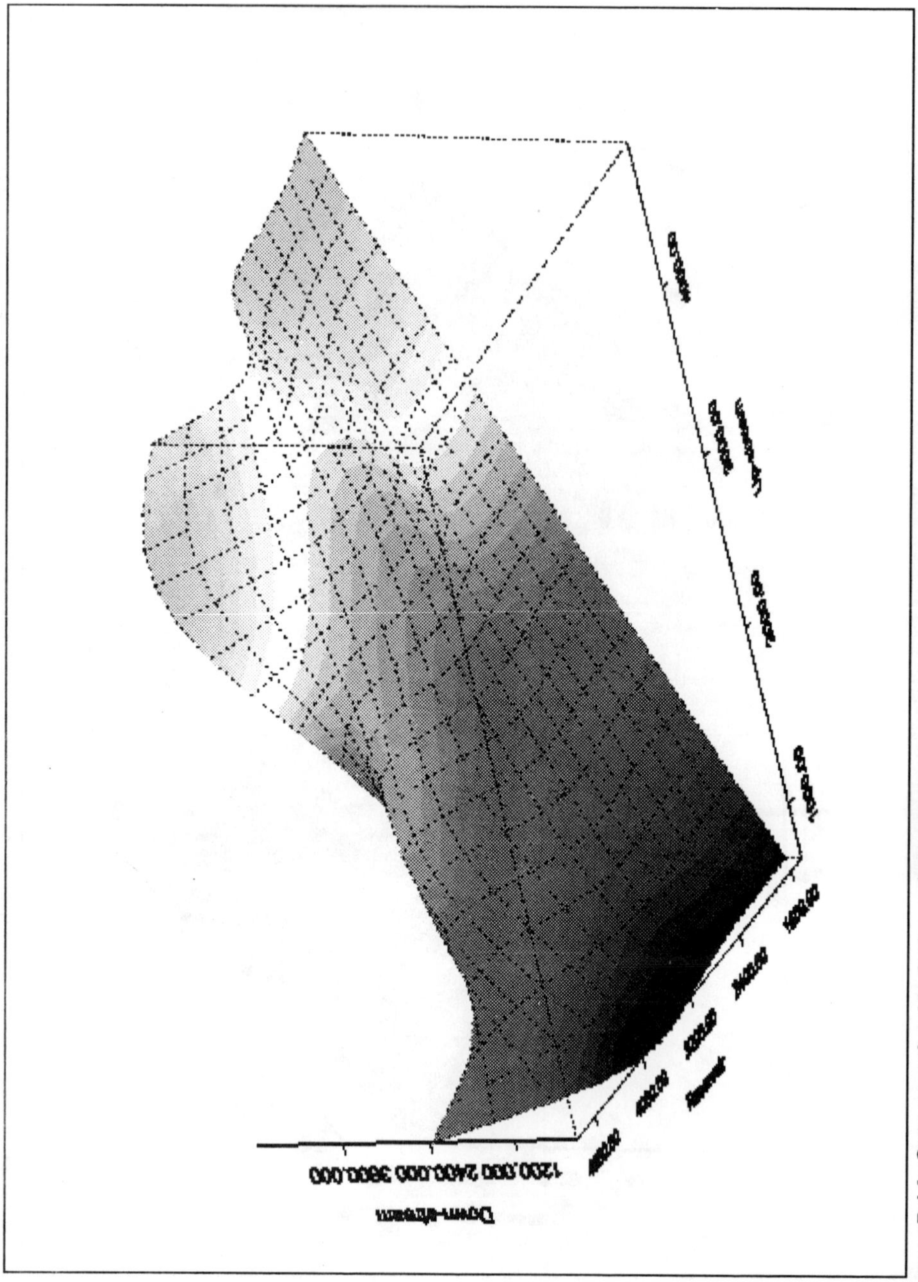

Figure 5.14: Comparative Variation of Phytoplankton between Upstream, Reservoir and Downstream during 2006–08

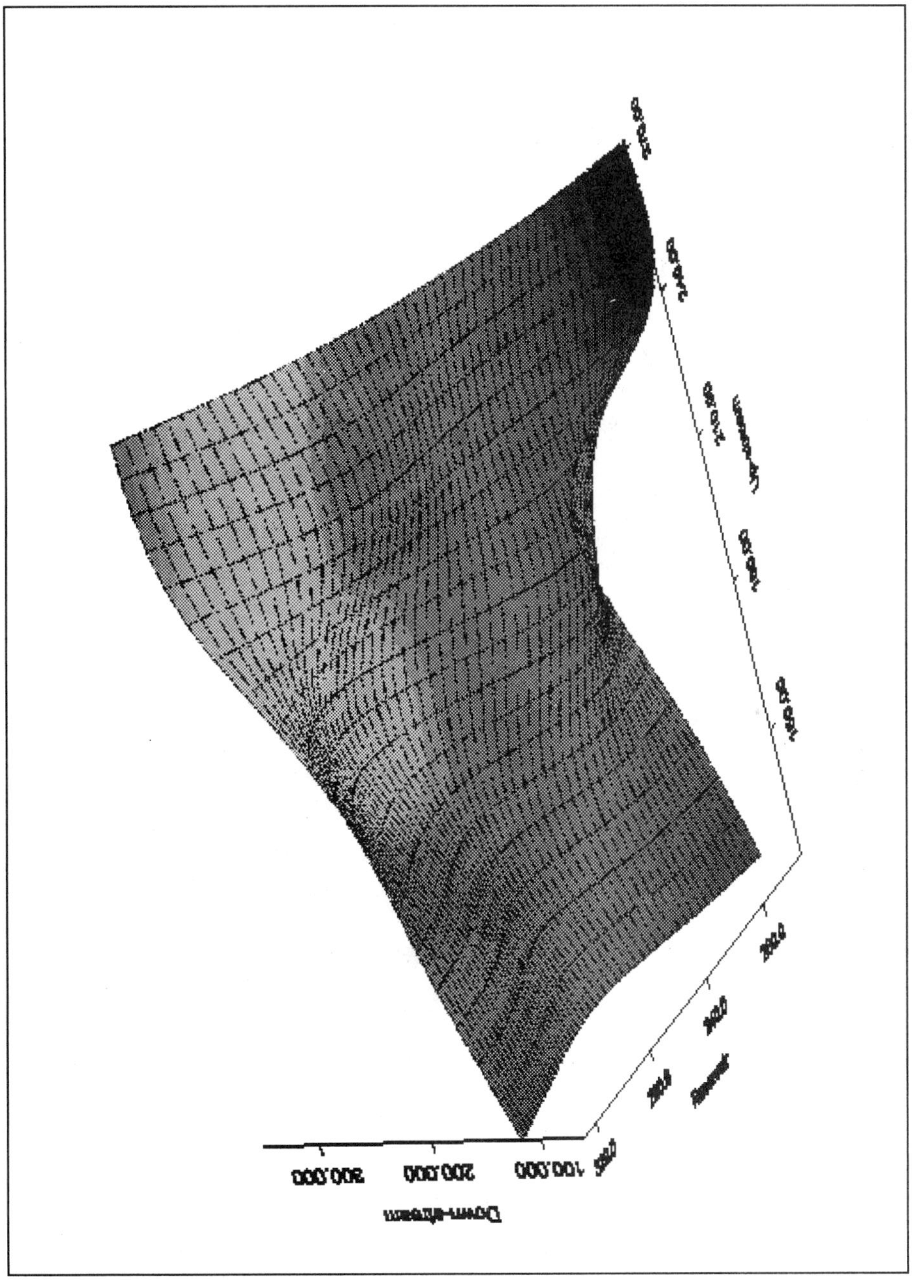

Figure 5.15: Comparative Variation of Zooplankton between Upstream, Reservoir and Downstream during 2006–08

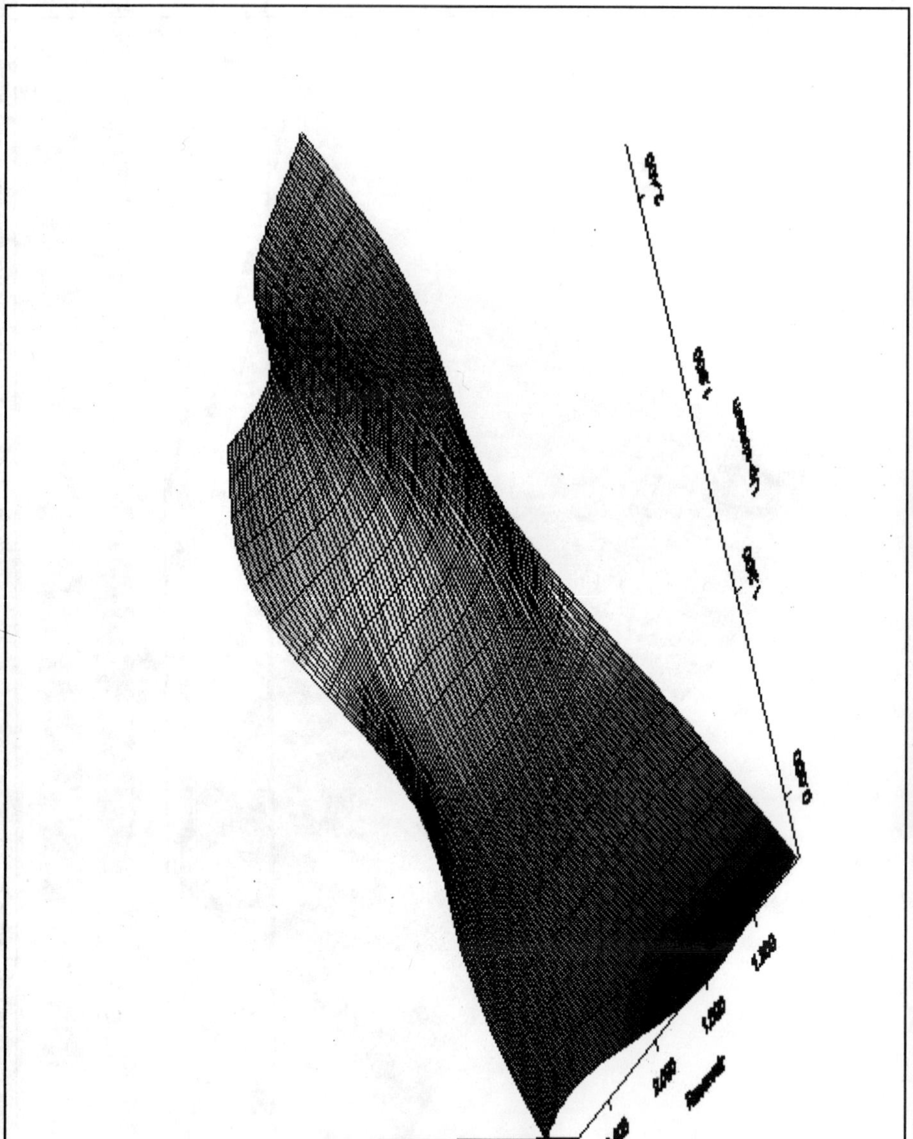

Figure 5.16: Comparative Variation of Biomass between Upstream, Reservoir and Downstream during 2006–08

get dissolved on the other hand due to turbidity photosynthetic activity decreases which results in decrease of dissolved oxygen. The regression equation between DO, temperature and velocity is DO = 12.3–0.111 temp–0.950 velocity, P =0.002.

In our study it is found that the DO and velocity are negatively related, it happened because the more velocity in river was observed during rainy season and rain made water highly turbid. On the other hand reservoir (surface water) has relatively more average DO than the River because of low turbidity and high number of phytoplankton in reservoir water. High level of DO in the surface water of lake was observed by the Randolph *et al.* (1984).

Brierley *et al.* (1987) recorded 6.9 mg/l in the Naivasha Main Lake and 7.1 mg/l in the Oloidien Bay near the Ol Kalia Gorge. Howard-Williams *et al.* (1989), recorded very low DO levels (0.3 to 5.1 mg/l) within a shallow floodplain oxbow lake in Sao Paulo, Brazil.

Higher surface DO level in Banglang reservoir was observed by Ariyadej *et al.* (2004). Minimum DO during winter months have been reported by Sehgal (1980) in Lake Sriunsar Jammu.

The chemical oxygen demand is used as a measure of the oxygen equivalent of the organic matter content of a system that is susceptible to oxidation by a strong chemical oxidant. Chemical oxygen demand is the indicator of industrial pollution. The maximum monthly average value of COD in upstream (sampling site I and II) for both the years was recorded 8.08 mg/l±0.55 in the month of July and minimum monthly average value of COD recorded was 3.21 mg/l±0.46 in the month of February. The maximum monthly average value of COD in reservoir (sampling site III and IV) for both the years was recorded 6.74 mg/l±1.51 in the month of July and minimum monthly average value of COD recorded was 2.79 mg/l±1.17 in the month of January. The maximum monthly average value of COD in downstream (sampling site V) for both the years was recorded 7.62 mg/l±1.06 in the month of July and minimum monthly average value of COD recorded was 3.18 mg/l±0.06 in the month of December. The annual average value of COD in upstream, reservoir and downstream in both the years of study fluctuated between 4.77±1.23 to 5.17±1.18.

The value of DO and COD are negatively correlated. The level of correlation between DO and COD fluctuates as follows in different study sites

Site I > Site II > Site IV > Site V > Site III. The regression equation for between DO and COD for the observations is, DO = 12.8–0.587 COD, (P= 0.00), it also clear that DO and COD are negatively related to one another.

Biochemical oxygen demand is an important parameter to judge the quality of water, it determine the extent of consumption of DO in aquatic system. The maximum monthly average value of BOD in upstream (sampling site I and II) for both the years was recorded 2.81 mg/l±0.21 in the month of July and minimum monthly average value of BOD recorded was 1.28 mg/l±0.40 in the month of January and February. The maximum monthly average value of BOD in reservoir (sampling site III and IV) for both the years was recorded 2.35 mg/l±0.53 in the month of July and minimum monthly average value of BOD recorded was 1.02 mg/l±0.45 in the month of January.

The maximum monthly average value of BOD in downstream (sampling site V) for both the years was recorded 4.84 mg/l±1.73 in the month of July and minimum monthly average value of BOD recorded was 2.26 mg/l±0.98 in the month of December.

A similar pattern for BOD was reported by Khanna (1993) and Prasad (2003). Raina *et al.* (1984) reported peak values during winter in river and Jena *et al.* (2003) during summer at Sagar Island. Khanna and Singh (2000) noticed peak values during summer in Suswa River and Khanna *et al.* (1997) observed peak values in monsoon season in river Ganga. Biochemical oxygen demand for any aquatic system is closely related with the DO and COD.

The value of BOD is negatively correlated with the DO and is positively correlated with the COD. Site I, site IV and site V showed equal level of correlation between BOD and DO and which was highest and than site II followed by site III. In case of BOD and COD correlation values varies as following:

Site I > Site III > Site II > Site IV > Site V. The regression equation obtained for BOD, DO and COD for this study is, DO = 13.5 + 0.0479 COD–2.08 BOD, (P= 0.00).

In case of dam the value of BOD was found relatively lower than the river system. The average value of BOD in upstream, reservoir and downstream for both the years varies from 1.77mg/l±0.41 to 3.52 mg/l±0.73. Fluctuation in the BOD was not very much throughout the year in reservoir and its value was less than the river's sites. Average amount of BOD in downstream found more than that of upstream and reservoir indicates to the decomposition activities in deeper water of reservoir. BOD and DO are negatively related.

Alkalinity of water is its acid-neutralizing capacity. The alkalinity of surface water is primarily due to carbonate, bi-carbonate and hydroxide content. Many studies also reveal that the alkalinity may increase due to decomposition of organic matter that has settled at the bottom. The maximum monthly average value of alkalinity in upstream (sampling site I and II) for both the years was recorded 87.19 mg/l±6.15 in the month of January and minimum monthly average value of alkalinity recorded was 32.64 mg/l±4.34 in the month of August. The maximum monthly average value of alkalinity in reservoir (sampling site III and IV) for both the years was recorded 92.28 mg/l±1.70 in the month of January and minimum monthly average value of alkalinity recorded was 30.83 mg/l±3.80 in the month of August.

The maximum monthly average value of alkalinity in downstream (sampling site V) for both the years was recorded 89.90 mg/l±1.40 in the month of January and minimum monthly average value of alkalinity recorded was 31.95 mg/l±0.75 in the month of August. The average value in upstream, reservoir and downstream for both the years varies from 51.38 mg/l±15.85 to 57.56 mg/l±18.39. The trend obtained in this study for the value of alkalinity in reservoir was similar to the findings of Harding (1992).

The trend of alkalinity observed for river in this study was also observed by Khanna and Chugh (2004), Holden and Green (1960) and Abdin (1948 a, b). In this study the correlation between alkalinity and pH was positive and level of correlation variation in different site was as follow:

Site III > Site V > Site I > Site II > Site IV. The regression equation between alkalinity and pH is, pH = 6.87 + 0.00930 Alkalinity, (P= 0.00). Mishra and Yadav (1978) found similar annual trend for alkalinity during their study of physico-chemical characteristics of river and lake water in central India.

Water hardness refers to the capacity to precipitate soap. Hardness in water is due to presence of chloride and sulphate of calcium and magnesium. In the present study the maximum monthly average value of total hardness in upstream (sampling site I and II) recorded was 64.60 mg/l±1.82 in the month of January and minimum value recorded was 38.58 mg/l±4.55 in the month of August. In the present study the maximum monthly average value of total hardness in reservoir (sampling site III and IV) recorded was 65.61 mg/l±2.59 in the month of January and minimum value recorded was 42.53 mg/l±6.17 in the month of August. In the present study the maximum monthly average value of total hardness in downstream (sampling site V) recorded was 63.81 mg/l±2.31 in the month of January and minimum value recorded was 49.84 mg/l±0.64 in the month of July. The annual average value of total hardness in upstream, reservoir and downstream for both the years of study varies from 52.08 mg/l±8.29 to 56.06 mg/l±3.95. Khanna *et al.* (1993, 2003) and Mishra and Joshi (2002) observed hardness in River Ganga at Haridwar and found more or less similar trends in their study.

Chloride in aquatic system is mainly present in the form of salt of sodium and potassium. In the present study maximum value of chloride in upstream (sampling site I and II) was recorded 3.86 mg/l±0.30 in the month of August and minimum value of chloride recorded was 1.44 mg/l±0.16 in the month of November.

Maximum value of chloride in reservoir (sampling site III and IV) recorded was 2.39 mg/l±0.39 in the month of July and minimum value recorded was 1.18 mg/l±0.14 in the month of November. Maximum value of chloride in downstream (sampling site V) recorded was 3.20 mg/l±0.00 in the month of July and minimum value recorded was 1.02 mg/l±0.03 in the month of November. The annual average value in upstream, reservoir and downstream varies from 1.74±0.38 to 2.36±0.71.

The regression equation hardness and chloride is given below:

Chloride = 4.88–0.0514 Hardness, (P= 0.017)

During this study it has been observed that the value of chloride in summer and monsoon is more than that of winter. The slightly higher value of chloride during summer and monsoon season was shown by Hutchinson (1957) and Sehgal (1980). Gonzalves and Joshi (1946) observed increased concentration of chloride during summer.

Algal density and biomass are the important indicator of their function in the aquatic ecosystem. The light penetration in water directly affects the survival and spatial distribution of aquatic vegetation. Because of unidirectional light supply, individuals which are most successful in capturing light do not have to suffer from the resultant shortage of light beneath. In contrast, pelagic primary producers such as freshwater and marine phytoplankton interact on a large scale. Plankton are subject

Table 5.13: Correlation between Different Parameter at Sampling Site I

	Temp.	*TS*	*TDS*	*TSS*	*Velocity*	*Turbidity*	*DO*	*COD*	*BOD*	*Alkalinity*	*pH*	*Hardness*
TS	0.751											
TDS	0.658	0.969										
TSS	0.761	0.999	0.959									
Velocity	0.751	0.914	0.860	0.917								
Turbidity	0.611	0.955	0.971	0.948	0.336							
DO	–0.565	–0.705	–0.732	–0.697	-0.703	–0.754						
COD	0.563	0.690	0.721	0.681	0.678	0.724	–0.969					
BOD	0.585	0.718	0.741	0.711	0.707	0.759	–0.989	0.985				
Alkalinity	–0.836	–0.680	–0.630	–0.684	–0.650	–0.582	0.564	–0.605	–0.588			
pH	–0.640	–0.534	–0.499	–0.537	–0.478	–0.472	0.351	–0.420	–0.389	0.562		
Hardness	–0.876	–0.827	–0.794	–0.827	–0.808	–0.786	0.738	–0.740	–0.750	0.801	0.562	
Chloride	0.317	0.643	0.724	0.627	0.417	0.721	–0.437	0.505	0.452	–0.484	–0.455	–0.559

Table 5.14: Correlation between Different Parameter at Sampling Site II

	Temp.	*TS*	*TDS*	*TSS*	*Velocity*	*Turbidity*	*DO*	*COD*	*BOD*	*Alkalinity*	*pH*	*Hardness*
TS	0.697											
TDS	0.605	0.966										
TSS	0.712	0.999	0.953									
Velocity	0.258	0.755	0.770	0.739								
Turbidity	0.586	0.947	0.959	0.940	0.665							
DO	–0.610	-0.674	-0.714	–0.664	–0.548	–0.601						
COD	0.529	0.641	0.689	0.630	0.597	0.551	–0.956					
BOD	0.550	0.645	0.688	0.633	0.569	0.558	–0.976	0.965				
Alkalinity	0.705	–0.526	-0.469	–0.534	–0.218	–0.474	0.626	–0.557	–0.539			
pH	–0.794	-0.549	-0.510	–0.557	–0.238	–0.465	0.549	–0.504	–0.501	0.526		
Hardness	–0.800	-0.818	–0.745	-0.827	–0.575	–0.732	0.634	–0.589	–0.565	0.730	0.716	
Chloride	0.391	0.537	0.620	0.514	0.392	0.619	–0.431	0.485	0.418	–0.444	–0.418	–0.439

Table 5.15: Correlation between Different Parameter at Sampling Site III

	Temp.	*TS*	*TDS*	*TSS*	*Velocity*	*Turbidity*	*DO*	*COD*	*BOD*	*Alkalinity*	*pH*	*Hardness*
TS	0.600											
TDS	0.534	0.907										
TSS	0.601	0.999	0.892									
Turbidity	0.611	0.962	0.952	0.954								
DO	–0.729	–0.700	-0.707	-0.693	–0.673							
COD	0.731	0.696	0.712	0.689	0.672	–0.924						
BOD	0.699	0.707	0.721	0.700	0.677	–0.934	0.970					
Alkalinity	–0.867	-0.525	–0.503	–0.523	–0.554	0.762	-0.710	–0.680				
pH	–0.742	-0.352	–0.350	–0.349	–0.382	0.687	-0.531	–0.531	0.752			
Hardness	–0.820	-0.667	-0.622	–0.665	–0.708	0.792	-0.690	–0.697	0.795	0.745		
Chloride	0.464	0.410	0.597	0.391	0.526	–0.575	0.548	0.502	–0.613	–0.542	–0.590	

Table 5.16: Correlation between Different Parameter at Sampling Site IV

	Temp.	*TS*	*TDS*	*TSS*	*Velocity*	*Turbidity*	*DO*	*COD*	*BOD*	*Alkalinity*	*pH*	*Hardness*
TS	0.539											
TDS	0.431	0.898										
TSS	0.544	0.999	0.880									
Turbidity	0.532	0.884	0.856	0.878								
DO	–0.782	–0.394	-0.208	-0.406	-0.367							
COD	0.766	0.363	0.216	0.373	0.366	–0.947						
BOD	0.783	0.383	0.219	0.393	0.381	–0.989	0.963					
Alkalinity	–0.833	-0.441	-0.349	-0.444	–0.477	0.608	-0.598	–0.599				
pH	–0.452	-0.311	-0.318	-0.308	–0.366	0.523	–0.479	–0.523	0.370			
Hardness	–0.760	-0.651	-0.547	–0.654	–0.611	0.584	–0.581	–0.593	0.866	0.457		
Chloride	0.553	0.536	0.503	0.534	0.415	–0.419	0.516	0.432	–0.576	–0.189	–0.624	

Table 5.17: Correlation between Different Parameter at Sampling Site V

	Temp.	*TS*	*TDS*	*TSS*	*Velocity*	*Turbidity*	*DO*	*COD*	*BOD*	*Alkalinity*	*pH*	*Hardness*
TS	0.546											
TDS	0.380	0.919										
TSS	0.560	0.999	0.900									
Velocity	0.126	0.665	0.656	0.659								
Turbidity	0.460	0.940	0.940	0.931	0.713							
DO	–0.259	-0.213	-0.209	-0.212	0.112	–0.259						
COD	0.368	0.288	0.241	0.291	–0.052	0.296	-0.933					
BOD	0.327	0.259	0.221	0.261	–0.101	0.280	-0.986	0.955				
Alkalinity	–0.816	-0.631	-0.522	-0.637	–0.239	0.648	0.344	-0.435	-0.414			
pH	–0.494	-0.410	-0.331	-0.415	–0.013	0.372	0.530	-0.513	-0.556	0.608		
Hardness	–0.769	-0.508	-0.390	–0.517	–0.340	0.561	0.334	-0.402	–0.376	0.793	0.412	
Chloride	0.287	0.148	0.235	0.136	0.085	0.247	-0.376	0.360	0.368	–0.488	–0.375	–0.474

to continuous turbulent mixing (Reynolds 1984, Jones *et al.*, 1996). In a well-mixed water column, individuals are, on average, exposed to the same availability of nutrients and light and are thus affected by the collective consumption of these resources through the entire plankton community (Huisman and Weissing, 1994; Grover, 1997). Turbidity affect the penetration of light in water column, higher turbidity cause penetration of light up to low depth causing low number of plankton (Khanna, 1993). This is one of the reasons for different number of plankton in different season.

In the present study the maximum monthly average number of diatoms 4360.15unit/l±549.98 were found at upstream (sampling site I and II) in the month of February while minimum number 631.68 unit/l ±182.43 was recorded in the month of July.

The monthly average value of diatoms in reservoir (sampling site III and IV) for both the years of study was obtained 4579.88±479.95 in the month of February while minimum value for diatom number was recorded 1209.97±465.89 in the month of July. In this way fluctuation of diatom number in reservoir is more than that of upstream. The maximum monthly average number of diatoms 3119.77 unit/l±835.47 were found at downstream (sampling site V) in the month of February while minimum number 476.41 unit/l ±49.60 recorded in the month of July. The average value fluctuated between 1848.30 unit/l±970.44 and 2851.42 unit/l±1103.64.

The maximum monthly average number of Green algae in upstream (sampling site I and II) was counted 451.57 unit/l±32.65 in the month of February and minimum number recorded 53.65 unit/l ±26.84 in the month of July. The maximum monthly average number of Green algae in reservoir (sampling site III and IV) was counted 448.32 unit/l±48.45 in the month of January and minimum number was recorded 109.13 unit/l ±8.88 in the month of June.

The maximum monthly average number of Green algae in downstream (sampling site V) was obtained 382.71 unit/l±2.12 in the month of February and minimum number was recorded 41.29 unit/l ±19.87 in the month of July. The annual average for both the years of study in upstream, reservoir and downstream fluctuated between 169.46 unit/l ±106.47 to 258.75 unit/l ±128.30.

The maximum monthly average number of blue green algae in upstream (sampling site I and II) was recorded 90.60 unit/l±11.90 in the month of January and minimum number was recorded 8.10 unit/l ±3.68 in the month of August. The maximum monthly average number of blue green algae in reservoir (sampling site III and IV) was counted 112.70 unit/l±13.25 in the month of February and minimum number was recorded 17.80 unit/l ±3.09 in the month of August. The maximum monthly average number of blue green algae in downstream (sampling site V) was counted 77.47 unit/l±14.04 in the month of January and minimum number was recorded 5.89 unit/l ±2.35 in the month of August. The annual averages in upstream, reservoir and downstream, for both the years of study fluctuated from 44.70 unit/l ±26.70 to 54.81 unit/l ±29.98.

In this study out of all the observed species the most abundant species was Diatoms and less abundant species was Blue green algae. Total phytoplankton is the sum of all the plankton observed during the study period. The maximum monthly average number of phytoplankton in upstream (sampling site I and II) was counted 4883.91 unit/l±535.94 in the month of February and minimum number was recorded 693.85 unit/l ±186.34 in the month of July. The maximum monthly average number of phytoplankton in reservoir (sampling site III and IV) was counted 5116.13 unit/l±492.01 in the month of February and minimum number was recorded 1429.74 unit/l ±303.43 in the month of June.

The maximum monthly average number of phytoplankton in downstream (sampling site V) was counted 3560.81 unit/l±840.46 in the month of February and minimum number was recorded 524.58 unit/l ±70.21 in the month of July. The annual average for both the years of study fluctuated from 2061.91unit/l ±1079.68 to 3185.92 unit/l ±1208.70.

Phytoplankton are the primary producers and their number is directly affected by water turbidity. The regression equation for the phytoplankton and turbidity for this study, Turbidity = 116–0.0289 Phytoplankton, (P= 0.01). From this regression equation it is clear that they are negatively related to one another.

Zooplankton are at second trophic level in aquatic food chain, they feed on phytoplankton. The maximum monthly average number of protozoa in upstream (sampling site I and II) was counted 72.50 unit/l±26.06 in the month of May and minimum number was recorded 17.50 unit/l ±9.57 in the month of August. The maximum monthly average number of protozoa in reservoir (sampling site III and IV) was counted 100.00 unit/l±42.29 in the month of December and minimum number was recorded 42.25 unit/l ±3.86 in the month of July. The maximum monthly average number of protozoa in downstream (sampling site V) was counted 75.50 unit/l±23.33 in the month of January and minimum number was recorded 18.50 unit/l ±6.36 in the month of July. The annual average for both the years of study fluctuated from 45.71 unit/l ± 15.38 to 66.54 unit/l ±14.36.

The maximum monthly average number of Rotifera in upstream (sampling site I and II) was counted 131.50 unit/l±41.73 in the month of February and minimum number was recorded 46.25 unit/l ±6.99 in the month of July. The maximum monthly average number of Rotifera in reservoir (sampling site III and IV) was counted 121.00 unit/l±24.10 in the month of October and minimum number was recorded 69.00 unit/l ±23.14 in the month of July. The maximum monthly average number of Rotifera in downstream (sampling site V) was counted 114.00 unit/l±63.64 in the month of January and minimum number was recorded 44.50 unit/l ±2.12 in the month of July. The annual average for both the years of study fluctuated from 74.54 unit/l ± 21.56 to 94.71 unit/l ± 17.42.

The maximum monthly average number of Cladocera in upstream (sampling site I and II) was counted 71.25 unit/l±43.18 in the month of January and minimum number was recorded 30.50 unit/l ±15.89 in the month of November. The maximum monthly average number of Cladocera in reservoir (sampling site III and IV) was counted 83.00 unit/l±25.02 in the month of January and minimum number was recorded 44.25 unit/l ±17.34 in the month of July. The maximum monthly average number of Cladocera in downstream (sampling site V) was counted 98.00 unit/l±15.56 in the month of January and minimum number was recorded 23.50 unit/l ±

3.54 in the month of July. The annual average for both the years of study fluctuated from 45.83 unit/l ±12.37 to 64.13 unit/l ±12.48. Minimum average value recorded in upstream while maximum in reservoir.

The maximum monthly average number of Copepoda in upstream (sampling site I and II) was counted 29.00 unit/l±13.29 in the month of January and minimum number was recorded 13.00 unit/l ±4.97 in the month of June. The maximum monthly average number of Copepoda in reservoir (sampling site III and IV) was counted 46.00 unit/l±20.82 in the month of January and minimum number was recorded 16.25 unit/l ±22.53 in the month of September. The maximum monthly average number of Copepoda in downstream (sampling site V) was counted 46.50 unit/l±2.12 in the month of January and minimum number was recorded 9.00 unit/l ±8.49 in the month of August. The annual average for both the years of study fluctuated from 20.38 unit/l ±5.61 to 32.69 unit/l±11.16. Maximum average for Copepoda observed in reservoir while minimum in upstream..

The maximum monthly average number of total zooplankton in upstream (sampling site I and II) was counted 271.75 unit/l±69.93 in the month of February and minimum number was recorded 125.25 unit/l ±31.51 in the month of July. The maximum monthly average number of zooplankton in reservoir (sampling site III and IV) was counted 329.50 unit/l±81.98 in the month of January and minimum number was recorded 173.50 unit/l ±32.88 in the month of July. The maximum monthly average number of zooplankton in downstream (sampling site V) was counted 334.00 unit/l±69.30 in the month of January and minimum number was recorded 99.50 unit/l ±0.71 in the month of July. The annual average of total zooplankton for both the years of study fluctuated from 193.21 unit/l ±59.01 to 258.06 unit/l ±41.98.

Total plankton is the sum of phytoplankton and zooplankton in this study the highest number of total plankton found was 6006.25 unit/l in the month of February in reservoir.The number of zooplankton was highly affected by the phytoplankton number in this study. The regression equation between zooplankton and phytoplankton is Phytoplankton =–2941 + 25.4 Zooplankton, (P= 0.00).

In the present study the maximum number of plankton was observed during winter season. Dobriyal and Singh (1988), Khanna (1993) and Khanna *et al.* (2000) observed high value of plankton during January- March. Khanna and Singh (2000) also reported high values of plankton during winter season. The study showed dominance of phytoplankton was also reported by Pahwa and Mehrotra (1966) in the Ganga River.

The maximum number of plankton was observed in winter similar result was also observed by Allen (1920), Ray (1955), Prasad (1956), Byars (1960). Venkateswarlu and Menon (1979) recorded maximum values of total plankton during winter and minimum during rainy season. Pahwa and Mehrotra (1966) observed maximum planktonic population in River Ganga both in summer and winter. Eddy (1934) and Chandler (1940) pointed out that the plankton production is mainly influenced by temperature. Poulickova *et al.* (2002) observed that diatom dominate the phytoplankton population in the reservoir which is similar to our findings. The distribution of plankton observed is related to the changing physico-chemical parameters such as temperature and dissolved oxygen.

The relationship of plankton with physico-chemical factors indicates that the peaks of plankton are not governed by any single factor. No exact relationship can be deduced between chemical conditions and plankton population at any specific time (Byras, 1960). Davis (1954) stated that a number of physico-chemical and biological environmental situation acting simultaneously must be taken into consideration in understanding the fluctuation of plankton populations.

The maximum monthly average amount of planktonic biomass in upstream (sampling site I and II) was counted 2.5818 mg/l ±0.2566 in the month of February and minimum amount was recorded 0.4233mg/l ±0.1214 in the month of July. The maximum monthly average amount of biomass in reservoir (sampling site III and IV) was recorded 2.6667 mg/l±0.2499 in the month of February and minimum amount was recorded 0.8343 mg/l ±0.1384 in the month of June. The maximum monthly average amount of biomass in downstream (sampling site V) was recorded 1.8729 mg/l±0.4432 in the month of February and minimum amount was recorded 0.3045 mg/l ±0.0286 in the month of July. The annual average for both the years of study fluctuated from 1.1262 mg/l ±0.5706 to 1.7105 mg/l ±0.6160. In this study maximum amount of planktonic biomass 2.9780 mg/l was observed at sampling site IV (reservoir) while minimum amount of it was obtained 0.2843mg/l at sampling site V (dam outlet). This trend reflects the change in plankton viability in reservoir.

Minimum amount of planktonic biomass in downstream reflects decreased plankton level at the deeper layers of water in reservoir. On the other hand more biomass in winter in reservoir is due to increased plankton population in reservoir Allen (1920), Ray (1955), Prasad (1956), Byars (1960) reported this trend of plankton population in different aquatic ecosystems.

During this study it was observed that average planktonic biomass of reservoir is more than that of river. Bays and Crisman (1983) found an increase in the relative biomass of micro-zooplankton in Florida Lake than that of river. Watson, (1992) reported increased phytoplankton biomass in dam reservoir than that of river in Oldman reservoir system.

Most of the world's large rivers are fragmented by dams. Fragmentation of the river ecosystem alters migration patterns among fish populations and converts free-flowing rivers to reservoir habitat (Jager *et al.*, 2001). Inundation of spawning grounds in the reservoir, irregular release of water from the dam and periodic inundation and drying out of spawning grounds affect the fishes directly. Fishes are affected indirectly due to modification in velocity, temperature and quality of water. Habitat change caused by dam construction modified the fish community, population density and are utilized by a particular species (Horvath and Munico, 1998).

During the course of study we found 21 species of fishes from upstream of Bhagirathi River (sampling site I and II). 6 species were observed in reservoir (sampling site III and IV) and only 11 species of fishes were observed in downstream (sampling site V). Bottom feeders were not observed in reservoir during the course of studies. This may be due to lack of suitable substrata (bottom) in the reservoir. Hill stream fishes of Bhagirathi River are not observed in reservoir and downstream during the course of study. *Cyprinus caprio* var *communis* and *Cyprinus caprio* var *specularis* are the two exotic fishes which were observed in reservoir in abundance due to their ability to survive in wide range varying aquatic environmental factors. Large dams dramatically reduce upstream abundances of migratory fishes in high-gradient systems. Second, migratory fish losses alter benthic resources (Great house *et al.*, 2006). Problem in the migration of migratory fishes due to construction of dam was also studied by Hora (1940, 1942), Linlokken (1993) in Glomma river system.). The relationship between fish life and suspended solids was the first to be considered by the Albaster (1972), Albaster and Lloyd (1980). Dam reduces the amount of sediments deposited downstream, creating an adverse effect on downstream aquatic species and their habitats (Newcomb and MacDonald, 1991).

Reduced DO in downstream of Tehri dam indicates the low oxygen content in the deeper layers of water in reservoir. The DO conditions in lakes and reservoirs have an important implication on aquatic life forms including fish. Generally, fish movements are constrained by lower oxygen content because fish prefer well-oxygenated water (Goldman and Horne 1983).

Chapter 6
General Summary

This thesis embodies the findings of study conducted under title "The impact of Tehri Dam on aquatic ecology of Bhagirathi River at Tehri Garhwal" during July 2006 to June 2008 and consists of 6 chapters. The text is supported by 63 tables, 111 figures, 3 maps and 5 photo-plates. In this study several aspects of the ecological factors which affect the water system through direct or indirect ways, are analyzed and evaluated in various aspects.

For this study, samples were collected from five sampling sites.

1. Sampling site I (Dharanshu)
2. Sampling site II (Chinyalisaur)
3. Sampling site III (Bhaldiyana)
4. Sampling site IV (Koti)
5. Sampling site V (Dam outlet near Chopra village)

The basic idea in selection of these five sites was based upon the ecological and hydrological variation in the water stream of River Bhagirathi due to construction of dam. Sampling site I and II were upstream of Bhagirathi River, sampling site III and IV were reservoir sites while sampling site V was downstream of Bhagirathi. The river bed in upstream and downstream sampling stations had rocks, sand and gravel covered by sand while sampling sites in reservoir region were dependent upon level of water in reservoir. In late winter and early summer water level in reservoir is lowered so we have to go down for sampling. Sampling site V was near most approachable site for sampling downstream to dam outlet. Sampling at this site was done near Chopra village (o-point bridge). Amount and velocity of water at this site was quite unexpected because it depends upon water release from reservoir.

The results which were observed during this study are summarized as below:

- ☆ A difference in the fluctuation of water temperature was calculated to be 6.49 °C±0.29 (minimum) in the month of January and 30.00 °C ±0.32 (maximum) in the month of July. The fluctuation of water temperature in reservoir region is minimum 9.20 °C±0.25 in the month of January and maximum 30.00 °C ±0.32 in the month of July. While that of upstream is 6.49 °C ±0.29 and 21.20 °C ±0.68 respectively and fluctuation in down stream is 8.50 °C ±0.29 and 20.10 °C ±0.22 respectively. This huge difference in the fluctuation in water temperature affects the population of different aquatic biota such as plankton, fish and other organisms. In this study temperature showed an upward trend from February to July (slight downward from May to July in upstream due to melting of glaciers) followed by downward trend from August to January.
- ☆ Turbidity is one of the most important variables pertaining to the optical characteristics of the river. In the month of winter and early summer some sampling sites recorded nil turbidity. Turbidity of reservoir remained almost constant while there was huge fluctuation in the turbidity of river during monsoon, winter and summer. Highest (362.70 JTU±1.55) was noted in the month of August.
- ☆ The velocity regulate the transport includes advection and dispersion of the water quality constituents. The maximum velocity in river was (2.30 m/s±0.12) in the month of August while minimum 0.14 m/s ±0.02 at sampling site II in the month of October (due to reaching of reservoir water level beyond this site). Velocity in reservoir noted nil throughout the year.
- ☆ In the present investigation it was noted that the total solids were maximum in the month of August (3069.91 mg/l±2.43), which were responsible for the turbidity in the river. The total solids were recorded minimum (67.00 mg/l±1.25) in the month of March. Due to gradual sedimentation of the filterable residue the average total solids in the reservoir are lesser than the upstream.
- ☆ During the course of study total dissolved solids (TDS) found minimum 46.28 mg/l±1.11 in the month of February and that of maximum 429.97 mg/l±1.15 in the month of August. Total suspended solids (TSS) found minimum 16.37mg/l±1.52 in the month of April and that of maximum 2641.24 mg/l±2.07 in the month of August.
- ☆ Dissolved oxygen (DO) is one of the most important chemical factors having a widespread effect upon the flora and fauna of any aquatic ecosystem. It also tells us about the phytoplanktonic status of the water. In our study we found DO maximum 12.70 mg/l±0.22 in the month of January and minimum 6.90 mg/l±0.21 in the month of August. The average DO in the surface water of reservoir is more than that of upstream and downstream.
- ☆ One of the well known water quality parameters BOD is very much responsible for the consumption of the DO in an aquatic ecosystem and for the degradation of water quality. The biochemical oxygen demand was

estimated maximum 3.45 mg/l±0.06 in the month of August and minimum 0.46 mg/l±0.12 in the month of January.

- ✰ The chemical oxygen demand is used as a measure of the oxygen equivalent of the organic matter content of a system that is susceptible to oxidation by a strong chemical oxidant. It was noted highest 9.20 mg/l±0.16 and lowest 1.37 mg/l±0.07 in the months of August and January respectively.
- ✰ Alkalinity of water is its acid neutralizing capacity. The total alkalinity was maximum 95.20 mg/l±0.16 in the month of January and minimum 27.20 mg/l±0.48 in the month of September In the period of lower rate of precipitation accompanied with warmer and humid climatic condition, the degradation of organic matter is enhanced.
- ✰ Total hardness was understood to be a measure of the capacity of water to precipitate the soap. Maximum values for total hardness were recorded 68.50 mg/l±1.49 in the month of January and minimum was recorded 31.20 mg/l±1.13 in the month of August.
- ✰ Chloride, in the form of chloride ion (Cl^-) is one of the major inorganic anions in water system. Micro as well as higher organisms present in the water system for nutrition uses this anion in the trace. The maximum values for chloride was found in the month of August (4.22 mg/l±0.21) and the minimum in the month of January (0.90mg/l±0.03) respectively.
- ✰ During the course of study the highest value for the total phytopklankton were 5670.25 unit/l found in the month of February and zooplankton were observed 435.00 unit/l in the month of January. The minimum concentration of total phytoplankton was recorded 471.83 unit/l in the month of July and zooplankton were observed 83.00 unit/l in the month of August. Average of plankton in reservoir was more than that of river. In the same way average of plankton in downstream was lesser than upstream. During the course of study the maximum value of planktonic biomass observed 2.9780 mg/l in the month of February. The minimum amount of planktonic biomass observed 0.2843 mg/l in the month of July. Average for planktonic biomass in reservoir was found more than river.
- ✰ The fish fauna of the River Bhagirathi from Dharanshu to Dam outlet near Chopra village was surveyed. As a result of this survey 21 species belonging to 4 families have been reported.

Conclusion

From the present study it may be concluded that:

1. There were minor differences in physico-chemical and biological parameters of Upstream (sampling site I and II) reservoir (sampling site III and IV) and downstream (sampling site V) recorded during the course of study.
2. Reduced fish fauna in reservoir, particularly bottom feeders due to depth and snow trouts due to increased average temperature of surface water of reservoir clearly indicates the impact of dam.

3. Migratory pathway of fishes especially Mahaseer and *Glyptothorax* etc. have been disturbed due to the construction of dam.
4. High plankton number in reservoir indicates its high potential for fish culture. Finally it has to be suggested that exotic fishes should be cultured in reservoir.
5. Beside this study, the regular monitoring of River Bhagirathi and Tehri Reservoir is very important to assess the impact of dam. Thus it is apparent that much attention should be paid on further studies of physico-chemical and biological parameters of River Bhagirathi at Tehri Garhwal.

References

Abdin, G., 1948a. Physical and chemical investigations relating to algal growth in river Nile. *Cairo. Bull. Inst. Egypt*, 29: 20-24.

Abdin, G., 1948b. Seasonal distribution of Phytoplankton and sessile algae in the river Nile. *Cairo. Bull. Inst. Egypt*, 29: 369-382.

Abraham, J.C., 1962. A survey of hydrobiology and fisheries of the Cooum River, Madras. *J. Fish*, 1: 50-59.

Adam, W., 2000. Downstream impacts of large dams, Thematic review I.1. World Commission on Dams. pp: 1-22.

Adams, W.M., 1985. 'The downstream impacts of dam construction: a case study from Nigería', *Transactions of the Institute of British Geographers N.S.*, 10: 292-302

Adams, W.M., 2000. *Social Impacts of Large Dams: equity and distributional issues*, Report to World Commission on Dams, Thematic Review II Social Impacts of Large Dams; Equity and Distributional Issues.

Aggarwal, D.K., Gaur, S.D., Tiwari, I.C., Narayanswami, M.S., and Marwali, S.M., 1976. Physico-chemical characters of the Ganga at Varanasi., *Indian J. Environ. Health*, 18: 201-206.

Akbay, N., Anul, N., Yerti, S., Soyupak, S. and Yurteri, C., 1999. Seasonal distribution of large phytoplankton in Keban Dam reservoir. *Plank. Res.*, 21(4): 771-787.

Alabaster, J.S. and Lloyd, R., 1980. Water quality criteria for fresh water fisheries. London, Butterworths.

Alabaster, J.S., 1972. Suspended solids and fisheries. *Proc. Royal Soc.* London,180, 395-406.

Allen, W.E., 1920. A quantitative and stastical study of the plankton on the San-Joaquin River and its tributaries in a near Stockton, California in 1913. *Univ. Calif. Pub. Zool.*, 22: 1-292.

APHA, AWWA, WPCF, 1998. Standard methods for the examination of water and wastewater, 20th ed., Washington D.C., New York.

Ariyadej, C., Tansakul, R., Tansakul, P. and Angsupanich, S., 2004. Phytoplankton diversity and its relationships to the physico-chemical environment in the Banglang Reservoir, Yala Province. *Songklanakarin J. Sci. Technol.*, 26(5): 595-607.

Arjaria, Amita, 2003. Physico chemical profile and plankton diversity of Ranital lake, Chhatarpur, M.P., *J. Nat. Env. and Poll. Tech.*, 2(3): 327-328.

Arunachalam, M., O. Divakaran, N. Balakrishnan Nair and Balasubramanian, N.K., 1982. Seasonal variation of zooplankton of the Veli Lake, South west coast of India. *Arch. Hydrobiol.*, 93: 359–374.

Atkin, W.R.G. and Harris, G.T., 1924. Seasonal changes in the water and heleoplankton of fresh water ponds., *Proc. Roly. Dub. Soc.*, VIII (N.S.): 1-21.

Atobatele, Oluwatosin Ebenezer and Ugwumba, O. Alex, 2008. Seasonal variation in the physicochemistry of a small tropical reservoir (Aiba reservoir, Iwo, Osun, Nigeria). *African Journal of Biotechnology*, 7(12): 1962-1971.

Badola, S.P. and Badola, Smita, 1999. Observations on destructive fishing methods and recommendations to protect fish fauna in Garhwal Himalaya. *Sus. Eco. Sys. Env. and Ed.* By D.R. Khanna, A. Gautam and A. Gautam. pp: 99-104.

Badola, S.P. and Pant, M.C., 1973. Fish fauna of Garhwal hills part I., *Ind J. zool* 14(1) 37-44.

Badola, S.P. and Singh, H.R., 1977a. Fish fauna of Garhwal hills part III (Chamoli district)., *Ind J. zool.*, 18(2): 119-122.

Badola, S.P. and Singh, H.R., 1977b. Fish fauna of Garhwal hills part IV (Tehri district)., *Ind J. zool.*, 18(2): 115-118.

Badola, S.P. and Singh, H.R., 1981. Fish and fisheries of River Alaknanda. Proc. Nat. Acad. Sci., India. B 51(2): 133-142.

Badola, S.P. and Singh, H.R., 1981. Hydrobiology of River Alaknanda at Garhwal Himalayas. *Indian J. Ecol.*, 8(2): 269-276.

Badola, S.P., 1975. Fish fauna of Garhwal hills part II., *Ind J. zool* 16(1) 57-70.

Badola, S.P., 1979. Limnological studies of Ichthyofauna of some fresh water resources of Garhwal region. D. Phil. Thesis, Garhwal University.

Bagnold, R.A., 1966. *An Approach to the Sediment Transport Problem from General Physics*, Geological Survey Professional Paper 422-I, GPO, Washington, DC.

Bahuguna, P.K., Joshi, H.K. and Dobriyal, A.K., 2007. Fecundity and sex ratio in *Puntius conchonius* (Pisces Cyprinidae) from Garhwal Himalayas, *Env. Cons. J.* vol. 8(1 and 2) 37-43.

Baig, N.A. and Khan, M.Y., 1976. Biological and chemical conditions of Manchar lake. (Distt:Dadu). *Pak. J. Zool.*, 5: 165-168.

Baloni, M.C. and Sarkar, H.L., 1965. Some of the pollution of Yamuna River at Okhla water works intake Delhi. *Indian J. Environ. Health*, 7: 84-86.

Banatwala, R.C., Dongre, S.D., Mishra, P.K. and Verma, R.G., 2004. Limnological study of Sapna reservoir of Betul M.P. *Env. Cons. J.*, 5(1-3): 71-75.

Baqai, I.U. and Ishrat, R., 1973. Seasonal fluctuation of fresh water copepods of Keenjhar lake, Sindh and its correlation with Physico-chemical factors. *Pak. J. Zool.*, 5: 165-168.

Basu, M. and Mitra, A.K., 2002. Studies on pollution of River Damodar at Burdwan with special reference to metals, COD and pH. *J. Nat. Env. and Poll. Tech.*, 1(4): 397-399.

Bays, J. S., and Crisman, T.L., 1983. Zooplankton and trophic state relationships in Florida lakes. *Can. J. Fish. Aquat. Sci.*, 40: 1813-1 8 19.

Berner, L.M., 1951. Linmnology at the lower Missourie River. *Ecology*; 32: 1-12.

Bhatt, S.D., Bisht, Y. and Negi, U., 1985. Ecology of the limnoflora in River Kosi of Kumaon, Himalaya. *UP Proc. Nat. Sci.*, 50(4): 395-405.

Bhowmick, B.N. and Singh, A.K., 1985. Phytoplankton population in relation to physico-chemical factors of River Ganga at Patna. *Indian J. Ecol.*, 12(2): 360-364.

Birge, E.A. and Judey, C., 1911. The Inland lake of Wisconis, the dissolved gases and their biological significance. *Bull. Wis. Great. Nat. Hist. Surv*, 22-259.

Biswas, A. and Tortajada, C., 2001. Development and Large Dams: A global perspective. *International Journal of Water Resources Development*, 9-21.

Biswas, A.K., 1970. History of hydrology, North- Holland, Amsterdam.

Biswas, K., 1928. A comparative study of the Chilka Lake and the Calcutta Salt lakes. Proc. Indian Sci. Congr. 15 (Pt. III): 227.

Brierley, B., Harper, D. and Thomas, R., 1987. Water Chemistry and phytoplankton studies at Lake Naivasha–Short term spatial and temporal variations. University of Leicester studies on the Lake Naivasha ecosystem. Final Report to the Government of Kenya.

Byars, J.A., 1960. A freshwater pond in New Zealand. *Aust. J. Mar. and Freshwater Res.*, 2: 222-240.

Chacko, P.I. and Srinivasan, R., 1955. Observations on the Hydrobiology of the major rivers of Madras state, South India., *Fresh Biol. Stn.* Madras, 13:1-14.

Chacko, P.I., Abraham, J.G. and Andal, R., 1953. Report on a survey of flora, fauna and fisheries of Pulicat lake, Madras state, India, 1951--52. Contrib. Fish. Biol. Station, Madras 8: 1-20.

Chakarbarty, R.D., Ray, P. and Singh, S.B., 1959. A quantitative study of plankton and physiological condition of River Jamuna at Allahabad in 1954-1955. *Indian J. Fish*, 6(1): 186-203.

Chakrapani, C.J. and Veizer, Jan, 2005. Dissolved inorganic carbon isotopic compositions in the Upstream Ganga River in the Himalayas. *Current Science*, 89(3): 553-556.

Chandler, D.S., 1940. Limnological studies of Western Eric I: Plankton and certain physico-chemical data of Basis Island region from September 1938 to November 1939. *Ohio., J. Sc.*, 40: 291-336.

Chatterjee, Ranjan, Pinaki and Raziuddin, M., 2003. A case study of evolution of water quality of lakes by pollution index method., *J. Nat,. Env. and Poll. Tech.*, 2(2): 171-172.

Chattopadhaya, S.N., Routh, T. Sharma, V.P., Arora, H.C., and Gupta, R.K., 1984. A short term study on the pollution status of River Ganga in Kanpur Region. *Indian J. Environmental Hlth.*, 26(3): 244-257.

Chopra, A.K. and Patrik, J.N., Nirmal, J., 1994. Effect of domestic sewage on self-purification of Ganga water at Rishikesh. *Ad. Bids.*, 13(1): 75-82.

Chopra, A.K., and Rahman, A., 1995. A study on self purification of physico-chemical properties of Ganga canal water at Jwalapur, Haridwar. *Him. Env. Zoo.*, 9: 11-13.

Clarke, F.W., 1924. The data geochemistery. *Sthed-Bull. U.S. Goel. Surv.* pp: 770.

Cordone, A.J. and Kelley, D.W., 1961. The influence of organic sediment on the aquatic life of stream. *Calif. Fish. Game*, 47:189-228.

Crayton, W.M. and Sommerfield, M.R., 1979. Composition and abundance of phytoplankton in tributaries in the lower Colorado River, Canyon region. *Hydrobiologia*, 66(1): 81-93.

Cristofano, E.A., 1965. *Method of Computing Erosion Rate for Failure of Earthfill Dams*, U.S. Bureau of Reclamation, Denver, Colorado.

Das, S.M. and Srivastava, V.K., 1956. Studies on freshwater plankton. I. Plankton of a fish tank in Lucknow, India. *Proc. Nat. Acad. Sci. India*, 26B: 85–92.

Das, S.M., 1989. Hand book of limnology and water pollution with practical methodology: South Asian Publishers, New Delhi.

Davis, C.C., 1954. A preliminary study of the plankton of the Cleveland harbour area, Ohi. II. The distribution and quantity of the phytoplankton. *Ecol. Monogr.*, 24: 21-47.

Dayananda, G.Y., Naveed, A., Naik, K.L. and Hosetti, B.B., 2007. Limnological studies of Gudavi Wetlands Sorab Shimoga Karnataka. *Env. Cons. J.* 8(3): 47-54.

De Silva, S.S., 1992. Reservoir Fisheries of Asia. Proceedings of the 2nd Asian Reservoir Fisheries Workshop held in Hangzhou, Peoples Republic of China, 15-19 October, 1990. IDRC, Ontario.

Deming, H.G., 1975. *Water the fundamental opportunity*. Oxford University Press, New York. pp: 942.

Dobriyal, A.K. and Singh, H.R., 1988. Ecological basis for icthyofaunal variation in two hill streams of Garhwal Himalaya. *Proc. First Indian Fisheries Forum*, Mangalore, 1987.

Dobriyal, A.K., 1985. Ecology of Limnofauna in the small streams and their importance to the village life in Garhwal Himalaya. Uttar Pradesh. *J. Zoo.*, 5(2): 139-144.

Dutta, S.P.S., Sharma, V. and Kour, S., 2007. Ecology of plankton in some paddy fields near Nagri bus stand Kathua (J&K state). *Env. Cons. J.*, 8(1&2): 45-67.

Dwivedi, H.N., Pandey, B.N., Manoj, Kumar and Alam, Naushad, 1995. Limnological studies in relation to fish production of two ponds of Patna. *Him. J. Env. Zool.*, 9: 7-10.

Dynesius, M. and Nilsson, C., 1994. Fragmentation and Flow Regulation of River Systems in the Northern Third of the World. *Science*, 266: 753–762.

Eddy, S., 1934. A study of the fresh water plankton communities Illinois. *Biol. Monogr.*, 12(4): 1-93.

Francis, A., Sikoki, F.D. and Ansa, E.J., 2007. Physico-chemical parameters of the Andoni River system- Niger delta, Nigeria. *Journal of Fisheries International*, 2(1): 27-31.

Galtsoff, P.S., 1924. Limnological observation in the Upper Mississippi, 1921. *Fish Bull. V. S.*, 39: 347-438.

Garner, J.H., Brown, F.W. and Lovett, M., 1936. Chemical and biological survey of the River Holme., *Rep. W. riding Riv.Bd.*

Gaudet, J.J. and Melack, J.M., 1981. Major ion chemistry in a tropical African lake basin. *Freshwater Biol.*, 11: 309- 333.

Gautam, A., Khanna, D.R. and Sarkar, P., 2000. Diurnal variation in the physico-chemical characteristics of Ganga water at Triveni Ghat, Rishikesh during winter season. *Indian J. Environ. and Ecoplan.*, 3(2): 369-371.

Gautam, Ashutosh., Sarkar, Praveen and Khanna, D.R., 2001. Quality deterioration in aquatic environment of the Ganges: A growing problem for future. *Ind. J. Env. Sc.*, 5(2): 113-117.

George, C.J., 1973. The role of the Aswan High Dam in changing the fisheries of the southeastern Mediterranean', pp. 159-78 in M.T. Farvar and J.P. Milton (eds) *The Careless Technology: ecologyand international development*, Stacey, London.

Gnilka, A., 1975. Some chemical and physical aspects of Center Hill Reservoir, Tennessee. *J. Tenn. Acad. Sci.*, 50(1): 7-11.

Goldman, C.R. and Horne, A.J., 1983. *Limnology*. McGraw-Hill, Tokyo.

Gonzalves, E.A. and Joshi, D.B., 1946. Fresh water algae near Bombay I. the seasonal succession of the algae in a tank at Bańdra. J. Bombay Nat. Hist. Soc., 46: 154-176.

Gopal, B. and Zutshi, D.P., 1998. Past, present and future of limnology in India. In B. Gopal and R.G. Wetzel (eds), Limnology in Developing Countries, vol. II. International Association for Limnology.

Greathouse, Effie A., Pringle, Catherine M., McDowell, William H. and Holmquist, Jeff G., 2006. Indirect upstream effects of Dams: Consequences of migratory consumer extirpation in Puerto Rico. *Ecological Applications*, 16(1): 339-352.

Greenfield, R.E., 1925. Comparison of chemical and bacteriological examination made on the Illinois River during a season of low and season of high water 1923-1924 Illinois State. *Wat. Surv. Bull.*, 10: 9-33.

Griffith, R.S., 1955. Analysis of plankton yield in relation to certain physical and chemical factors of Lake Michigan. *Ecology*, 36(4): 543-552.

Grover, J.P., 1997. *Resource competition*. Chapman and Hall.

Haldar, G.C., Mazid, M.A. and Ahmed, K.K., 1992. Limnology and primary production of Kaptai Lake, Bangladesh. In: de Silva, S.S., 1992. Reservoir Fisheries of Asia. Proceedings of the 2nd Asian Reservoir Fisheries Workshop held in Hangzhou, Peoples Republic of China, 15-19 October, 1990. IDRC, Ontario.

Hannan, H.H., Fuchs, I.R. and Whitenberg, D.C., 1979. Spatial and temporal patterns of temperature, alkalinity, dissolved oxygen and conductivity in an oligo-mesotrophic, deep-storage reservoir in central Taxas. *Hydrobiologia*, 66(3): 209-221.

Harding, Jon S., 1992. Physico-chemical parameters and invertebrate faunas of three lake inflows and outlet in Westland, New Zealand. *New Zealand Journal of Marine and Freshwater Research*, 26: 95-102.

Holden, S.M. and Green, J., 1960. Hydrology and plankton at River Sokoto. *J. Anim. Ecol.*, 29(1): 65-84.

Hora, S.L., 1921. Fish and fisheries of Manipur with some observations on those of the Naga hills. Rec. Indian Museum 27.

Hora, S.L., 1940. Dams and the problem of migratory fishes. *Curr. Sci.*, 9: 406-407.

Hora, S.L., 1942. The effect of dams on the migration of the Hilsa fish in Indian waters. *Curr. Sci.*, 11: 470-471.

Horvath, E. and Munico, M.A.T., 1998. Impacts of dams on fish fauna: Feasibility of mitigation measures. 2nd Int. PhD Symposium in Civil Engineering.

Howard-Williams, C., Esteves, F., Santos, J.E. and Downes, M.T., 1989. Short term nitrogen dynamics in a small Brazilian wetland (Lago Infernao, Sao Paulo). *Journal of Tropical Ecology*, 5: 323-335.

Huisman, J. and Weissing, F.J., 1994. Light-limited growth and competition for light in well-mixed aquatic environments: an elementary model. *Ecology*, 75: 507-520.

Hutchinson, G.E., 1937. Ecological observations on the fishes of Kashmir and Indian Tibet. *Ecol. Monogr.*, 9: 146–182.

Hutchinson, G.E., 1957. *A treaties on limnology*. Vol. 1 part 2, Chemistry of lakes. John Willey and Sons, New York.

ICOLD, 2000. Dams and the Environment International Commission on Large Dams, Boulevard Haussmann, Paris, France. pp: 1-7.

Idowu, Rachel Toyosi, Inyang, Nicholas Matthias and Eyo, Joseph Effiong, 2004. The physico-chemical parameters of an African arid zone man made lake. *Animal Research International*, 1(2): 113-119.

Imevbore, A.M.A., 1967. Hydrology and plankton of Eleyele reservoir. *Hydrobiologia*, 30: 134-176.

IRN, 2002. Tehri dam. IRN fact sheet. pp: 1-4

Jadhav, S.S., Surve, P.R., Biradar, R.G. and Ambore, N.E., 2006. Seasonal variation of physico-chemical characteristics of Sonkhed Dam water. *Aquacult.*, 7: 117-122.

Jager, Henriette I., Chandler, James A., Lepla, Kenneth B. and Winkle, Webb Van, 2001. A Theoretical study of river fragmentation by dams and its effects on white sturgeon populations. *Environmental Biology of Fishes*, 60: 347-361.

Jain, Renu and Sharma, Dushyant, 2000. Water quality of Rampur Reservoir of Guna district Madhya Pradesh, India. *Env. Conserv. J.*, 1 (2 and 3): 99-102.

Jakher, G.R., Bhargava, S.C. and Sinha, R.K., 1990. Comparative limnology of Sambhar and Didwana lakes (Rajasthan), India. *Hydrobiologia*, 197: 245-256.

Jana, B.B., Dey, U.K. and Das, R.N., 1980. Environmental factors affecting the seasonal changes of net plankton in two tropical fish ponds in India. *Schweiz. Z. Hydrol.*, 42: 225-246.

Jekatierynczuk-Rudczyk, E., Gorniak, A., Zielinski, P. and Dziemian, J., 2002. Daily dynamics of water chemistry in a lowland polyhumic dam reservoir. *Polish Journal of Environmental Studies*, 11(5): 521-526.

Jena, B., Sudarshan, R. and Choudhary, S.B., 2003. Studies on water quality parameters around sagar island, Sunderbans. *J. Nat. Env. and poll Tech.*, 2(3): 329-332.

Jhajharia, Anita, 2002. Limnobiotic studies of Sekhawat Pond, Jodhpur with special reference to plankton. *J. Nat. Con.*, 14(2): 215-224.

Jhingran, C.G., 1975. *Fish and Fisheries of India.* Hindustan Publishing Corporation (India). pp: 1-954.

Jones, R.I., Young, J.M., Hartley, A.M. and Bailey-Watts, A.E., 1996. Light limitation of phytoplankton development in a oligotrophic lake-Loch Ness, Scotland. *Freshwater Bio.*, 35: 533-543.

Joshi, C.B., 1996. Hydrobiological profile of River Sutlej in its middle stretch in western Himalaya. *U.P. J. Zool.*, 16(2): 9-103.

Juettner, I., Sharma, S., Dahal, B.M., Ormerod, S.J., Chimonides, P.J. and Cox, E.J., 2003. Diatoms as indicators of stream quality in the Kathmandu Valley and middle hills of Nepal and India. *Freshwater Biol.*, 48: 2065-2084.

Kala, Rakhi and Sharma, R.C., 2002. Effect of physico chemical factors on phytoplankton population in lotic environment of Alaknanda River, Garhwal Himalaya. *Indian J. Ecol.*, 29(2): 221-226.

Kant, S. and Anand, V.K., 1978. Inter relationship of phytoplankton and physical factors in Mansar Lake, Jammu, (J&K). *Indian J. Ecol.*, 5(2): 134-140.

Keshre, Vivek, Mudgal, L.K., Khanna, D.R., Matta, Gagan and Kumar, Dheeraj, 2007. Study of physico-chemical parameters for a reservoir at Khandwa district (M.P.). *Env. Cons. J.*, 8(3): 127-132.

Khan, M.A., Fagbemi, T. and Ejike, C., 1983. Diurnal variations of physicochemical factors and planktonic organisms in Jos plateau (West Africa) Water reservoir. *Jap. J.Limnol.*, 44: 65-71.

Khanna, D.R. and Bhutiani, R., 2003. Limnological Characteristics of the River Ganga at Haridwar (Uttaranchal). *U.P. J. Zool.*, 23(3): 179-183.

Khanna, D.R. and Bhutiani, R., 2003. Limnological status of Satikund Pond at Haridwar (UA). *Indian J. Env. Sc.*, 7(2): 131-136.

Khanna, D.R. and Bhutiani, R., 2004. *Fish and their ecology of River Ganga at Gohri Ghat Garhwal (Uttaranchal). Pro. Nat. Sem,* Fish diversity in protecterd habitats, Ed. By S. Ayappan, D.S.Malik, Rani Dhanze and R.S.Chauhan. pp: 291-298.

Khanna, D.R. and Chugh, T., 2004. Microbial Ecology: A study of River Ganga. Discovery publishing house, New Delhi. pp: 1-277.

Khanna, D.R. and Singh, R.K., 2000. Seasonal fluctuation in the plankton of Suswa River at Raiwala (Dehradun). *Env. Cons. J.*, 1(2 and 3): 89-92.

Khanna, D.R. and Singh, S., 2002. Pond Fish Ecology and Economics, Daya Publishing House, Delhi. pp: 1-816.

Khanna, D.R., 1993. *Ecology and pollution of Ganga River*. Ashish Publication House, Delhi. pp: 1-241.

Khanna, D.R., and Bhutiani, R., 2005. Benthic fauna and its ecology of river Ganga from Rishikesh to Haridwar (Uttaranchal) India. *Env. Cons. J.*, 6(1): 33-40.

Khanna, D.R., and Bhutiani, R., 2005. *Water analysis at a glance*, ASEA publications, 1-115.

Khanna, D.R., Badola, S.P. and Dobriyal, A.K., 1993. Plankton ecology of the River Ganga at Chandighat, Haridwar. Advances in limnology, Ed. by H.R. Singh, Narendra publishing House, New Delhi. pp: 171-174.

Khanna, D.R., Gautam, Ashutosh, Chugh, Tarun and Sarkar, Praveen, 2000. Impact of abiotic factors on the phytoplanktonic population of a pond. *Env. Con. J.*, 1(1): 41-46.

Khanna, D.R., Malik, D.S. and Rana, D.S., 1997. Phytoplanktonic communities in relation to certain physico-chemical parameter of Ganga canal at Hardwar. *Him. J. Env. Zool.*, 12: 193-197.

Khanna, D.R., Malik, D.S., Seth, T.R. and Rupendra, 1999. Co-relation between abiotic factors and planktonic population in river Ganga at Rishikesh (UP). *Sus. Eco. Sys. and Env.*, Ed. By D.R. Khanna, A. Gautam and A. Gautam, Published By ASEA, Rishikesh. pp: 69-76.

Khanna, D.R., Pathak, S.K., Bhutiani, R. and Chandra, K.S., 2006. Study of water quality of River Suswa Near Raiwala, Uttaranchal. *Env. Con. J.*, 7(3): 79-84.

Khanna, D.R., Seth, T.R., Malik, D.S. and Singh, R.K., 1999. Fishes and their ecology of River Suswa at Raiwala Dehradun. *Sus. Eco.Sys. and Env.*, Ed. By D.R. Khanna and A. Gautam. pp: 105-110.

Khanna, D.R., Singh, Vikas, Bhutiani, R., Chandra, K.S., Matta, Gagan and Kumar, Dheeraj, 2007. A study of biotic and abiotic factors of Song River at Dehradun, Uttarakhand. *Env.Cons. J.*, 8(3): 117-126.

Khanna, D.R., Ashraf, J., Chandan, Beena and Bhutiani, R., 2005. Physico chemical analysis of River Panvdhoi at Saharanpur U.P. *Env. Cons. J.*, 6(2): 89-94.

Khare, Bharti, 2007. Hydrobiological studies in the upstream of River Kunda at khargone, Madhya Pradesh. *Env. Cons. J.*, 8(1 and 2): 81-83.

Khulbe, R.D., Bisht, G.S. and Joshi, Chandra, 1995. An ecological study of water moulds of some rivers of Kumaon Himalayas. *J. Indian Bot. Soc.*, 74: 61-64.

Kofoid, C.A., 1903. Plankton studies-IV. The plankton of Illinois River, 1894- 99 with introductory notes upon the Hydrography of the Illinois River and its basin. Part 1. Quantitative investigations and general results. *Bull. Ill. Nat. Hist. Sury.*, 6(2): 95-624.

Kumar, Neeraj and Sharma, R.C., 2002. Water quality of River Krishni {Part–I Physicochemical characteristics}. *J. Nat. Con.*, 14(2): 273-297.

Kumar, Neeraj and Sharma, R.C., 2002. Water quality of River Krishni{ part –II Physicochemical characteristics}. *J. Nat. Con.*, 14(2): 299-333.

Kuuppo, P., 1994. Annual variation in the abundance and size of heterotrophic nanoflagillates on the SW coast of Finland, the Baltic Sea. *J. Plankton Res.*, 16: 1525-1542.

Lakshminarayana, J.S.S., 1965. Studies on the phytoplankton of the River Ganges, Varanasi, India. I–IV. *Hydrobiologia*, 25: 119-175.

Lazaras, Asha, D. and Mishra, P.K., 2002. Physico-chemical analysis of Machna River water and related fish ecology at Betul, M.P., India. *Env. Cons. Cons. J.*, 3(3): 75-76.

Linlokken, A., 1993. Efficiency of fishways and impacts of dams on the migration of Grayling and Brown trout in Glomma River system, South-Eastern Norway. *Regulated River*, 8: 145-153.

Mahajan, S.K., Garde, Yogesh, Bhawsar, Rakesh and Sharma, Sharad, 2002. Physico-chemical and biological characterisation of the River Kunda at down stream of Khargone (Madhya Pradesh), India. *Env. Cons. J.*, 3(2): 53-55.

Mathur, R.P., 1982. Water and waste water testing, Neemchand and Bros. Publishers, Roorkee. pp: 1-54.

McCully, P., 2001. Silenced Rivers: The Ecology and Politics of Large Dams. Zed

Michael, R.G., 1964. Diurnal variation of the plankton correlated with physico-chemical factors in three different ponds. Ph.D. thesis, University of Calcutta.

Minns, C.K., 1989. Factors affecting fish species richness in Ontaria lake. *Transaction of American Fisheries Society*, 118: 533-545.

Mir, A.M. and Kachroo, P., 1990. Limnological studies in Kashmir Lakes VIII Dynamics of Phytoplankton in Dal Lake. *J. Indian Bot. Soc.*, 69: 435-441.

Mishra, G.P. and Yadav, A.K., 1978. A comperative study of physico-chemical characteristics of river and lake water in central India. *Hydrobiologia*, 59(3): 275-278.

Mishra, Sarika and Joshi, Namita, 2002. Generation of solid waste during festive occassions in Hardwar city. *Him. J. Env. Zool.*, 16(2): 263-266.

Moehl, J.F. and Davies, W.D., 1993. Fishery intensification in small water bodies. A review for North America. FAO Fisheries Technical Paper No. 333, Rome.

Mohanty, P.K., 2006. Water: The Elixir of Life and Wonderful Gift. *Orissa Review* (April) pp: 59-62.

Monichan, K.K. and Sharma, R.K., 2003. Study on the distribution status of Fish fauna in Etawah District, U.P. *Indian J. Env. Sc.*, 7(2): 115-117.

Mruthunjaya, T.B. and Hosmani, S.P., 2002. Comparative account of physico-chemical parameters, enzymatic activity and plankton population in three water bodies of Mysore city. *J. Nat. Env. and Poll. Tech.*, 1(3): 156-169.

Mwaura, F., 2006. Some aspects of water quality characteristics in small shallow tropical man-made reservoirs in Kenya. *African Journal of Science and Technology*, 7(1): 82-96.

Nautiyal, P., 1990. *Ecology of the Ganga River system in the upland of Garhwal Himalaya.* Environment Resource and development. pp: 69-73.

Newcombe, C.P. and MacDonald, D.D., 1991. Effects of suspended sediments on aquatic ecosystems. *North Mmerican Journal of Fisheries Management*, 11: 72-82.

Nicholas, W.R. and Gordon, J.A., 1972. Effects of impoundments on water quality. Observations of several mechanisms of D.O. depletion. A.S.C.E. Annual and National Environmental Meeting, Houston, Texas

Nix, J.F., 1974. Distribution of trace metals in a warm water release impoundment. Arkansas Water Resources Res. Center Publ. 12. pp: 377.

Oliff, W.O., 1959. Hydrological studies on the Tugela River system. Part I the main tugela River. *Hydrobiologia*, 14: 281-370.

Paasche, E., 1980. The physiological ecology of phytoplankton. Editor, I. Morris, University of California Press, Berkeley, California, USA. pp: 259-284

Pahwa, D.V. and Mehrotra, S.M., 1966. Observations on fluctuations in abundance of plankton in relation to certain hydrological conditions of River Ganga. *Proc. Nat. Acad. Sci.*, 36B(2): 157-189.

Pandey, B.N. Kumar, K. and Das, P.K.L., 1993. A preliminary study on the physicochemical quality of water of River Kosi. *J. Ecobiol.*, 5: 237-339.

Pandey, B.N., Ambasta, O.P.A., Jha, A.K. and Shambhu, K., 2007. Seasonal variation in physico chemical and biological properties of River Panar (Bihar). *Env. Cons. J.*, 8(3): 133-153.

Parashar, Nagendra, Bhusan, Kaushik, Purroshottam and Pandey, Sachindra, 2003. Physico-chemical and microbiological studies of Ganga canal at Haridwar during Kumbh period. *Him. J. Env. Zool.*, 17(2): 167-171.

Patel, Pushpa., Mahajan, S.K. and Pathak, S.K., 2002. Hydrobiological observations on Virla Reservoir at Khargone (Madhya Pradesh), India. *Env. Cons. J.* 3(2): 57-59.

Pathak, J.K., 1991. Some aspects of the ecology of the River Sarju and Gomti of Kumaon Himalayas, Ph.D. Thesis, Kumaon University Nainital.

Pathak, S.K. and Mudgal, L.K., 2005. Limnology and biodiversity of fish fauna in Virla Reservoir, M.P. India. *Env. Cons. J.*, 6(1): 41-45.

Pathak, S.K. and Mudgal, L.K., 2006. Corelation of Physico chemical parameters in Virla Reservoir M.P.*Env. Cons. J.*, 7(1-2): 79-82.

Pearce, F., 1992. The Dammed: Rivers, Dams, and the Coming World Water Crisis.

Peerapornpisal, Y., Sonthichai, W., Somdee, T., Mulsin, P. and Rott, E., 1999. Water quality and phytoplankton in the Mae Kuang Udomtara Reservoir, Chiang Mai, Thailand. *J. Sci. Fac. Cmu.*, 26(1): 25-43.

Perry County, 2008. Multi-Jurisdictional hazard mitigation plan. Appendix-C, Hazard profiles; Draft, 2008.

Poff, N.L. and Hart, D.D., 2002. How dams vary and why it matters for the emerging science of dam removal. *BioScience*, 52: 659-668.

Poff, N.L., Allan, J.D., Bain, M.B., Karr, J.R., Prestegaard, K.L., Richter, B.D., Sparks, R.E. and Stromberg, J.C., 1997. The natural flow regime: A paradigm for conservation and restoration of river ecosytems. *BioScience*, 47: 769-784.

Poulickova, Aloisie, Belohlavek, Jiri, Sykorova, Palvla and Baresova, Milada, 2002. The influence of reservoirs on stream phytobenthos. *Biol.*, 39-40: 27-39.

Prapurna, N. and Shashikanth, K., 2002. Pollution level in Hussain Sagar lake of Hyderabad- A case study. *Poll. Res.*, 21: 187-190.

Prasad, B.B. and Singh, R.B., 2003. Composition, abundance and distribution of phytoplankton and zoobenthos in a tropical water body. *J. Nat. Env. and Poll. Tech.*, 2(3): 255-258.

Prasad, G., Shankar, V. and Chopra, A.K., 2003. Environmental state of river Ganges in Rishikesh-Haridwar and its management. *Env. Cons. J.*, 4(1-3): 51-88.

Prasad, Guru B., 2003. Physico-chemical and bacteriological quality of ground waters at Tadepalli Mandal of Guntur district, Andhra Pradesh. *Nature Environment and Pollution Technology*, 2(2): 173-178.

Prasad, R.R., 1956. Further studies on the plankton of the inshore waters off Mandapam. *Indian J. Fish.*, 3: 1-42.

Pruthi, H.S., 1933. Studies on the bionomics of fresh water of India I Seasonal changes in the Indian Museum Compound Int. Rev. *Hydrobiologia*, 28:46-67.

Raghvendra and Hosmani, S.P., 2002. Hydrobiological study of Mandakally Lake, a polluted water body of Mysore. *J. Nat. Env. and Poll. Tech.*, 1(3): 291-293.

Raina, V., Sah, A.R. and Ahmed, S.R., 1984. Pollution studies on River Jhelum-I:An assessment at water quality. *Indian J. Env. Health.*, 26: 187-210.

Randolph, James C. and Wilhm, Jerry, 1984. Seasonal variation in the phytoplankton and the trophic state of a southern great plains reservoir. *Proc. Okla. Acad. Sci.*, 64: 57-62.

Rao, K.L., 1975. India's water wealth. Orient Longman, N. Delhi. pp: 225.

Rautela, Piyoosh, Rakshit, Rahul, Jha, V.K., Gupta, Rajesh Kumar and Munshi, Ashish, 2002. GIS and remote sensing-based study of the reservoir-induced land-use/ land-cover changes in the catchment of Tehri Dam in Garhwal Himalaya, Uttaranchal (India). *Current Science*, 83(3): 308-311.

Ray, H.K., 1955. Plankton ecology of the River Hoogly at Patna. *Ecology*, 36(2): 169-175.

Reddy, P.M. and Venkateshwarlu, V., 1987. Assessment of water quality and pollution in the River Tungbhadra near Kurnool, (A.P.). *J. Environ. Biol.*, 8(2): 109-119.

Reynolds, C.S., 1984. *The ecology of freshwater phytoplankton*. Cambridge University Press, UK.

Rhee, G.Y. and Gotham, I.J., 1981. The effect of environmental factors on phytoplankton growth temperature and the interaction of the temperature with nutrient limitation. *Limnol. Oceanogr.*, 26: 635-648.

Richardson, R.E., 1921. Changes in bottom and shore fauna of the middle Illinois River and its connecting lakes Since 1913-1915, as a result of the increased southward of sewage pollution Illinois, *Nat. Hist.Surv.Bull.*, 14: 33-75.

Rodgher, S., Espindola, E.L.G., Rocha, O., Fracacio, R., Pereira, R.H.G. and Rodrigues, M.H.S., 2005. Limnological and ecotoxicological studies in the cascade of reservoirs in the Tiete River (Sao Paulo, Brazil). *Braz. J. Biol.*, 65(4): 697-710.

Sakalauskiene, G., Hansen, F.T. and Raulinaitis, M., 2004. Assessment of Hydro Power Plants Effect on River Water Quality, *Environmental research, engineering and management*, 3(29): 14-20.

SANDRP, 2008. The disappearing Ganga: Is there any hope for this holy river? Dam, Rivers and People.

Sarkar, A.U. and Karagoz, S., 1995. Sustainable development of hydroelectric power.

Saxena, K.K. and Chauhan, R.R.S., 1993. Physico chemical aspects of pollution studies of River Ganga near Kanpur, India. *J. Env. Health*, 8: 270-285.

Saxena, K.L., Chakaraborty, R.N., Khan, A.Q. and Chattopdhyay, S.N., 1966. Pollution studies of River Ganga near Kanpur, Indian. *J. Env. Health*, 8: 270-385.

Sehgal, H.S., 1980. Limnology of lake Sruinsar, Jammu with reference to zooplankton and fisheries prospectus. *Ph.D. Thesis*, University of Jammu.

Selva Raj, D. Jeyraj, Thampi, C. and Chithra, G., 2003. Studies on hydrology and fish species diversity of Thamiraparani River in Kanyakumari district.

Senger, R.M.S., Sharma, K.D. and Pathak, P.D., 1985. Studies on distribution of algal flora in polluted and non polluted regions in Yamuna River at Agra (U.P.). *J. Indian Bot. Soc.*, 64: 365-376.

Shanthi, K., Ramasamy, K. and Lakshmanaperumalsamy, P., 2002. Hydrobiological study of Singanallur Lake at Coimbatore, India. *J. Nat. Env. and Poll. Tech.*, 97-101.

Sharma, Dushyant and Jain, Renu, 2001. Diurnal variation in some aspects of limnology of Gopalpura Tank of district Guna (Madhya Pradesh), India. *Env. Cons.J.*, 2(1): 41-44.

Sharma, R.C., 1986. Effect of physico chemical factors on benthic fauna of Bhagirathi River Garhwal Himalaya. *Indian J. Ecology*, 13(1): 133-137.

Sharma, R.C., 2004. Protection of endangered fish *Tor tor* and *Tor putitora* population impacted by transportation network in the area of Tehri dam project, Garhwal Himalaya, India. In proceedings of the 2003 International Conference of Ecology and Transportation, Eds. Irwin CL, Garrett P, Mc Dermott K.P. Center for Transportation and the Environment, North Carolina State University, Raleigh, NC: pp: 83-90.

Sharma, S. Bajracharya, R.M., Sitaula, B.K. and Merz, J., 2005. Water quality of the central Himalaya. *Current Science*, 89(5): 774-786.

Sharma, S.D. and Pandey, K.S., 1998. Pollution status of Ramganga River at Moradabad, *Poll. Res.*, 17(2): 201-209.

Sharma, Shailendra, Mudgal, L.K., Sharma, Praveen and Belsare, D.K., 2007. Biodiversity of fish fauna of Kishanpura lake, Indore (M.P.) *Env. Cons. J.*, 8 (1 and 2): 109-113.

Shivakumar, A.A., Arunadevi, P. and Aruchami, M., 2003. Studies on water quality of the river Ambarampalayam, Coimbatore Tamilnadu., *J. Nat.Env. and Poll.Tech.*, 2(3): 305-308.

Singh, A.K. and Ahmed, S.H., 1990. A comparative study of the phytoplankton of the River Ganga and Pond of Patna (Bihar), India. *J. Indian. Bot. Soc.*, 69: 153-158.

Singh, H.R. and Dobriyal, A.K., 1982. First report on the occurrence of *Botla geto* (Ham) in the River Alaknanda of Garhwal Himalaya. *Proc. Nat. Acad. Sci., India.* B 52(2):137-139.

Singh, J.P., Yadava, P.K. and Singh, L., 1988. Pollution status at sangam and its adjoining rivers before the Kumbh mela at Allahabad. *Indian. J. Env. Port*, 8(11): 839-842.

Singh, K.P., 1960. The Chlorophyceae of Kumaon hills U.P. India I, Agra University *J. Res. (Science)*, 9(2): 211-216.

Singh, N., Bahuguna, S.N. and Bhatt, K.C., 1993. The profile of river ecosystem, food and feeding habits of hill-stream fishes and consequences of recent environmental degradation in Garhwal Himalaya. *Acta Ichthyologica Et Piscatoria*, 23: 3-30.

Singh, R.K., 1985. Limnological observations on Rihand Reservoir (Uttar Pradesh) with reference to the physical and chemical parameters of its water. *Int. Revue Ges. Hydrobiol.*, 70: 857-875.

Snoeijs, P., 1994. Distribution of epiphytic diatom species composition diversity and biomass of different macroalgal hosts along seasonal and salinity gradients in the Baltic Sea. *Diatom Res.*, 9: 189-211.

Subramanyam, B. and Subrahmanya, Kumar, C.S.V., 2003. A case study on evaluation of water quality of lakes by pollution index method. *J. Nat. Env. and Poll. Tech.*, 2(1): 69-72.

Suravit, S., 1996. Relation between phytoplankton and water quality in Rachapapa Dam reservoir, Surathanee province, Master thesis of Fisheries, Kasetsart University, Thailand.

Talling, J.F., 1957. The phytoplankton population as a compound phoyosynthetic system. *New Phytologist*, 56: 133-149.

Talling, J.F., 1965. The photosynthetic activity of Phytoplankton in east African Lakes.

Tilzer, M.M., Gieskes, W.W. Heusel, R. and Fenton, N., 1994. The impact of phytoplankton on spectral water transparency in the southern Ocean: implications for primary productivity. *Polar Biology*, 14: 127-136.

Tirath, R. and Bhargawa, D.S., 1982. Water quality of Gomti River at Lucknow (UP). J. *Ind.Water Works. Assoc.*, 14: 299-304.

Trivedi, R.K. and Goel, P.K., 1984. Chemical and biological methods for water pollution studies, Karad: *Environmental Publication*. pp: 1-251.

Turaga, U., 2000. Damming waters and wisdom: protest in the Narmada River Valley.

UN, 2003. Water for People, Water for Life. The United Nations Water Development Report.

Unni, K.S., 1982. Limnological studies of Sampna reservoir, Betul, M.P. India. *Proc. Nat. Acad. Sci.*, 52B: 365-372.

Unni, K.S., 1993. Phytoplankton distribution in reservoirs and lakes of Indian subcontinent – a review. *Verh. int. Ver. Limnol.*, 25: 1262-1265.

Venkateshwarlu, T. and Menon, A.G.K., 1979. A list of fishes of River Ganges and its branches. *Hydrobiologia*, 9: 46-70.

Venkateshwarlu, T. and Jayanti, T.V., 1968. Hydrobiological studies of the River Sabarmati to evaluate water quality. *Hydrobiologia*, 33(3-4): 442-448.

Verma, P. K. and Datta Munshi, J., 1987. Plankton community structure of Badua reservoir, Bhagalpur (Bihar). *Trop. Ecol.*, 28: 200–207.

Verma, P.K., 2007. Pond and riverine algae of Khargone, M.P. *Env.Cons. J.* 8(1-2): 17-20.

Verma, S.R. and Shukla, G.R., 1969. Pollution in a perennial stream Khala, By the sugar factory effluent near Laksar (Distt. Saharanpur). U.P., India. *J. Env. Health.*, 11:145-162.

Verma, S.R., 1978. Pollution studies of few rivers of western Uttar Pradesh with reference to biological indices. *Proc. Indian. Acad. Sci.*, 87B: 121-131.

Verma, S.R., Sharma, P., Tyagi, A., Rani, S., Gupta, A.K. and Dalel R.C., 1984. Pollution of saprobic status of Eastern Kalindi. *Limnologica*, 15(1): 69-133.

Vishnoi, Unnati, Chandra, K.S. and Matta, Gagan, 2007. Abiotic staus of Song River and its relation to zooplankton and phytoplankton at Nepali Farm, Dehradun. *Env. Cons. J.*, 89(1-20): 69-72.

Vishnoi, Unnati, Kaur, Jagjeet and Pathak, S.K., 2007. Water quality of River Ganga in respect to physico-chemical characteristics at Shyampur district Haridwar. *Env. Cons. J.*, 8(3): 23-26.

Voros, L. and Padisak, J., 1991. Phytoplankton biomass and chlorophyll a in some shallow lakes in Central Europe. *Hydrobiologia*, 215: 111-119.

Vyas, L. N., 1968. Studies in phytoplankton ecology of Pichhola Lake, Udaipur. In R. Misra and B. Gopal (eds), Proceedings Symposium on Recent Advances in Tropical Ecology, Int. Soc. Trop. Ecol., Banaras Hindu Univ., Varanasi: 334–347.

Wagh, N.S., 1998. Hydrobiological parameters of Harsul Dam, in relation to pollution. *Ph. D. Thesis*, Dr. B.A. Univ. Aurangabad.

Watson, S., 1992. The phytoplankton biomass of the Oldman reservoir system. Report submitted to surface water assessment branch technical services and monitoring division Alberta environmental protection. pp: 1-12.

WCD, 2000. *Dams and Development: A new Framework for Decision-making*. The Report of the World Commission on Dams. World Commission on Dams, Cape Town, South Africa.

Welch, P.S., 1948. Limnological methods, *The Blakiston. Co. Philadelphia*. pp: 1-381.

Whalen, S.C., Leathe, S.A., Gregory, R.W. and Wright, J.C., 1982. Physicochemical limnology of the Tongue River reservoir, Montana. *Hydrobiologia*, 89: 161-176.

Yakubo, A.F., Sikoki, F.D. and Horsfall, J.R.M., 1998. An investigation into the physico-chemicasl conditions and planktonic organisms of the lower reaches of the Nun River, Nigeria. *J. Applied Sci. Env. Manag.*, 1: 38-41.

Zade, S.B. and Sitre, S.R., 2007. Water quality assessment of a polluted urban lake with respect to zooplankton. *Env. Cons. J.*, 8(3): 67-71.

Zafer, A.R., 1964. On the ecology of algae in certain fish ponds of Hyderabad, India. I Physico-chemical Complexes, *Hydrobiologia*, 23: 179-195.

Zdankus, Narimantas, Vaikasas, Saulius and Sabas, Gintautas, 2008. Impact of a hydropower plant on the downstream reach of a river. *Journal of Environemntal Engineering and Landscape Management*, 16(3): 128-134.

Zingde, M.D. Narvekar, P.V. Sharma, R.V. and Desai, B.N., 1980. Water quality of the River Damanganga (Gujrat). *Indian J. Mar. Sci.*, 9: 94-99.